Lecture Notes in Physics

Edited by H. Araki, Kyoto, J. Ehlers, München, K. Hepp, Zürich
R. L. Jaffe, Cambridge, MA, R. Kippenhahn, München, D. Ruelle, Bures-sur-Yvette
H. A. Weidenmüller, Heidelberg, J. Wess, Karlsruhe and J. Zittartz, Köln
Managing Editor: W. Beiglböck

374

L. Ting R. Klein

Viscous Vortical Flows

Springer-Verlag
Berlin Heidelberg GmbH

Authors

Lu Ting
Courant Institute of Mathematical Sciences
New York University, 251 Mercer Street
New York, NY 10012, USA

Rupert Klein
Institut für Technische Mechanik, RWTH Aachen
Templergraben 64, D-5100 Aachen, BRD

This book was processed by the authors using the T_EX macro package from Springer-Verlag

ISBN 978-3-662-13850-2 ISBN 978-3-540-47009-0 (eBook)
DOI 10.1007/978-3-540-47009-0

Originally published by Springer-Verlag Berlin Heidelberg New York in 1991
Softcover reprint of the hardcover 1st edition 1991

2153/3140-543210 – Printed on acid-free paper

Preface

These lecture notes are based upon a series of ten lectures given by L. Ting in commemoration of the 75th anniversary of the founding of the Aerodynamisches Institut, Rheinisch-Westfälische Technische Hochschule Aachen, in April 1988. Besides making some amplifications here and there and the addition of a few recent results, we retain the lecture style, the sequence and the spirit delivered there. No attempt was made to hide the authors' biases and interests.

The lecture notes are composed of four chapters preceded by an extensive introduction in which the contents of each chapter are summarized section by section.

In the first chapter we formulate the mathematical problem of a vortex-dominated viscous flow as an initial value problem for the incompressible Navier-Stokes (N-S) equations in an unbounded domain with an initial vorticity field. Some general properties of the solutions are presented, namely, the consistency conditions on the moments of vorticity, and their relationships to the far field representations of the vector and scalar potentials of velocity.

In the second chapter we study the motion of slender vortex filaments with small diffusive core structures, using the method of matched asymptotics. Details of the analyses were reported in a sequence of four papers which began in 1965 with the relatively simple two dimensional problem and ended in 1978 with the three dimensional problems of slender vortex filaments with large axial and circumferential velocities in their cores. Here we emphasize the scalings, or rather the intuition, leading to the general scheme of the asymptotic analysis and the physical interpretation of the solutions. Also we present some recent studies on the practical region of validity of the asymptotic solutions.

In the third chapter we study the merging of the core structures of vortices by numerical solution of the N-S equations. We explain our numerical schemes and how the asymptotic and numerical methods can complement each other in the solution for the entire flow field. In the last chapter we mention several new problem areas.

Finally we wish to thank Professor Egon Krause, director of the Aerodynamisches Institut, for his invitation to the commemoration lectures and his constant encouragement, without which this book could never have materialized. We are especially indebted to Professors Joseph B. Keller of Stanford University and Andrew Majda of Princeton University who read our preliminary manuscript and made many useful suggestions that were incorporated in the final text. We gratefully thank Dr. Frances Bauer of the Courant Institute of Mathematical Sciences for her continuing help to bring the manuscript to its final form.

We wish to acknowledge the support of this research area by NASA Langley Research Center under grant No. NCCI-58 over the past seven years and the continuing support of Air Force Office of Scientific Research under grant No. AFOSR-90-0022, personified by Dr. C. H. Liu and Dr. Arji Nachman respectively. R. Klein wishes to acknowledge the support of the Deutsche Forschungsgemeinschaft for a two year post-doctoral fellowship at the Department of Mathematics, Princeton University, during which these notes were prepared. We also express our thanks to Prof. W. Beiglböck for inviting us to publish in this Springer-Verlag series.

New York, NY — L. Ting

November 9, 1990 — R. Klein

Contents

Introduction

Flow fields with vorticity concentrated in a distinct region have been observed in many fluid dynamic, astrophysical and geophysical problems. Extensive collections of photographs of such flow fields and references to experimental and theoretical investigations can be found in two recent books by van Dyke (1982) and Lugt (1983). Many of those flow fields are very complex, involving body forces, energy sources, compressibility and real gas effects. We shall mention only a few classical examples. When we first studied fluid dynamics we saw photographs of vortical flows in the textbook of Prandtl and Tietjens (1957). They showed the formation and shedding of eddies or vortices in the wake of circular cylinders, blunt bodies and airfoils, the formation of von Kármán vortex streets, the effect of rotation of a circular cylinder on the eddies shedding off its surface, the trapping of vortices in a concave corner, the roll up of an interface of a separated flow into eddies, and the organized eddies in a turbulent flow. Those phenomena have been re-examined theoretically, experimentally and numerically in recent years, see van Dyke (1982) and Lugt (1983), and references therein.

Vortical flow can frequently be observed or experienced in our daily life. We often see smoke rings generated from orifices, a pair of vortex filaments trailing a commercial airplane and encounter the gusty flow around the corner of a tall building. Vortical flows have also been of great practical interest to aeronautical engineers. They have analyzed, for example, the trailing wakes of aircrafts during landing and take off for the sake of flight safety and studied the generation of leading edge vortices and their stability for high performance configurations. Several special conferences or workshops dealing with these two problems were held in recent years, see for example, Olsen et al (1971), Wendt (1982), Young (1983) and Staufenbiel (1985).

A comprehensive theoretical study of inviscid vortical flows can be found in Chapters III and VI of the treatise by Lamb (1932) and the effects of viscosity in his Chapter XI. Classical theories, or results which can be found in those three chapters, will often be quoted here without reference. For the method of matched asymptotics and that of multiple scales our standing references are van Dyke (1975), Schneider (1978) and Kevorkian and Cole (1981).

Now we shall identify the class of problems which we will focus on and define the terminology and symbols to be used in here.

We consider the flow field to be incompressible and viscous, unless stated otherwise. The velocity field $\mathbf{v}(t, \mathbf{x})$ is composed of the velocity induced by the vorticity field $\Omega(t, \mathbf{x})$ and of a background potential flow. Note that all vector quantities will be denoted by boldface Roman letters except for the vorticity vector whose symbol is Ω. All unit vectors will be denoted by symbols with a hat accent. For example, we use $\mathbf{x}$ to denote the position vector with Cartesian coordinates x_i, $i = 1,2,3$, $\mathbf{v}$ the velocity with components v_i, Ω the vorticity with components ω_i and $\hat{\imath}$, $i = 1,2,3$ the unit vectors along the Cartesian coordinate axes.

In **Chapter 1** we study general properties of a viscous vortical flow with a typical length scale ℓ and velocity scale U. Unless specified otherwise, the time is scaled by ℓ/U, vorticity and velocity gradients by U/ℓ and the kinematic viscosity ν by $U\ell$. From the last scaling, we define the Reynolds number of the flow field by

$$R_e = U\ell/\nu \ . \tag{1}$$

For an incompressible fluid we use its density ρ as the density scale, i.e., we set $\rho = 1$. The typical mass scale is therefore $\rho\,\ell^3$. We assume that the initial vorticity distribution is centered near the origin of a suitable reference frame and decays rapidly at large distance, so that at any finite time $t \geq 0$ and for some sufficiently large N we have

$$|\Omega|\ell/U = o\big([\ell/|\mathbf{x}|]^N\big) \quad \text{as} \quad |\mathbf{x}|/\ell \to \infty \ . \tag{2}$$

Condition (2) is certainly fulfilled if the vorticity distribution is of bounded support or decays exponentially. The latter is a realistic condition for a vortical flow and implies that condition (2) holds for all N. Note that we do not distinguish between the order of magnitude of ℓ , the effective size L of the vorticity distribution and a typical vorticity decay length ℓ_d at this point.

In **Sec.1.1** we state the governing equations for the vortical flow induced by any initial vorticity field in free space. In doing so, we assume that there is no background flow and that the distance from the vorticity field to a boundary or body is much larger than ℓ. Some general results for such flow fields are then presented in **Sec. 1.2**. We recount, in particular, the consistency and time invariance conditions on the moments of vorticity due to Truesdell (1951, 1954) and Moreau (1948, 1949), since these conditions are usually not reported in textbooks on Fluid Dynamics. Also, we provide the corresponding results for axi-symmetric and two-dimensional flows. The two-dimensional relations will be employed in **Sec. 2.2** to identify the physical meaning of an optimum similarity solution for a viscous vortex, called an optimum Lamb vortex, and in **Sec.3.2** to generate rules for merging of viscous vortices into one single approximating vortex.

The conditions of Truesdell and Moreau were employed by Ting (1983) to derive an asymptotic description of the far field behavior of a vortical flow. In the far field the vector potential of velocity, $\mathbf{A}$, is represented by a series in inverse powers of $|\mathbf{x}|$,

$$\mathbf{A}(t,\mathbf{x}) = \sum_{n=0}^{m} A^{(n)}(t,\mathbf{x}) + O(|\mathbf{x}|^{-m-2}) \ , \tag{3}$$

where $\mathbf{A}^{(n)}$ is proportional to $|\mathbf{x}|^{-n-1}$. As we shall see later, $\mathbf{A}^{(n)}$ is defined because of (2) when $m+3 < N$. Using Truesdell's consistency conditions, we show that the series for the vector potential begins with $n = 1$ instead of $n = 0$ and that for each $n \geq 1$, $\mathbf{A}^{(n)}$ is defined by a linear combination of $n(n+2)$ instead of all the $3(n+2)(n+1)/2$ nth moments. Using Moreau's conditions, we show, in addition, that the first term $\mathbf{A}^{(1)}$ represents three time invariant doublets and that the second term $\mathbf{A}^{(2)}$ represents eight quadrupoles, three of which are time invariant. This far field behavior will be employed in **Sec.3.1** to provide higher order approximate boundary conditions on a finite computational domain for the numerical simulation of a viscous vortical flow.

In **Sec.1.3** we show that for $n > 2$ the nth term $\mathbf{A}^{(n)}$ depends on only $4n$ linear combinations of the nth moments of vorticity and that only $2n + 1$ of them contribute to the far field velocity of order $O(|\mathbf{x}|^{-n-2})$. The latter can be expressed as the gradient of a scalar potential, $\nabla\phi^{(n)}$, for which we provide an explicit expression in terms of $2n + 1$ linear combinations of the nth moments of vorticity, see Klein and Ting (1990).

In **Sec.1.4** we identify the solution for an incompressible vortical flow as the zeroth order solution of a compressible flow at a low Mach number, $M \ll 1$. The leading unsteady far field velocity of the incompressible vortical flow is matched with the near field solution of acoustic quadrupoles. The far field contribution of the global compressibility effect of the next order solution of the vortical flow, $O(M^2)$, is matched to the near field solution of an acoustic monopole. This global dilatation effect is related directly to the rate of energy dissipation of the incompressible vortical flow. Thus we obtain the formulae for the leading acoustic field induced by the vortical flow (Crow 1970, Möhring 1978, Obermeier 1985 and Ting and Miksis 1990). These results are applied to turbulent flows and their applications to the mean flow are discussed. The formulae are also employed in **Sec.2.3** to compute the sound generation due to the motion of slender vortex filaments and due to the evolution of their core structures.

In **Chapter 2** we study a special class of vortical flows which are characterized by multiple length scales. This includes, in particular, flows induced by a finite number of vortex filaments submerged in a background potential flow with length scale ℓ and velocity scale U. We call it a vortex filament or in short a filament or a vortex, when the bulk of vorticity is concentrated in a slender tube-like region. We introduce a reference line, called the center line $\mathcal{C}$, which essentially is the line of maximum vorticity in cross sections normal to the vortex tube. This curve is described by a vector function $\mathbf{X}(t, s)$ of time t and of a tangential parameter s. In general, the effective core size of the filament, or the size of the cross section of the tube, δ, is also a function of s and t. By slenderness, we mean that the core size is much smaller than a typical radius of curvature of $\mathcal{C}$ which is assumed to be on the order of the length scale ℓ. Thus, the flow field has two distinct length scales given by a typical core size δ^* and an overall length scale ℓ of either the outer flow or the filament geometry. Their ratio defines a small parameter,

$$\delta^*/\ell = \epsilon \ll 1 . \tag{4}$$

Now we can specify the order of magnitude of other dimensionless quantities, namely the velocities scaled by U, the lengths scaled by ℓ, the time t scaled by ℓ/U etc., in terms of the small parameter ϵ as it approaches zero. For example, the statements that the radius of curvature R of $\mathcal{C}$ is of the order of ℓ and that it is one order larger than its effective core size are equivalent to

$$R(t,s)/\ell = O(1) \tag{5}$$

and

$$R(t,s)/\ell = O(\epsilon^{-1}\delta(t,s)/\ell)) \ . \tag{6}$$

Based on this, we replace the qualitative characterization of a vortex filament, saying that the bulk of vorticity is concentrated in a slender tube-like region of effective core size δ, by the following two properties. We require that 1) the effective core size δ remains on the order of δ^* and 2) the vorticity at a point $\mathbf{x}$ decays rapidly as the distance r between $\mathbf{x}$ and C becomes large relative to δ^*. The second condition means that

$$|\Omega(t,\mathbf{x})|\ell/U = o([\delta^*/r]^N) \qquad \text{as} \qquad r/\delta^* \to \infty \ , \tag{7}$$

for some sufficiently large N and it implies that the vorticity decay length ℓ_d is on the order of the core size,

$$\ell_d = O(\delta^*) \quad \text{or} \quad \ell_d/\ell = O(\epsilon) \ll 1 \ . \tag{8}$$

This filament structure, described by (4-8), will survive the effect of viscous diffusion at least for a finite time, only if the Reynolds number based on the length ℓ is large, i. e., $R_e \gg 1$. The order of magnitude of $1/Re$ relative to ϵ will be specified later in (13). Under these circumstances, the flow field far away from C is basically irrotational and we can define the strength of the vortex filament by the circulation Γ along a circuit around C with the minimum distance between C and the circuit much larger than δ^*. We assume that the strength Γ is finite in the sense that ,

$$\Gamma/(U\ell) = O(1) \ . \tag{9}$$

From (4) and (9), we see that the velocity and the vorticity (or the velocity gradients) near C are large, of the order of ϵ^{-1} and ϵ^{-2} respectively, which means

$$|\mathbf{v}|/U = O(\Gamma/[U\delta^*]) = O(\epsilon^{-1}) \tag{10a}$$

and

$$|\Omega|\ell/U = O(\Gamma\ell/[\delta^{*2}U]) = O(\epsilon^{-2}) \qquad \text{for} \qquad r/\delta^* = O(1) \ . \tag{10b}$$

In contrast, the velocity at a distance far away from C (relative to δ^*) is of order one, so that

$$|\mathbf{v}|/U = O(1) \qquad \text{for} \qquad r/\ell = O(1) \ , \tag{11}$$

while the vorticity vanishes in the sense of (7).

In general, there are other typical lengths in the flow field, e. g., the length of a filament, the distance between two filaments, the distance from a filament to a rigid surface, etc. We consider all those typical lengths to be $O(\ell)$ unless stated otherwise. In case that the flow field is composed of vortex filament(s) without a background flow, we can use (5) and (9) to define the length scale ℓ by a typical radius of curvature of C and the velocity scale U by $|\Gamma|/\ell$.

In the classical inviscid theory, the effective size δ is reduced to zero and the filament becomes a vortex line along $\mathcal{C}$. The velocity $\mathbf{Q}$ at a point $\mathbf{x}$ induced by the vortex line $\mathcal{C}$ is defined by the Biot-Savart integral along $\mathcal{C}$,

$$\mathbf{Q}(t,\mathbf{x}) = \frac{\Gamma}{4\pi}\int_{\mathcal{C}} \frac{\mathbf{X}' - \mathbf{x}}{|\mathbf{X}' - \mathbf{x}|^3} \times d\mathbf{X}' \ . \tag{12}$$

From this equation it is evident that the inviscid theory suffers from the two serious defects that 1) the fluid velocity on the vortex line, $\mathbf{x} = \mathbf{X}$, becomes infinite and that 2) the velocity of the vortex line itself is undefined.

Often such flow fields are modelled by an irrotational flow outside a slender tube-like surface, and a rotational flow with a prescribed vorticity distribution inside of it. In that situation the velocity defined by the Biot-Savart formula will be finite everywhere and the velocity of the center line of the vortex tube will be defined and depend on the vorticity distribution. It is important to note, however, that the vorticity distribution inside the tube, no matter how slender it is, cannot be arbitrarily assigned. The reason is that the rotational flow has to fulfill the continuity and Euler equations inside the tube and on the interface its velocity and pressure have to match with those of the potential flow outside in order to satisfy the kinematic interface conditions. Whether we consider the flow in the vortical core to be viscous or not, in either case we may carry out a matched asymptotic analysis using the slenderness ratio ϵ as the small expansion parameter to derive an approximate description of the core flow. Since the velocity gradients in the the vortical core are large of order ϵ^{-2} according to (10b), we opt to retain viscous effects in the inner solution by assuming the distinguished limit

$$Re = \frac{U\ell}{\nu} = O(\epsilon^{-2}) \qquad \text{or} \qquad \bar{\nu} = \frac{1}{\epsilon^2 Re} = \frac{\nu}{U\ell\epsilon^2} = O(1), \tag{13}$$

while the Reynolds number based on the typical core size is

$$Re^* = \frac{U\delta^*}{\nu} = O(\epsilon^{-1}). \tag{14}$$

Thus, the asymptotic solution of the Navier-tokes equations will be valid for any fixed value of $\bar{\nu}$ and will be free of an artificial interface confining the highly rotational flow.

In **Sec.2.1** we describe in detail the deficiencies of the classical inviscid theory. It is well known that in the inviscid theory of two-dimensional flow, a point vortex is assumed to move with the local velocity of the background flow (without the vortex). For a real fluid, a point vortex will diffuse immediately into a vortex with a Gaussian vorticity distribution, called a Lamb vortex. We will show that the inviscid theory is a good approximation only on the normal time scale, ℓ/U, and may not be applicable on a much smaller time scale. To bring home this point, we first reproduce the solution of the oscillatory motion of a small spinning disc in a uniform stream, as described by Milne-Thompson (1973). The period and the amplitude of the oscillation decrease towards zero as the disc radius a, scaled by U/Γ, vanishes, where Γ denotes the circulation around the disc. We then perform a perturbation analysis for the oscillatory motion of a small rotating blob of fluid, called a Rankine vortex, in an inviscid uniform stream. We identify the correspondence between the motions of a spinning disc and the center of a Rankine vortex and show the similarity of the effect of their size on the amplitude and period of their trajectories. In both cases, the respective centers of disc and vortex are drifting along with the uniform stream on the normal time scale. These two simple examples illustrate the necessity for, and also the physical meaning of, the two-time matched asymptotic analysis of a viscous vortex to be presented in **Sec.2.2**. In the last subsection **2.1.2** we explain the defects of the inviscid theory for a curved vortex line in space. We derive the singular behavior of the flow field near the vortex line defined by the Biot-Savart formula (12). These singular terms will be identified with the far field behavior of an inner solution obtained through a matched asymptotic analysis in **Sec.2.2** and **Sec.2.3** for two and three-dimensional problems, respectively.

In **Sec.2.2** we present the matched asymptotic solutions of the N-S equations of Ting and Tung (1965) for a two-dimensional vortex submerged in a background potential flow. By comparing the one-time solution, in the normal time t, and the two-time solution, in t and a short time variable τ, we show that the average of the two-time solution over the short time is equivalent to the one-time solution to leading order and to first order. In particular, the trajectory of the vortex center defined by the two-time solution and its average path differ from the trajectory defined by the one-time solution by merely $O(\epsilon^2)$. The latter is independent of the core structure and differs from the classical inviscid result by the same order. The leading order core structure, which is axi-symmetric in the inner variables, depends only on the normal time and its solution is defined in terms of the initial data. It is not necessarily a similarity profile. The corresponding optimum similarity solution, or the solution for an optimum Lamb vortex, is defined by the condition that it is the best one term truncation of a series solution in terms of powers of $1/t$ for $t \to \infty$. By applying the consistency conditions, to be discussed in **Sec.1.2**, the optimum Lamb vortex turns out to be the one which not only has the same strength and center but also the same second polar moment as that of the non-similar solution for $t \geq 0$.

Similar to the two-dimensional solutions, the correspondence between the two-time and one-time asymptotic solutions of a circular vortex ring submerged in a background axi-symmetric potential flow were obtained by Tung and Ting (1966) and Ting (1971). There, however, the velocity of the center line of the vortex ring does depend on the core structure. Similar correspondence for a general three-dimensional problem is to be expected albeit the two-time analysis will be extremely tedious and has not yet been attempted. In **Sec.2.3** we present the one-time asymptotic analysis of Callegari and Ting (1978) for a curved slender vortex filament in space. We emphasize the essential differences between the three and two-dimensional problems, namely, 1) the dependence of the velocity of the center line on the core structure, 2) the compatibility conditions between large axial and circumferential velocity components in the vortical core and 3) the evolution of the core structure due to both stretching of the filament and viscous diffusion. The general results may then be applied to the case studied by Ting (1971), where the axial velocity in the core is only order one, and to the axi-symmetric case of Tung and Ting (1966).

It is well known that the leading order asymptotic solution in terms of a small expansion parameter $\epsilon \ll 1$, is often accurate enough for practical purposes, even when ϵ is not very small. In order to establish such a practical region of validity for our asymptotic description of a vortex filament, the solution was employed to study the interaction of vortex filaments by Liu , Tavantzis and Ting (1986). Numerical examples were presented for filaments which were initially slender and far apart from each other relative to a typical size of their vortical cores. The motion and deformation of the filaments were compared with a finite difference solution of the N-S equations with the same initial data. It was found that the asymptotic solution remains in good agreement with the finite difference solution even when the ratio between the typical core size δ and the reference length ℓ is almost 1/2. Here the reference length is either the minimum half distance between two filaments or the minimum radius of curvature of the filament centerline. The numerical comparison, which will be reported in the last section of **Chapter 2**, suggests that the practical region of validity for the asymptotic solution is $\epsilon < 1/2$. The methods used for the numerical solution of the N-S equations, suitable for the study of viscous vortical flows with rapidly decaying vorticity, will be described in **Chapter 3**.

In **Chapter 3**, we present efficient numerical schemes for the simulation of viscous vortical flows, i.e., for solving the initial value problem formulated in **Sec.1.2**. In particular, we study the merging or intersection of vortex filaments. We assume that the initial data are either specified or provided by the matched asymptotic solutions when these solutions are still valid, i.e., when ϵ is within its region of validity, say $\epsilon \leq 1/4$. A numerical scheme is considered to be efficient when the finite computational domain $\mathcal{D}$ is as small as possible under the side condition that the error in the boundary conditions imposed along $\partial\mathcal{D}$ is of the same order as that of the finite difference is within $\mathcal{D}$. Thus, it is essential to specify approximate boundary conditions along $\partial\mathcal{D}$ which are consistent with the unbounded domain problem and to assess their accuracy.

In **Sec. 3.1**, we explain the basic concepts in the choice of a computational domain, $\mathcal{D}$, and the formulation of boundary conditions tailored for a specified type of problem. We explain in **Sec. 3.1.1** our classification of different types of merging problems, such as global merging of filament(s) or local merging of small segments of filament(s).

In **Sec. 3.1.2** we consider global merging problems for which the decay length ℓ_d can be of the order of the size, $L = O(\ell)$, of the initial vorticity distribution. For example, global merging is taking place when the core radius δ of a filament becomes comparable to its radius of curvature. In this case the size of the computational domain $\mathcal{D}$ has to be much larger than L, say $4L$, and then the volume of $\mathcal{D}$ is about 64 times the volume of the effective vorticity distribution. Approximate boundary data along $\partial\mathcal{D}$ are obtained from the far field behavior of the vortical field described in **Sec.1.2**.

In **Sec. 3.1.3** we consider global merging problems for which the decay length ℓ_d is much smaller than the size ℓ. For example, in a nearly head on collision of two slender filaments, the core size becomes comparable to the distance between the filaments but is still much smaller than their radii of curvature. In this case, the size of the computational domain may be only several core sizes larger than L and the approximate boundary data are obtained by a suitable adaptation of the far field expansion of the vortical field.

In **Sec.3.1.4** we consider the case of local merging, which does not occur in two-dimensional or axi-symmetric problems. For example, when two slender filaments intersect each other at a finite number of non-overlapping local regions, then small segments of the filaments merge with each other in each of these regions. In this case, the computational domain for a local region has to cover only the merged segments and thus may be much smaller than the overall size L of a filament. The numerical solution then must be supplemented by the asymptotic solution for the motion and diffusion of the filament(s) outside the computational domains.

In **Sec.3.2** and **3.3** we study the merging of vortices in two and three dimensions. The numerical schemes formulated in **Sec. 3.1** are employed to study the merging of two-dimensional vortices, coaxial vortex rings and vortex filaments in **Sec. 3.2.1, 3.3.1** and **3.3.2** respectively.

Recall that integral invariants for the moments of a two-dimensional vorticity distribution are discussed in **Sec.1.2.4** and that they are employed to identify the physical meaning of the optimum similarity solution for a viscous vortex. Using those results, we present in **Sec.3.2.2**, rules for merging of two-dimensional viscous vortices into a single one with an optimum similarity core structure, i. e., an optimum Lamb vortex (Ting (1986)). In continuation of these physical concepts, an approximate solution to an initial value problem for a two-dimensional viscous vortical flow was formulated by Ting (1986) and Ting and Liu (1986). The vorticity field is represented by a finite number of Lamb vortices which may overlap each other. The velocities of the vortex centers are defined by the condition that the N-S equations are satisfied approximately such that the integral of the square of the error is a minimum. For a vortex which is not merging or overlapping with the others, the velocity of the vortex center defined by the minimum principle agrees with the single time asymptotic solution and hence with the classical inviscid theory. Numerical examples for the merging of vortices based on the minimum principle are presented and are shown to be in good agreement with the corresponding finite difference solutions.

In the last Chapter, we mention additional topics of vortical flows to demonstrate that there are many interesting and challenging problems which can be analyzed by modifications or extensions of the method of matched asymptotic expansions and/or the numerical methods discussed in **Chapters 2** and **3**. The topics are assembled according to the types of modifications required into three categories : 1) vortex filaments with multiple axial length scales, 2) vortices in a background rotational flow with order one vorticity and 3) the interaction of a filament with a moving surface. They are described in **Sec. 4.1, 4.2** and **4.3** respectively.

1. Vortex Dominated Flows and General Theory

In **Sec.1.1**, we formulate the mathematical problem of a vortical flow in general as an initial-value problem for the incompressible N-S equations in space. In **Sec.1.2** we recount the consistency and time invariance conditions on the moments of vorticity obtained by Truesdell (1951) and Moreau (1948) and their application in a far-field approximation to the vortical flow. We represent the vector potential of velocity in the far field by a series $\sum \mathbf{A}^{(n)}$, with $\mathbf{A}^{(n)}$ proportional to r^{-n-1}, $n = 0, 1, \ldots$. Because of the $(n+3)(n+2)/2$ consistency conditions for the $3(n+2)(n+1)/2$ nth moments of vorticity, we find $\mathbf{A}^{(0)} = 0$ and show that $\mathbf{A}^{(n)}$ for $n = 1, 2, \ldots$ can be represented by a linear combination of $J_n = n(n+2)$ vector potentials of nth order whose coefficients are nth moments of vorticity. Using the time invariants of Moreau we show that the first term in the series, $\mathbf{A}^{(1)}$, is represented by three doublets of constant strengths, while the second term, $\mathbf{A}^{(2)}$, is represented by $J_2 = 8$ quadrupoles, three of which are constant. The corresponding results for the axisymmetric and two-dimensional problems are stated as well. The results for the latter were attributed to Poincaré (1892) by Truesdell (1954). In **Sec.1.3**, we identify $n^2 - 1$ curl-free linear combinations of the J_n vector potentials of nth order which do not contribute to the velocity field. The far field velocity induced by the remaining $2n + 1$ linear combinations is identified as the gradient of a scalar potential of nth order. The coefficients of the $2n + 1$ spherical harmonics in the scalar potential $\phi^{(n)}$ are then related directly to linear combinations of nth moments of vorticity. In **Sec.1.4**, we identify the solution for an incompressible vortical flow as the leading solution of a low Mach number flow and include in its far field behavior the effect of density fluctuations. We apply the method of matched asymptotics to obtain the outer acoustic field induced by the vortical flow. The leading acoustic field is composed of five quadrupoles and a monopole. The strengths of the quadrupoles are defined by the retarded values of five linear combinations of the second moments of vorticity while the strength of the monopole is defined by the retarded value of the energy dissipation rate of the inner vortical flow.

1.1 Governing Equations for Viscous Vortical Flows

We consider an incompressible viscous flow field induced by an initial vorticity distribution in unbounded space. The velocity $\mathbf{v}(t,\mathbf{x})$ and pressure $p(t,\mathbf{x})$ obey the continuity equation,*

$$\nabla \cdot \mathbf{v} = 0 \tag{1.1.1}$$

and the momentum equation,

$$\mathbf{v}_t + (\mathbf{v} \cdot \nabla)\mathbf{v} = -\nabla p + \nu \Delta \mathbf{v} \ . \tag{1.1.2}$$

They are known as the incompressible Navier Stokes equations or in short N-S equations. Here we have chosen the unit of mass such that the density of the fluid is equal to unity and the subscript t denotes partial differentiation with respect to time t. By applying the curl operator to (1.1.2), we eliminate the pressure p and obtain the vorticity evolution equation,

$$\Omega_t + \nabla \cdot (\mathbf{v}\ \Omega) - \nabla \cdot (\Omega\ \mathbf{v}) = \nu \Delta \Omega \ , \tag{1.1.3}$$

where

$$\Omega = \nabla \times \mathbf{v} \tag{1.1.4}$$

denotes the vorticity vector. It follows from (1.1.4) that the vorticity is divergence free, i. e.,

$$\nabla \cdot \Omega = 0 \ . \tag{1.1.5}$$

The incompressible velocity field can be expressed as the sum of a potential flow and a divergence free field,

$$\mathbf{v} = \nabla\phi + \nabla \times \mathbf{A} \ , \tag{1.1.6}$$

where ϕ and $\mathbf{A}$ are known as the scalar and vector potentials, respectively. On the right-hand side of the vector equation (1.1.6), there are four unknown scalar functions, therefore, an additional condition is needed. Since the addition of the gradient of any scalar function, $\nabla\chi$, to the vector potential will not change the velocity field, we may suppose χ to be chosen such that

$$\nabla \cdot \mathbf{A} = 0 \ . \tag{1.1.7}$$

This in turn serves as the additional condition.

By applying the divergence operator to (1.1.6) and using the condition of incompressibility (1.1.1), we obtain a Laplace equation

$$\Delta\phi = 0 \tag{1.1.8}$$

for the scalar potential. The solution ϕ is defined when its far field behavior is specified. If the latter is independent of t, we have a steady solution. In particular,

* With the exception of those in the **Introduction**, equation numbers will be preceded by the chapter and section numbers.

if the flow is uniform upstream, ϕ represents the potential of a uniform flow or that of a uniform flow around a body provided that the distance from the vorticity field to the body is much larger than the size of the vorticity field. This restriction is introduced so that we can consider the vorticity field to be in free space. It will be removed later in **Chapter 4** when we study the interaction of vortex filaments and bodies. The scalar potential will be called the background velocity potential because it is defined independent of the vector potential **A**, i. e., independent of the flow field induced by the vorticity field.

By applying the curl operator to (1.1.6) and using (1.1.4) and (1.1.7), one obtains a vector Poisson equation

$$\triangle \mathbf{A} = -\Omega \tag{1.1.9}$$

for the vector potential with the vorticity as a source term. Equations (1.1.3) and (1.1.9) are the governing equations for **A** and Ω with **v** defined by the auxiliary equation (1.1.6). Because of the contribution of $\nabla\phi$ to **v**, the vortical flow field **A** depends on the background potential ϕ.

For vortical flows in unbounded space, it is convenient to use Ω and **A** as the state variables. They are governed by the vorticity evolution equation (1.1.3) and the Poisson equation (1.1.9), supplemented by (1.1.7) and (1.1.6), which relates v to **A**. Since Ω_t is the only t-derivative appearing in the governing equations, an initial condition on Ω or **A** is needed. To be consistent with (1.1.5), we assign a divergence free initial vorticity profile, Υ, i. e.,

$$\Omega(0, \mathbf{x}) = \Upsilon(\mathbf{x}) \ , \tag{1.1.10a}$$

with

$$\nabla \cdot \Upsilon \equiv 0 \ . \tag{1.1.10b}$$

The initial distribution, Υ, is usually assumed to be of bounded support or to decay exponentially in $r = |\mathbf{x}|$. For our purposes, it is sufficient to require that Υ decays rapidly in the sense of (2). Since the vorticity $\Omega(t, \mathbf{x})$ is convected at finite speed while it diffuses and initially decays exponentially in $r = |\mathbf{x}|$, it will continue to decay rapidly in r, thereby satisfying condition (2) for $t \geq 0$. Thus, the far field flow is irrotational to the order $(\ell/r)^N$.

We now consider a flow induced solely by a rapidly decreasing vorticity distribution without a background potential flow, i. e.,

$$\phi = 0 \qquad \text{and} \qquad \mathbf{v} = \nabla \times \mathbf{A} \ . \tag{1.1.11}$$

Then the far field condition for the vortical flow is

$$|\mathbf{v}| \to 0 \quad \text{and hence} \quad \mathbf{A} \to 0 \qquad \text{as} \qquad r \to \infty \ . \tag{1.1.12}$$

Thus we have completed the mathematical formulation of the viscous vortical flow problem. The differential equations (1.1.3) and (1.1.9), the auxiliary condition (1.1.11), the initial condition (1.1.10) and the far field conditions (2) and (1.1.12) define an initial-value problem in an unbounded domain for Ω, **A** and **v**. It remains to be shown that all three of them will remain divergence free, i. e., they fulfill (1.1.1), (1.1.5) and (1.1.7). From (1.1.6) and (1.1.8) or from (1.1.11), we see that $\mathbf{v}$ is divergence free. By applying the divergence operator to (1.1.3), we obtain a simple diffusion equation for $\nabla \cdot \Omega$. By using the homogeneous initial data (1.1.10b) and far field condition (2), we show that Ω will remain divergence free. Finally we apply the divergence operator to (1.1.7) to obtain a Laplace equation for $\nabla \cdot \mathbf{A}$ and then use the homogeneous far field condition (1.1.12) to show that $\mathbf{A}$ is divergence free.

The solution of the Poisson equation (1.1.6) under the far field condition (1.1.12) is given by the Poisson integral,

$$\mathbf{A}(t,\mathbf{x}) = \frac{1}{4\pi} \int\int\int_{-\infty}^{\infty} \frac{\Omega(t,\mathbf{x}')}{|\mathbf{x}'-\mathbf{x}|} d^3\mathbf{x}' , \tag{1.1.13}$$

where $d^3\mathbf{x}'$ stands for $dx_1'\, dx_2'\, dx_3'$. With the velocity $\mathbf{v}$ related to Ω through (1.1.11) and (1.1.13), (1.1.3) becomes an integro-differential equation for Ω subject to the initial condition (1.1.10).

Numerical schemes for the finite difference solution of this initial-value problem will be presented in **Chapter 3** . We shall see how to improve the efficiency of a numerical scheme by reducing the size of the computational domain without loss of accuracy. This is achieved by employing the far field description of $\mathbf{A}$ to increase the accuracy of the boundary data. Such a far field description is given by the series expansion of the Poisson integral (1.1.13) in powers of r^{-1}, which reads

$$\mathbf{A}(t,\mathbf{x}) = \sum_{n=0}^{m} \mathbf{A}^{(n)}(t,\mathbf{x}) + O(r^{-m-2}) , \tag{1.1.14}$$

where

$$\mathbf{A}^{(n)}(t,\mathbf{x}) = \frac{1}{4\pi\ r^{n+1}} \int\int\int_{-\infty}^{\infty} \Omega(t,\mathbf{x}')[(r')^n P_n(\hat{\mathbf{x}}\cdot\hat{\mathbf{x}}')] d^3\mathbf{x}' .$$

Equation (1.1.14) says that if $\mathbf{A}$ is approximated by the $(m+1)$th partial sum of the series the error is order r^{-m-2}. As we shall see later, the number m has an upper bound $N-3$, i. e., $m \leq N-3$, because of the far field condition (2) on Ω. In the integral representation for $\mathbf{A}^{(n)}$, $\hat{\mathbf{x}}$ and $\hat{\mathbf{x}}'$ denote unit vectors in the directions of $\mathbf{x}$ and $\mathbf{x}'$ respectively, P_n is a Legendre polynomial and, as a consequence, $(r')^n P_n$ is a homogeneous polynomial in x_i' of degree n. Likewise $r^n P_n$ and hence $r^{2n+1}\mathbf{A}^{(n)}$ are homogeneous polynomials in x_i of degree n. The coefficients of the latter polynomial are linear combinations of nth moments of vorticity. The first three terms of (1.1.14) are

$$\mathbf{A}^{(0)} = \frac{1}{4\pi} < \Omega > (\frac{1}{r}) , \tag{1.1.15}$$

$$\mathbf{A}^{(1)} = \frac{1}{4\pi} \sum_{j=1}^{3} < x_j \Omega > \partial_j (\frac{-1}{r}) , \tag{1.1.16}$$

$$\mathbf{A}^{(2)} = \frac{1}{8\pi} \sum_{j,k=1}^{3} < x_j x_k \Omega > \partial_j \partial_k (\frac{1}{r}) , \tag{1.1.17}$$

where $< >$ denotes the volume integral over the entire space and ∂_j denotes the partial derivative with respect to x_j.

In general, an nth moment of the kth component of vorticity is defined by,

$$< \omega_k \prod_{i=1}^{3} x_i^{j_i} > \qquad \text{with} \quad k = 1,2,3, \quad j_i \geq 0 \quad \text{and} \quad \sum_{i=1}^{3} j_i = n \,. \tag{1.1.18}$$

All the $\frac{3}{2}(n+2)(n+1)$ nth moments exist for any $n \leq m \leq N-3$, on account of the far field condition (2) on Ω.

From (1.1.14), we see that the far field description of $\mathbf{A}$ to the order r^{-m-1} is defined by the nth moments of vorticity for $n \leq m$. The moments of vorticity in general are time dependent. On the other hand, it is known that these moments are not linearly independent and that some of them are time invariant so long as the initial data Υ is divergence free and decays sufficiently rapidly. These consistency and time invariance conditions are the topics of the following section.

1.2 Consistency Conditions and Time Invariance of Moments of Vorticity

Using the condition that Ω is divergence free, (1.1.5), and that it decays rapidly, (2), Truesdell (1951, 1954) showed that $\frac{1}{2}(n+3)(n+2)$ linear combinations of nth moments of Ω must vanish for $t \geq 0$. These conditions, called consistency conditions, are valid for any vector function that is divergence free and decays rapidly in r, but does not have to be associated with a flow field. Using the fact that the vorticity of flow fields has to satisfy the vorticity evolution equation (1.1.3), Moreau (1948, 1949) obtained six additional time invariant combinations of the moments, namely three for the first moments and three for the second moments. Since these consistency conditions and time invariants are usually not available in textbooks of fluid dynamics but will be needed in **Chapter 1** and **3**, we rederive them in **Sec.1.2.1** and **Sec.1.2.2** respectively. The results for the general three-dimensional case are then reduced to the cases of axisymmetric flow in **Sec.1.2.3** and to the two-dimensional case with appropriate modifications in **Sec.1.2.4**.

1.2.1 Derivation of the Consistency Conditions

To derive Truesdell's consistency conditions on the $\frac{3}{2}(n+2)(n+1)$ nth moments we will first show that an nth coaxial moment of Ω along an axis parallel to any vector $\mathbf{b}$ should vanish:

$$I^{(n)}(t, b_1, b_2, b_3) = \int\int\int_{-\infty}^{\infty} (\mathbf{x} \cdot \mathbf{b})^n \; \Omega \cdot \mathbf{b} \; d^3\mathbf{x} = 0 \,, \tag{1.2.1}$$

for $t \geq 0$, $n = 0, 1, 2, \ldots$, and for all b_i, $i = 1, 2, 3$, which are the components of $\mathbf{b}$.

A simple proof of (1.2.1) follows from integration by parts and by using the far field conditions (2) and (1.1.5):

$$< (\mathbf{x} \cdot \mathbf{b})^n \, \Omega \cdot \mathbf{b} > = \frac{1}{n+1} < (\mathbf{x} \cdot \mathbf{b})^{n+1} \, \nabla \cdot \Omega > = 0 \,. \tag{1.2.2}$$

Since $I^{(n)}$ is a homogeneous polynomial in b_i of degree $n+1$ and the components b_i are arbitrary, (1.2.1) holds if and only if all the coefficients in the polynomial are equal to zero. There are $\frac{1}{2}(n+3)(n+2)$ coefficients, which are linearly independent combinations of nth moments of vorticity. They are proportional to the left-hand side of the following equation,

$$i < x_1^{i-1} x_2^j x_3^k \omega_1 > + j < x_1^i x_2^{j-1} x_3^k \omega_2 > + k < x_1^i x_2^j x_3^{k-1} \omega_3 > = 0 \,, \tag{1.2.3}$$

for $i + j + k = n + 1$ and $i, j, k \geq 0$. Consequently, the number of linearly independent combinations of the nth moments is

$$J(n) = \frac{3}{2}(n+2)(n+1) - \frac{1}{2}(n+3)(n+2) = n(n+2) \,. \tag{1.2.4}$$

In particular for $n = 0$, we have $J(0) = 0$ and three consistency conditions,

$$< \omega_i > = 0, \quad i = 1, 2, 3, \qquad i.e., \quad < \Omega > = \mathbf{0} \,. \tag{1.2.5}$$

For $n = 1$, we have $J(1) = 3$ and six consistency conditions,

$$< x_i \omega_j > + < x_j \omega_i > = 0, \qquad (i, j = 1, 2, 3 : \; j \geq i) \,. \tag{1.2.6}$$

It says that the 3×3 matrix $< x_i \omega_j >$ is skew symmetric and only three combinations of the first moments, which are the three components of $< \mathbf{x} \times \Omega >$, remain to be defined by the initial data and through the evolution of Ω in time. For $n = 2$, the following ten linear combinations of the second moments vanish,

$$< {x_i}^2 \Omega > + 2 < x_i \omega_i \mathbf{x} > \; = 0 \,, \quad (i = 1, 2, 3) \tag{1.2.7a}$$

and

$$< x_1 x_2 \omega_3 > + < x_2 x_3 \omega_1 > + < x_3 x_1 \omega_2 > \; = 0 \,. \tag{1.2.7b}$$

and eight linear combinations remain to be defined.

As a result of (1.2.5), we get from (1.1.15) that $\mathbf{A}^{(0)} \equiv 0$ and then from (1.1.14) the following:

$$\mathbf{A}(t, \mathbf{x}) = \mathbf{A}^{(1)}(t, \mathbf{x}) + O(r^{-3}) \tag{1.2.8}$$

and

$$|\mathbf{v}| = |\nabla \times \mathbf{A}| = O(r^{-3}) \qquad \text{as} \quad r \to \infty \,. \tag{1.2.9}$$

1.2.2 Time Invariance and the Far-field Behavior of the Vector Potential

To derive temporal relationships for the moments of vorticity, we rewrite the vorticity evolution equation (1.1.3) as

$$\Omega_t = -\nabla \times [\nabla \cdot (\mathbf{v} \, \mathbf{v})] + \nu \Delta \Omega \,. \tag{1.2.10}$$

Note that each term on the right-hand side of the equation is differentiated twice with respect to the space variables.

We multiply (1.2.10) by x_i, $i = 1, 2, 3$ and integrate both sides over the entire space. The left side becomes the rate of change of the corresponding first moment. We show that each term on the right-hand side of the equation vanishes by carrying out integration by parts twice and making use of the far-field behavior of Ω and $\mathbf{v}$, (2) and (1.2.9). The result is that the first moments of vorticity are time invariant,

$$< x_i \, \Omega >_t = 0 \, , \qquad i = 1, 2, 3 \, . \tag{1.2.11}$$

Combining this with the consistency conditions (1.2.6), we see that there are only three non-trivial linear combinations of the first moments, identified as the three components of the vector $< \mathbf{x} \times \Omega >$, and this vector is time invariant, i. e.,

$$< \mathbf{x} \times \Omega > = < \mathbf{x} \times \Upsilon > = \mathbf{E} \, , \qquad \text{for} \quad t \geq 0 \, , \tag{1.2.12}$$

where $\mathbf{E}$ denotes the constant vector specified by the initial data.

To evaluate the rate of change of second moments, we multiply (1.2.10) by ${x_i}^{n_i} {x_j}^{n_j}$ with n_i and $n_j \geq 0$ and $n_i + n_j = 2$. After integration by parts twice, the right-hand side of the equation contains integrals of products of velocity components, which are unknown prior to the solution of the initial value problem. We seek linear combinations of rates of change of second moments so as to eliminate those unknown terms. In addition to the ten consistency conditions (1.2.7a, b), we obtain three time invariants,

$$< r^2 \, \Omega(\mathbf{x}, t) > = < r^2 \, \Upsilon(\mathbf{x}) > = \mathbf{D} \, . \tag{1.2.13}$$

This equation says that the polar moment of vorticity with respect to the origin is time invariant. No additional invariants for $n \geq 3$ have been found from integrations of the vorticity evolution equation (Howard 1957).

The remaining five linearly independent combinations of second moments can be chosen as:

$$F_i(t) = < \omega_i ({x_j}^2 - {x_k}^2) > \tag{1.2.14a}$$

and

$$H_i(t) = < 2\omega_i x_j x_k - \omega_j x_k x_i - \omega_k x_i x_j >, \tag{1.2.14b}$$

where $i = 1, 2, 3$ and i, j, k are in cyclic order. Here the H_i obey the constraints

$$H_1 + H_2 + H_3 \equiv 0 \tag{1.2.15}$$

due to the consistency condition (1.2.7b).

The above consistency conditions and time invariants, (1.2.5–7) and (1.2.11–14), imply that $\mathbf{A}^{(0)}$, the first term in the far field expansion (1.1.14) of $\mathbf{A}$, vanishes and that the next two terms become

$$\mathbf{A}^{(1)}(\mathbf{x}) = - \, \frac{1}{8\pi} \mathbf{E} \times \nabla \Big[\frac{1}{r}\Big] \tag{1.2.16}$$

and

$$\mathbf{A}^{(2)}(t,\mathbf{x}) = -\frac{1}{16\pi}\nabla(\mathbf{D}\cdot\nabla)\left[\frac{1}{r}\right] + \frac{1}{16\pi}\sum_{i=1}^{3} F_i(t)[\hat{\imath}\,(\partial_j^2-\partial_k^2)-\hat{\jmath}\,\partial_j\partial_i+\hat{k}\,\partial_k\partial_i]\left[\frac{1}{r}\right] + \frac{1}{12\pi}\sum_{i=1}^{3}\hat{\imath}\,H_i(t)\,\partial_j\partial_k\left[\frac{1}{r}\right]. \tag{1.2.17}$$

Here $\hat{\imath}$ represents the unit vector along the ith axis and i, j, k are in cyclic order. We observe that the first term in (1.2.17) is curl free and therefore the constant second moment $\mathbf{D}$ does not contribute to the far-field velocity.

In **Sec. 1.3**, we will show that the far field velocity $\mathbf{v}^{(n)}$ induced by $\mathbf{A}^{(n)}$ is equivalent to the gradient of a potential $\Phi^{(n)}$, i. e., $\mathbf{v}^{(n)} = \nabla\times\mathbf{A}^{(n)} = \nabla\Phi^{(n)}$. Here we point out the equivalence for $n=1$ in order to identify the physical meaning of the constant vector $\mathbf{E}$. From (1.2.16), we have

$$\mathbf{v}^{(1)} = \nabla\times\mathbf{A}^{(1)} = \frac{\mathbf{E}}{2}\cdot\nabla\left[\frac{-1}{4\pi r}\right]. \tag{1.2.18}$$

Therefore, $\mathbf{v}^{(1)}$ represents the velocity of a doublet with strength $|\mathbf{E}|/2$ located at the origin and oriented in the direction of $\mathbf{E}$. In short, we call $\mathbf{E}/2$ the doublet vector.

To conclude this subsection, we mention several identities in Lamb (1932) which we need later in **Sec. 1.4** and in **Chapter 3**. Those identities relate the rate of change of total kinetic energy, $\frac{1}{2} < \mathbf{v}\cdot\mathbf{v} >_t$, the total rate of energy dissipation, $< \Theta >$, and the integral of the scalar product of vorticity and vector potential $< \Omega\cdot\mathbf{A} >$. They are :

$$\frac{1}{2} < \mathbf{v}\cdot\mathbf{v} >_t = \frac{1}{2} < \Omega\cdot\mathbf{A} >_t = - < \Theta >, \tag{1.2.19}$$

where Θ is the dissipation function,

$$\Theta = \frac{\nu}{2}\sum_{j,k=1}^{3}\left(\frac{\partial v_j}{\partial x_k}+\frac{\partial v_k}{\partial x_j}\right)^2.$$

The results presented in this subsection for the general case of a three-dimensional vortical flow are reduced to the special cases of axisymmetric and two-dimensional flows in the following two subsections respectively.

1.2.3 Axisymmetric Flow

In this subsection we consider axisymmetric flows with zero circumferential velocity. Let the axis of symmetry be the x_3-axis and $\hat{\imath}, \hat{\jmath}$ and $\hat{k}$ denote the basic unit vectors. All the vortex lines are coaxial circles and the vorticity vector can be expressed in terms of its circumferential component ϖ,

$$\Omega = \hat{\theta}\;\varpi(t,\sigma,x_3) = [-\sin\theta\,\hat{\imath}+\cos\theta\,\hat{\jmath}]\varpi(t,\sigma,x_3), \tag{1.2.20}$$

where σ, θ, x_3 denote the cylindrical coordinates: the radial, circumferential and axial coordinates. For Ω to decay rapidly we require ϖ to decay rapidly in $r = \sqrt{\sigma^2 + x_3^2}$. It is clear that Ω is divergence-free and it follows that all the consistency conditions are fulfilled automatically.

For such an axisymmetric vortical field, we define its total strength or circulation by

$$\Gamma(t) = \int_{-\infty}^{\infty} v_3(t, 0, x_3) dx_3 = \int_{-\infty}^{\infty} \int_0^{\infty} \varpi(t, \sigma, x_3) \, d\sigma \, dx_3 \, . \tag{1.2.21}$$

From the vorticity evolution equation in cylindrical coordinates, and from the condition of axial symmetry, implying $\varpi = 0$ along the x_3-axis, we get

$$\Gamma'(t) = -2\nu \int_{-\infty}^{\infty} \varpi_\sigma(t, 0, x_3) \, dx_3 \, . \tag{1.2.22}$$

In case that ϖ is of the same sign everywhere, say $\varpi \geq 0$ and hence $\Gamma > 0$ and $\varpi_\sigma(t, 0, x_3) \geq 0$, we conclude from (1.2.22) that $\Gamma'(t) \leq 0$, saying that the total strength will decrease.

Among the three time invariants of first moments (1.2.12) the only non-trivial one is the axial component of $< \mathbf{x} \times \Omega >$,

$$\begin{aligned} < x_1\omega_2 - x_2\omega_1 > = 2 < x_1\omega_2 > &= 2 \int_{-\infty}^{\infty} \int_0^{\infty} \pi\sigma^2 \, \varpi(\sigma, x_3, t) \, d\sigma \, dx_3 \\ &= E_3 = \text{const.} \end{aligned} \tag{1.2.23}$$

From (1.2.18) and (1.2.23), we conclude that the axisymmetric vortical flow behaves in the far field like a doublet located at the origin with constant strength $E_3/2$ and oriented along the x_3-axis. This result can be obtained directly by using the equivalence between the flow field induced by a slender circular vortex ring and by a jump of velocity potential across the circular disk bounded by the ring† (See Lamb 1932, pp. 237 -239, in particular, the first two equations for the velocity potential and stream function in p. 239). Consider an elementary circular vortex ring of cross-sectional area $d\sigma \, dx_3$ lying in the plane $x_3 = c$, with radius $\sigma = R$ and strength $d\Gamma = \varpi d\sigma dx_3$. This elementary vortex ring induces a constant jump of velocity potential, $d\Phi$, by the amount $-d\Gamma$ across the circular disk, $\sigma < R, x_3 = c$. Since the induced potential $d\Phi$ is axisymmetric and an odd function of $\bar{x}_3 = x_3 - c$, we find the values of $d\Phi$ on the two faces of the disk to be $d\Phi(x_3 = c^+) = -d\Phi(x_3 = c^-) = \frac{1}{2} d\Gamma$ for $\sigma < R$. Note that the jump of potential across a surface is equivalent to a doublet oriented normal to the surface. Consequently the axisymmetric vorticity distribution is equivalent to an axisymmetric doublet

† The velocity potential here is discontinuous across the disc and hence can not be a part of the background potential mentioned in **Sec.1.1** because the latter is assumed to be regular in the entire flow field.

distribution oriented along the x_3-axis and (1.2.23) expresses the conservation of the total doublet strength, $E_3/2$.

In order to illustrate the meaning of (1.2.22) and (1.2.23), we consider a vortex ring originated at $t = 0$ with zero cross-sectional area, ring radius R_0 and total strength or circulation $\Gamma(0)$. As time increases, the total strength $\Gamma(t)$ decreases, the effective core size increases due to diffusion and eventually becomes comparable to the ring radius and then the distinctive structure of a vortex ring disappears. On the other hand the total strength of the doublet distribution maintains the constant value $E_3\,/2 = \pi R_0^2 \Gamma(0)$.

Due to axisymmetry and the absence of circumferential or swirling flow, only two combinations of the eighteen second moments of vorticity (1.2.20), say H_1 and H_2, are non-trivial. They are related to an integral,

$$\begin{aligned} < x_3 x_1 \omega_2 > &= \frac{1}{3} H_2(t) = - < x_2 x_3 \omega_1 > \\ &= -\frac{1}{3} H_1(t) = \int_{-\infty}^{\infty} x_3 \Big[\int_0^{\infty} \pi \sigma^2 \varpi \, d\sigma \Big] \, dx_3 \; . \end{aligned} \tag{1.2.24}$$

The term inside the square brackets denotes the area integral of the equivalent doublet distribution in the constant x_3 plane. Therefore, the right-hand side of (1.2.24) represents the first moment of the doublet distribution with respect to x_3.

Using the two preceding equations for the moments, we obtain the far-field vector potential for an axisymmetric flow field:

$$\mathbf{A} = \hat{\theta} \sigma^{-1} \psi(t, \sigma, x_3) \; , \tag{1.2.25a}$$

with

$$\psi = \frac{\sigma^2}{4\pi} \Big[\frac{E_3}{2} \, r^{-2} + H_2(t) \, \frac{x_3}{r} \, r^{-3} + O(r^{-4}) \Big] \; .$$

Here ψ can be identified as the stream function for the axisymmetric non-swirling flow, which defines the velocity through

$$\sigma \, \mathbf{v} = -(\partial_3 \psi) \hat{\sigma} + \big[\psi / \sigma + \partial_\sigma \psi \big] \hat{k} \; ,$$

where $\hat{\sigma} = \cos\theta \, \hat{\imath} + \sin\theta \, \hat{\jmath}$. Note that the second term inside the square bracket on the right-hand side of (1.2.25a) can be absorbed into the first term by moving the coordinate system along the x_3-axis with the origin located at $x_3 = X_3(t)$ so that (1.2.25a) becomes

$$\begin{aligned} \psi &= \frac{\sigma^2}{4\pi} \Big[\frac{E_3}{2} \, (r')^{-2} + O((r')^{-4}) \Big] \; , \qquad \text{where} \\ r' &= \sqrt{(x_3 - X_3)^2 + \sigma^2} \quad \text{and} \quad X_3(t) = H_2(t)/E_3 \; . \end{aligned} \tag{1.2.25b}$$

Thus we could locate the instantaneous center of the axisymmetric vorticity field at the point $\mathbf{X}(t) = (0, 0, X_3(t))$ so that the far field behaves as a doublet of constant strength located at $\mathbf{X}$ without the quadrupole term.

1.2.4 Two-dimensional flow

For a two-dimensional flow in the x_1x_2 plane, we use (σ, θ) to denote polar coordinates and define the far-field by $\sigma \gg 1$. Both the vorticity and the vector potential are normal to the plane :

$$\Omega = \omega_3(t, x_1, x_2)\, \hat{\mathbf{k}} \qquad \text{and} \qquad \mathbf{A} = \psi(t, x_1, x_2)\, \hat{\mathbf{k}}\ . \tag{1.2.26}$$

The stream function ψ obeys the two-dimensional Poisson equation $\Delta\psi = -\omega_3$. With the vorticity decaying rapidly in σ, we obtain the far-field behavior of $\mathbf{v}$ from the Poisson integral for the stream function:

$$|\mathbf{v}| = O(\sigma^{-1})\ , \qquad \text{for} \quad \sigma \gg 1\ . \tag{1.2.27}$$

When the vorticity field (1.2.26) in two spatial variables is considered as a three-dimensional field, it does not decay rapidly in $r = |\mathbf{x}|$ because it is independent of x_3. Consequently, the results obtained in **Sec. 1.2.1** and **1.2.2** for the three-dimensional field decaying rapidly in r are not applicable here. Since Ω given by (1.2.27) is divergence free, we seek only time invariance of the moments of vorticity in the x_1x_2 plane. The invariance of the total strength and the first moments as well as a decay law for the polar moment were first derived by Poincaré (1898) directly from the two-dimensional N-S equations. They are

$$< \omega_3 > = \Gamma\ , \quad < x_2\omega_3 > = C_1\ , \quad < x_1\omega_3 > = C_2 \tag{1.2.28a}$$

and

$$< \sigma^2\omega_3 > = 4\nu\Gamma t + D_3\ , \tag{1.2.28b}$$

where Γ, C_1, C_2 and D_3 are constants defined by the initial data and $< >$ denotes double integration over the x_1x_2 plane. The preceding equations (1.2.28a,b) say that

- *If the strength, first moments and polar moment of two vorticity distributions are equal initially, they remain equal for all* $t \geq 0$.

This result will be employed in **Sec.2.2** to give physical meaning to the "optimum one term similarity solution" and in **Sec.3.2** to formulate rules for the merging of two-dimensional vortices to a single approximate one.

Besides the polar moment, we choose the other two time dependent linear combinations of the second moments to be:

$$F_3(t) = < (x_1^2 - x_2^2)\omega_3 > \quad \text{and} \quad H_3(t) = < 2x_1x_2\omega_3 >\ . \tag{1.2.29}$$

We then arrive at the following far-field behavior of the stream function:

$$\begin{aligned}\psi(t, \sigma, \theta) = &-\frac{\Gamma}{2\pi}\ln\sigma + \frac{1}{2\pi\sigma}\,[C_1\cos\theta + C_2\sin\theta] \\ &+ \frac{1}{4\pi\sigma^2}\,[F_3(t)\cos 2\theta + H_3(t)\sin 2\theta] + O(\sigma^{-3})\ .\end{aligned} \tag{1.2.30}$$

When $\Gamma \neq 0$ we can use (1.2.28) to show that the center of gravity of the vorticity distribution, $\mathbf{X} = [C_1\hat{\imath} + C_2\hat{\jmath}]/\Gamma$, is stationary. We can then choose the center of gravity as the origin of a new reference frame and obtain $C_1 = C_2 = 0$. In this case the far-field behavior of ψ in (1.2.30) represents that of a point vortex and two quadrupoles located at the new origin, the center of gravity of the vorticity distribution.

The far-field representations derived in this section for three-dimensional, axisymmetric and two-dimensional vortical flows will be employed in **Chapter 3** to provide approximate boundary data for numerical solutions of the N-S equations in a finite computational domain.

Following (1.2.16) and (1.2.17), we pointed out that the first term in (1.2.17) for $\mathbf{A}^{(2)}$ is curl free and hence does contribute to the far field velocity $\mathbf{v}^{(2)}$. We show via (1.2.18) that the velocity $\mathbf{v}^{(1)}$ induced by $\mathbf{A}^{(1)}$ is equivalent to that of a doublet. These properties are not coincidences. In fact they can be generalized for all $\mathbf{A}^{(n)}$. The identification of the curl free terms in $\mathbf{A}^{(n)}$ and the equivalence of the far field velocity induced by $\mathbf{A}^{(n)}$ to that of a potential flow singular at the origin are the topics of the following section.

1.3 Far Field Potential Flow Induced by a Rapidly Decaying Vorticity Distribution

We recall that in the far field the vector velocity potential, $\mathbf{A}(\mathbf{x})$,† has a power series expansion in r^{-1}, according to (1.1.14). Let us consider the n-th term, $\mathbf{A}^{(n)}$. We note that each component of the vector $r^{2n+1}\,\mathbf{A}^{(n)}$, say the kth component, is a homogeneous polynomial in x_1, x_2 and x_3 of degree n and its $(n+2)(n+1)/2$ coefficients are n-th moments of ω_k. We let $\mathcal{S}_n$ denote the set of all the $\frac{3}{2}(n+1)(n+2)$ nth moments of ω_k , $k = 1,2,3$. In **Sec. 1.2**, we employed the consistency conditions of Truesdell to identify which $\frac{1}{2}(n+3)(n+2)$ linear combinations of the nth moments have to vanish and which $J_n = n(n+2)$ nth moments can be assigned or determined from the numerical solutions of the initial value problem formulated in **Sec. 1.1.** We denote the set of those J_n nth moments by $\bar{\mathcal{S}}_n$.

From the definition of $\mathbf{A}^{(n)}$, (1.1.14), we note that $r^{n+1}\mathbf{A}^{(n)}$ is a linear combination of spherical harmonics of order n while the far field behavior (2) of Ω insures the existence of its nth moments appearing as the coefficients in $\mathbf{A}^{(n)}$. Therefore, we get $\Delta\mathbf{A}^{(n)} = 0$ for $r > 0$ and certainly for large r. Another way of arriving at this conclusion is by substituting the power series (1.1.14) into the vector Poisson equation (1.1.9) and then equating the coefficients of like powers of r^{-1} while observing (2). Similarly, we conclude $\nabla\cdot\mathbf{A}^{(n)} = 0$ either by direct evaluation or by substituting the power series (1.1.14) into to (1.1.7). We then have $\nabla\times(\nabla\times\mathbf{A}^{(n)}) = 0$ and hence the far field velocity $\mathbf{v}^{(n)}$ induced by $\mathbf{A}^{(n)}$ is irrotational and can be expressed as the gradient of a scalar velocity potential $\Phi^{(n)}(\mathbf{x})$. The above statements imply :

---

† In this section, the dependence on t shall be suppressed because we derive consistency relationships with t treated as a parameter.

$$\mathbf{v} = \sum_{n=1}^{m} \mathbf{v}^{(n)} + O(r^{-m-3}) = \sum_{n=1}^{m} \nabla\Phi^{(n)} + O(r^{-m-3}), \tag{1.3.1}$$

with

$$\mathbf{v}^{(n)} = \nabla \times \mathbf{A}^{(n)} = \nabla\Phi^{(n)}, \tag{1.3.2}$$

and

$$\Phi^{(n)} = Y_n(\theta, \phi)\, r^{-n-1}. \tag{1.3.3}$$

In (1.3.2), Y_n is a Laplace spherical harmonic of order n while θ and ϕ denote the spherical angles. Since there are only $2n+1$ linearly independent spherical harmonics of order n, only $2n+1$ coefficients can be assigned to define Y_n (Courant and Hilbert (1953) pp 511-521). Consequently, only $2n+1$ linear combinations of the nth moments in $\bar{\mathcal{S}}_n$ will contribute to the potential flow in the far field. The following questions arise: what are those $2n+1$ linear combinations and what happened to the remaining n^2-1 linear combinations of the nth moments in $\bar{\mathcal{S}}_n$? We shall outline the answers to these two questions in the following sub-sections and refer for details to Klein and Ting (1990).

In **Sec.1.3.1** we identify n^2-1 curl free terms in $\mathbf{A}^{(n)}$, associated with n^2-1 linear combinations of nth moments in $\bar{\mathcal{S}}_n$. Those terms do not contribute to the velocity in the far field. In **Sec.1.3.2** we relate the $2n+1$ coefficients in the spherical harmonics Y_n directly to linear combinations of the nth moments of vorticity.

1.3.1 Reduction of the Vector Velocity Potential in the Far-field to the Corresponding Scalar Potential

Here we shall introduce Maxwell's representation of spherical harmonics as used by Courant and Hilbert (1953), in the far field expansion (1.1.14) of the vector potential $\mathbf{A}$. We use the consistency conditions to identify the $J_n = n(n+2)$ terms in $\mathbf{A}^{(n)}$ associated with the nth moments in $\bar{\mathcal{S}}$ and show that each term corresponds to a potential flow. Each term, denoted by $\mathbf{A}_j^{(n)}$ for $j = 1, \ldots, J_n$, is a solution of Laplace's equation proportional to r^{-n-1} and will be called a divergence-free vector potential of order n. The set of these J_n vector potentials will be denoted by $\mathcal{A}^{(n)}$. Any divergence-free vector potential of order n is a linear combination of the elements of $\mathcal{A}^{(n)}$. Finally, we identify the n^2-1 vector potentials in this set which are curl free.

We represent the mth component of vector potential $\mathbf{A}^{(n)}$ by

$$A_m^{(n)} = H_m^{(n)}(\xi, \eta, \zeta)\, \frac{1}{r}, \tag{1.3.4a}$$

$$\text{with} \qquad H_m^{(n)} = \sum_{i+j+k=n} C_{i,j,k,m}^{(n)}\, \xi^i \eta^j \zeta^k. \tag{1.3.4b}$$

Here $H_m^{(n)}$ is a homogeneous polynomial of degree n and its variables, ξ, η and ζ, stand for ∂_1, ∂_2 and ∂_3 respectively. From here on we suppress the superscript (n) until it is necessary. A Taylor series expansion of the Poisson integral (1.1.13) for $|\mathbf{x}'|/r \ll 1$, shows each coefficient in (1.3.4b) to be related to an nth moment by

$$C_{i,j,k,m} = \frac{(-1)^n}{4\pi\, i!\, j!\, k!} < x_1^i\, x_2^j\, x_3^k\, \omega_m > \qquad m = 1,2,3 \quad \text{and} \quad i+j+k = n \quad (1.3.5)$$

so that the consistency conditions (1.2.3) yield $\frac{1}{2}(n+3)(n+2)$ constraints on the coefficients $C_{i,j,k,m}$:

$$C_{i-1,j,k,1} + C_{i,j-1,k,2} + C_{i,j,k-1,3} = 0 \qquad \text{for} \qquad i+j+k = n+1, \qquad (1.3.6)$$

with the understanding that a coefficient is equal to zero when one of its subscripts is negative.

Consider first the three cases that either i, j or k is equal to $n+1$ in (1.3.6) while the remaining two are equal to zero. Then (1.3.6) reduces to

$$C_{n,0,0,1} = 0, \qquad C_{0,n,0,2} = 0 \qquad \text{and} \quad C_{0,0,n,3} = 0. \qquad (1.3.7)$$

These three equations are equivalent to the condition that the nth coaxial moments along the three axes are equal to zero, i.e., $I(t, \mathbf{b}) \equiv 0$ for $\mathbf{b} = \hat{h}_l$, where $\hat{h}_l$, $l = 1,2,3$ stand for the three basic unit vectors.. Had we set $n = 0$, we would have recovered the conditions $C_{(0,0,0,m)} = 0$, $m = 1,2,3$, showing that, indeed, the series in (1.3.2) should begin with $n = 1$.

Next we consider the cases with only one of the coefficients in (1.3.6) equal to zero, say the first one, so that $i = 0$ and $n \geq j, k \geq 1$. Equation (1.3.6) then relates the second coefficient to the third and the corresponding two terms in (1.3.4b) combine to

$$\mathbf{U}_{01} = C_{0,j,k-1,3}\, [-\hat{h}_2\, \zeta + \hat{h}_3\, \eta]\, \eta^{j-1} \zeta^{k-1}\, \frac{1}{r}, \qquad (1.3.8a)$$

with $\quad j = 1, \ldots, n, \quad k = n+1-j.$

Similarly, we find n vector potentials for $j = 0$ and n more for $k = 0$. They are

$$\mathbf{U}_{02} = C_{i,0,k-1,3}\, [-\hat{h}_1\, \zeta + \hat{h}_3\, \xi]\, \xi^{i-1} \zeta^{k-1}\, \frac{1}{r}, \qquad (1.3.8.b)$$

with $\quad i = 1, \ldots, n, \quad k = n+1-i,$ and

$$\mathbf{U}_{03} = C_{i,j-1,0,2}\, [-\hat{h}_1\, \eta + \hat{h}_2\, \xi]\, \xi^{i-1} \eta^{j-1}\, \frac{1}{r}, \qquad (1.3.8c)$$

with $\quad i = 1, \ldots, n, \quad j = n+1-i.$

Finally, we consider the cases with $1 \leq i, j, k \leq n-1$, when neither one of the three coefficients in (1.3.6) vanishes. We can then express the first one in terms of the second and the third. The three terms in (1.3.4b) associated with the three coefficients in (1.3.6) combine to two terms. They are

$$\mathbf{U}_2 = C_{i,j,k-1,3}\, [-\hat{h}_1\, \zeta + \hat{h}_3\, \xi]\, \xi^{i-1} \eta^j \zeta^{k-1}\, \frac{1}{r}. \qquad (1.3.9a)$$

and

$$\mathbf{U}_3 = C_{i,j-1,k,2}\, [-\hat{h}_1\, \eta + \hat{h}_2\, \xi]\, \xi^{i-1} \eta^{j-1} \zeta^k\, \frac{1}{r}, \qquad (1.3.9b)$$

for $1 \leq i, j, k \leq n-1$ and $i+j+k = n+1$.

There are $\frac{1}{2}n(n-1)$ vector potentials of the type $\mathbf{U}_2$ and the same number of the type $\mathbf{U}_3$. All together from (1.3.8) and (1.3.9), we have a set of $3n + n(n-1) = n(n+2) = J_n$ divergence free vector potentials of order n. We denote the set of those J_n vector potentials by $\mathcal{A}^{(n)}$.

Had we expressed the second coefficient in (1.3.6) in terms of the third and the first we would have obtained vector potentials which are linear combinations of those in (1.3.9a,b). For later reference, we write down the combination of the vector potentials associated with the second and the third coefficients,

$$\mathbf{U}_1 = C_{i,j,k-1,3}\,[-\hat{h}_2\zeta + \hat{h}_3\eta]\;\xi^i\eta^{j-1}\zeta^{k-1}\,\frac{1}{r}, \tag{1.3.9c}$$

for $1 \le i,j,k \le n-1$ and $i+j+k = n+1$.

Now we are ready to show that the flow field defined by each vector potential in $\mathcal{A}^{(n)}$ is a potential flow. We see that the velocity corresponding to a vector potential $\mathbf{U}_3$ is

$$\begin{aligned}\nabla\times\mathbf{U}_3 &= C_{i,j-1,k,2}\;[\hat{h}_3\,(\xi^2+\eta^2) - \hat{h}_1\,\xi\zeta - \hat{h}_2\,\eta\zeta]\;\xi^{i-1}\eta^{j-1}\zeta^k\,\frac{1}{r}\\ &= \nabla\varphi_3,\end{aligned} \tag{1.3.10a}$$

where

$$\varphi_3 = -C_{i,j-1,k,2}\;\xi^{i-1}\eta^{j-1}\zeta^{k+1}\,\frac{1}{r}. \tag{1.3.10b}$$

for $\quad n \ge i \ge 1,\; n-1 \ge k \ge 0,\; j = n+1-i-k \ge 1.$

Here we have absorbed $\mathbf{U}_{03}$ into $\mathbf{U}_3$ by allowing $k=0$ and $i=n$. Similarly we have,

$$\nabla\times\mathbf{U}_2 = \nabla\varphi_2 \qquad \text{with} \qquad \varphi_2 = C_{i,j,k-1,3}\;\xi^{i-1}\eta^{j+1}\zeta^{k-1}\,\frac{1}{r}, \tag{1.3.11}$$

for $n \ge i \ge 1,\; n-1 \ge j \ge 0,\; k = n+1-i-j \ge 1$. For the remaining vector potentials $\mathbf{U}_{01}$, we have

$$\nabla\times\mathbf{U}_{01} = \nabla\varphi_1 \qquad \text{with} \qquad \varphi_1 = -C_{0,j,k-1,3}\;\xi\eta^{j-1}\zeta^{k-1}\,\frac{1}{r}, \tag{1.3.12}$$

for $n \ge j \ge 1,\; k = n+1-j \ge 1$. Thus, each vector potential in $\mathcal{A}$ defines an irrotational velocity. Let the set of these J_n velocities be denoted by $\mathcal{V}^{(n)}$. Since there are only $2n+1$ linearly independent potential functions of the order r^{-n-1}, there can be only $2n+1$ non-trivial linear combinations of the velocities in $\mathcal{V}^{(n)}$. We should be able to find n^2-1 linear combinations of the velocities in $\mathcal{V}^{(n)}$, or combinations of the curl of the vector potentials in $\mathcal{A}^{(n)}$, which are equal to zero. In fact, we can construct them as pairwise sums of appropriate vectors in the sets of $\mathbf{U}_{01}\ldots\mathbf{U}_3$:

We find $(n-1)(n-2)/2$ such combinations of vectors $\mathbf{U}_2$ and $\mathbf{U}_3$. They are

$$\mathbf{L}_{23} = \frac{C_{i,j,k,3} + C_{i,j+1,k-1,2}}{2}[-\hat{h}_1(\eta^2+\zeta^2) + \hat{h}_2\,\xi\eta + \hat{h}_3\,\xi\zeta]\xi^{i-1}\eta^j\zeta^{k-1}\,\frac{1}{r}$$
$$= \frac{C_{i,j,k,3} + C_{i,j+1,k-1,2}}{2}\,\nabla\{\,\xi^i\eta^j\zeta^{k-1}\,\frac{1}{r}\,\}, \qquad (1.3.13a)$$

for $1 \le i,j,k \le n-2$ and $i+j+k = n$. The lower bounds for i,j and k are chosen such that they are consistent with those for both $\mathbf{U}$'s. We include the coefficients in the combinations in order to identify the corresponding vector potentials in $\mathcal{A}^{(n)}$. It is obvious that $\mathbf{L}_{23}$ is not only curl-free but also divergence-free since $[\xi^2+\eta^2+\zeta^2](1/r) = 0$. Similarly, the linear combinations of $\mathbf{U}_1$ and $\mathbf{U}_3$ yield

$$\mathbf{L}_{13} = -\,\frac{C_{i,j,k,3} - C_{i+2,j-1,k-1,2}}{2}\,\nabla\{\,\xi^i\eta^j\zeta^{k-1}\,\frac{1}{r}\,\}, \qquad (1.3.13b)$$

for $1 \le i,j,k \le n-2$ and $i+j+k = n$.

We recall that only two of the three sets of vector potentials, $\mathbf{U}_1, \mathbf{U}_2$ and $\mathbf{U}_3$ are linearly independent, so that combinations of the $\mathbf{U}_1$ and $\mathbf{U}_2$ will not provide additional independent vectors. However, from the linear combinations of $\mathbf{U}_{01}$ and $\mathbf{U}_2$ and also $\mathbf{U}_{02}$, we get further $(n-1)$ gradients of scalar potentials

$$\mathbf{L}_{012} = -\,\frac{C_{0,j+2,k,3} + C_{2,j,k,3}}{2}\,\nabla\{\,\eta^j\zeta^{k+1}\,\frac{1}{r}\,\}, \qquad (1.3.14a)$$

for $0 \le j,k \le n-2$, $j+k = n-2$. Likewise, we obtain $2(n-1)$ additional ones,

$$\mathbf{L}_{023} = \frac{C_{i+1,0,k+1,3} + C_{i+1,1,k,2}}{2}\,\nabla\{\,\xi^{i+1}\zeta^k\,\frac{1}{r}\,\}, \qquad (1.3.14b)$$

for $0 \le i,k \le n-2$, $i+k = n-2$ and

$$\mathbf{L}_{031} = -\,\frac{C_{i+2,j,0,2} - C_{i,j+1,1,3}}{2}\,\nabla\{\,\xi^i\eta^{j+1}\,\frac{1}{r}\,\}, \qquad (1.3.14c)$$

for $0 \le i,j \le n-2$, $i+j = n-2$. Together (1.3.13a,b) and (1.3.14a, b, c) define the desired n^2-1 curl-free linear combinations.

An alternate procedure of finding these n^2-1 linear combinations is to utilize the following facts: i) there are $J_{n-1} = n^2-1$ independent vector potentials of the order $n-1$, $\mathbf{A}_j^{(n-1)}$, in $\mathcal{A}^{(n-1)}$, ii) $\nabla \times \mathbf{A}_j^{(n-1)} = \nabla\varphi_j$ is divergence-free and curl-free and iii) $\nabla \times \mathbf{A}_j^{n-1}$ is a vector potential of order n. Hence it represents a linear combination of the J_n vector potentials in $\mathcal{A}^{(n)}$ and such a combination is curl free. Thus we find the n^2-1 combinations corresponding to $j = 1, \ldots, n^2-1$. For details refer to Klein and Ting (1990).

In this subsection, we identified n^2-1 curl free terms in $\mathbf{A}^{(n)}$, which do not contribute to the far field velocity $\mathbf{v}^{(n)}$. In principle, we can find the remaining $2n+1$ terms in $\mathbf{A}^{(n)}$ and the corresponding $2n+1$ linear combinations of nth moments and then define $\mathbf{v}^{(n)}$. It is easier to define it directly from $\mathbf{A}^{(n)}$ by making use of (1.3.2) and (1.3.3), which say that $\mathbf{v}^{(n)}$ is proportional to r^{-n-2} and can be expressed as the gradient of a scalar potential of nth order. This direct method is carried out in the next subsection.

1.3.2 Relating the Coefficients in the Scalar Potentials of nth Order Directly to nth moments of Vorticity

At the beginning of **Sec. 1.3**, we conclude from (1.3.2) and (1.3.3) that the far field velocity $\mathbf{v}^{(n)}$ induced by the vector potential $\mathbf{A}^{(n)}$ can be expressed as the gradient of a scalar potential $\Phi^{(n)} = Y_n(\theta,\phi)r^{-n-1}$. We now relate the $2n+1$ coefficients of a basis of spherical harmonics Y_n directly to the nth moments of vorticity by making use of (1.3.3) and the radial component of (1.3.2). The latter says that the radial velocity is equal to $\hat{\mathbf{x}}\cdot(\nabla\times\mathbf{A}^{(n)}) = \partial\Phi^{(n)}/\partial r$. Using (1.1.14) and (1.3.3), we evaluate the left- and right-hand side of the equality. It becomes

$$\begin{aligned}\hat{\mathbf{x}}\cdot\nabla\times\mathbf{A}^{(n)} &= \frac{1}{4\pi\, r^{2n+2}}\int\!\!\int\!\!\int_{-\infty}^{\infty} P_n'(\mathbf{x}\cdot\mathbf{x}')\ \mathbf{x}\cdot[\mathbf{x}'\times\Omega(\mathbf{x}')]\ d^3\mathbf{x}' \\ &= \frac{1}{4\pi\, r^{n+2}}\int\!\!\int\!\!\int_{-\infty}^{\infty} P_n'(\mathbf{x}'\cdot\hat{\mathbf{x}})\ \hat{\mathbf{x}}\cdot[\mathbf{x}'\times\Omega(\mathbf{x}')]\ d^3\mathbf{x}' \\ &= \Phi_r^{(n)}(r,\theta,\phi) = -(n+1)Y_n(\theta,\phi)r^{-n-2},\end{aligned} \tag{1.3.15}$$

where $P_n'(u)$ stands for the derivative of $P_n(u)$. Since $\hat{\mathbf{x}} = \mathbf{x}/r$ is a function of the spherical angles θ and ϕ only, the same is true for the last volume integral in the above equation. The corresponding spherical harmonic function of nth order is

$$Y_n(\theta,\phi) = \frac{-1}{4\pi(n+1)}\int\!\!\int\!\!\int_{-\infty}^{\infty} P_n'(\mathbf{x}'\cdot\hat{\mathbf{x}})[\mathbf{x}'\times\Omega]\cdot\hat{\mathbf{x}}d^3\mathbf{x}'\ . \tag{1.3.16}$$

Since the integrand is a homogeneous polynomial of x_i' and x_i/r, $i=1,2,3$ of degree n, the $2n+1$ coefficients in Y_n are linear combinations of nth moments of vorticity. We now show the explicit results for the leading two terms including their dependence on t which was suppressed in this subsection. For $n=1$, (1.3.16) becomes

$$Y_1(\theta,\phi) = -\frac{\hat{\mathbf{x}}\cdot\mathbf{E}}{8\pi}\ , \tag{1.3.17}$$

where $\mathbf{E} = <\mathbf{x}'\times\Omega>$ is independent of time on account of condition (1.2.12). For $n=2$, we have,

$$\begin{aligned}Y_2(t,\theta,\phi) &= -\frac{1}{8\pi}\sum_{i=1}^{3}\hat{\mathbf{x}}\cdot\left[<\mathbf{x}'\times\Omega\, x_i'>\frac{x_i}{r}\right] \\ &= \frac{1}{24\pi}\sum_{i=1}^{3}\left[9F_i(t)\frac{x_jx_k}{r^2} + 2\Big(H_j(t)-H_k(t)\Big)\left(\frac{x_i}{r}\right)^2\right],\end{aligned} \tag{1.3.18}$$

with i,j,k in cyclic order. Here the three components, x_i/r, $i=1,2,3$, of the unit radial vector are functions of the spherical angles θ and ϕ. The coefficients F_i and H_i are defined by (1.2.14) as linear combinations of the second moments of vorticity and are time dependent. There are only five linearly independent combinations since $H_1+H_2+H_3\equiv 0$. In deriving the last equation in (1.3.18), we made use of the consistency condition (1.2.7). Now we shall express the first two terms of the far field potential, Φ, in terms of doublets and quadrupoles. They are

$$\Phi^{(1)}(\mathbf{x}) = \frac{1}{8\pi}\mathbf{E}\cdot\nabla\frac{1}{r} \tag{1.3.19}$$

and

$$\Phi^{(2)}(t,\mathbf{x}) = \frac{1}{8\pi}\sum_{i=1}^{3}\partial_j\partial_k\left[\frac{F_i(t)}{r}\right] + \frac{1}{36\pi}\sum_{i=1}^{3}\partial_i^2\left[\frac{H_j(t)-H_k(t)}{r}\right], \tag{1.3.20}$$

with i,j,k in cyclic order.

Additional results regarding the far field representation of the vector potential and the corresponding scalar potential can be found in Klein and Ting (1990). For example, it is shown that $\mathbf{A}^{(n)}$ can be expressed as a combination of M_n linearly independent divergence free vector potentials of order n. The number M_n for $n=1$ is equal to $J_1=3$ and for $n\geq 2$ is equal to $4n\leq J_n$. The equality only holds when $n=2$. This means that the J_n vector potentials in $\mathcal{S}(n)$ are not linearly independent for $n>2$. In the paper, derivations of the aforementioned results along two different paths are presented. The first path, which follows the one outlined in this section, goes from the motivations to the conjectures and then to their confirmations. Knowing what to prove, the authors arrive at the same results in a straighter fashion using symbolic Cartesian tensor calculus. This approach, furthermore, reveals that Truesdell's consistency conditions restrict the moment of vorticity tensors of any order to have zero symmetric part.

As mentioned before, the investigations presented in this section were partially motivated by the need of the leading terms of the far field scalar potential induced by a vortical flow with rapidly decaying vorticity field. Knowing the leading terms we shall derive the acoustic field generated by such a vortical (low Mach number) flow in the next section.

1.4 Sound Generation by an Unsteady Vortical Flow

In this section, we study the acoustic field induced by an unsteady viscous vortical flow at low Mach number. We define the reference Mach number M by

$$M = U/C \ll 1\ , \tag{1.4.1}$$

where C denotes the ambient speed of sound. In the vortical flow field scaled by length ℓ and velocity U, compressibility effects are of the order M^2. Therefore, the leading order solution of the vortical flow is an incompressible viscous flow as described in **Sec.1.1** and its far field behavior is defined by (1.3.1), (1.3.19) and (1.3.20). In the far field, the leading unsteady term of the velocity decays as $O(\ell^4/r^4)$. When r becomes much larger than ℓ by an order $O(M^{-1})$, the scaled velocity contributions decrease to $O(M^4)$. At such distances, effects of compressibility are no longer negligible. This is a typical problem amenable to matched asymptotic analysis. We consider the flow field of length scale ℓ to be the inner region, the acoustic field of length scale

$$L_a = \ell/M \tag{1.4.2}$$

to be the outer region and we use the Mach number M as a small parameter in an asymptotic expansion and for the matching of inner and outer solutions.

Before describing the analysis, we note that sound generation due to turbulence has been studied by several authors. In the pioneering work of Lighthill (1952), he found that the leading order acoustic field should be dominated by quadrupoles. Later on, Ribner (1962) pointed out that there could be an acoustic monopole due to fluid dilatation. Crow (1970) carried out an asymptotic analysis for an inviscid *isentropic* flow and concluded that the leading order acoustic field could only be composed of quadrupoles. Extension of Crow's analysis to sound generation by a vortical flow induced by a vorticity distribution of bounded support, in particular for vortex rings, was reported by Möhring (1978). Viscous effects were taken into account by Obermeier (1985). A different derivation of the results, essentially equivalent to those given by Möhring and Obermeier, was obtained by Ting and Miksis (1990) albeit the recent derivation is valid under the more general condition of a vorticity distribution which decays rapidly in r. Here we will highlight the analysis of Ting and Miksis (1990), because it is is a follow-up of the analyses in the preceding sections for the far field behavior of an incompressible viscous vortical flow.

Let the fluid be an ideal gas so that the ambient speed of sound is related to the thermodynamic state variables of the gas by

$$C^2 = \gamma P_0/\rho_0, \tag{1.4.3}$$

where γ denotes the ratio of specific heats and P_0 and ρ_0 stand for the ambient pressure and density. Generally, the gas is assumed to be at rest at infinity, such that

$$|\mathbf{v}| \to 0, \quad p \to P_0 \quad \text{and} \quad \rho \to \rho_0 \qquad \text{as} \quad r \to \infty. \tag{1.4.4}$$

In **Sec.1.4.1** we introduce expansion schemes similar to those for an inviscid turbulent flow (Crow 1970) and obtain the governing equations for the inner viscous vortical flow and for the outer acoustic field. In **Sec. 1.4.2**, we study the inner flow field and identify its leading solution with that of an incompressible vortical flow as analyzed in the preceding sections. Using the far field behavior of the vortical flow we show that the leading unsteady terms, which all contribute to the acoustic field, are five quadrupoles defined by (1.3.20). We then study the pressure variation $(p - P_0)$ and relate the total pressure fluctuation, $< p - P_0 >_t$, to the rate of energy dissipation, $< \Theta >$, where Θ is the dissipation function of the incompressible viscous flow as defined in (1.2.19). From the leading energy equation, we obtain the total density fluctuation, $< \rho >_t$, which in turn yields a measure of the global dilatation effect of the next order inner solution and defines the strength of an unsteady source , or monopole, in the far field. The source strength is two orders smaller, $O(M^2)$, than that of the quadrupoles of the leading order incompressible flow, but it decays two orders slower; $O(r^{-2})$ instead of $O(r^{-4})$. Therefore, the corresponding acoustic source and quadrupole terms are of the same order on the acoustic length scale L_a. In **Sec.1.4.3** we show that the leading solution of the acoustic field is composed of five quadrupoles and one monopole and that their strengths are related to the retarded values of the five time dependent linear combinations of the second moments of vorticity and to the retarded value of the rate of energy dissipation, respectively. These results will be applied in **Sec. 2.3** to evaluate sound generation due to slender vortex filaments after the presentation of their asymptotic description. In **Sec. 1.4.4**, we apply our results on the sound generation due to unsteady vortical flows to turbulent flows and show that the strengths of the quadrupoles are defined by the moments of the mean vorticity field while the effect of turbulence appears explicitly only in the strength of the monopole. We note that only when the rate of energy dissipation is negligible, the acoustic field will be dominated by the quadrupoles in agreement with Lighthill (1952) and Crow (1970).

1.4.1 Expansion Schemes and the Governing Equations

As mentioned in the **Introduction**, in a vortical flow field the position vector $\mathbf{x}$ and velocity $\mathbf{v}$ are scaled respectively by the reference length ℓ and velocity U. Time t and vorticity Ω are then scaled by ℓ/U and U/ℓ. For low speed flow, the pressure variation is of the order of $\rho_0 U^2$. Therefore, the pressure variation scaled by $\rho_0 C^2$ and the density variation scaled by ρ_0 are of the order $O(M^2)$. We expand the scaled variables in the vortical flow field, the inner region, in power series in M^2 as follows :

$$\begin{aligned} \mathbf{v}/U &= \mathbf{v}^{(0)} + M^2\mathbf{v}^{(1)} + \cdots, \\ \frac{p - P_0}{\rho_0 C^2} &= \qquad M^2 p^{(1)} + \cdots, \\ \frac{\rho - \rho_0}{\rho_0} &= \qquad M^2 \rho^{(1)} + \cdots. \end{aligned} \tag{1.4.5}$$

This is the well known expansion of Janson and Rayleigh (see Sears 1954 and also Majda 1984). Since we shall not carry out the matching procedure to higher order, we can only rule out the terms linear in M but not all higher odd powers of M in the expansions in (1.4.5). Therefore, we may have to admit M^3 terms in the expansions in (1.4.5) if it is necessary for the matching of higher order terms.

In the acoustic field, the outer region, we scale $\mathbf{x}$ by the reference length $L_a = \ell/M$. We define the outer space variable

$$\tilde{\mathbf{x}} = M\mathbf{x}/\ell \tag{1.4.6}$$

and add a tilde to all variables in the outer region to indicate that they are functions of $\tilde{\mathbf{x}}$. The scaled variables are now expanded in power series of M as follows,

$$\begin{aligned} \frac{\mathbf{v}}{U} &= \epsilon M^{-1}[\tilde{\mathbf{v}}^{(0)} + M\tilde{\mathbf{v}}^{(1)} + M^2\tilde{\mathbf{v}}^{(2)} \cdots], \\ \frac{p - P_0}{\rho_0 C^2} &= \epsilon[\tilde{p}^{(0)} + M\tilde{p}^{(1)} + M^2\tilde{p}^{(2)} + \cdots], \\ \frac{\rho - \rho_0}{\rho_0} &= \epsilon[\tilde{\rho}^{(0)} + M\tilde{\rho}^{(1)} + \tilde{\rho}^{(2)} + \cdots]. \end{aligned} \tag{1.4.7}$$

Later, ϵ shall be identified as M^3 in order to match with the inner solution.

After substituting the series (1.4.5) into the compressible N-S equations and equating the coefficients of like powers of M, we obtain the governing equations for the inner solution. Since we shall deal with only the first two terms in the series (1.4.5), we shall drop the superscript (0) and replace the superscript (1) by a prime, e. g., $p^{(1)}$ by p'. The leading continuity and momentum equations are the incompressible N-S equations for $\mathbf{v}$ and p'. They are :

$$\nabla \cdot \mathbf{v} = 0 \tag{1.4.8}$$

and

$$\mathbf{v}_t + \mathbf{v} \cdot \nabla \mathbf{v} = -\nabla p' + \frac{1}{Re} \Delta \mathbf{v}. \tag{1.4.9}$$

The leading energy equation becomes an equation for the density fluctuation ρ'.

$$(p' - \rho')_t + \nabla \cdot [(p' - \rho')\mathbf{v}] = \frac{1}{Re\ Pr} \Delta(\gamma p' - \rho') + (\gamma - 1)\Theta, \tag{1.4.10}$$

where Pr is the Prandtl number and Θ is the dissipation function,

$$\Theta = \frac{1}{2\,Re} \sum_{j,k=1}^{3} \left(\frac{\partial v_j}{\partial x_k} + \frac{\partial v_k}{\partial x_j}\right)^2. \tag{1.4.11}$$

The next order continuity equation shows the density variations acting as sources for the next order velocity field,

$$\nabla \cdot \mathbf{v}' = -{\rho'}_t - \nabla \cdot (\rho' \mathbf{v}). \tag{1.4.12}$$

Similarly, we obtain the governing equation for the first two terms of the outer solution (1.4.7). They are :

$$\begin{aligned} \tilde{\rho}_t^{(i)} &= -\nabla \cdot \tilde{\mathbf{v}}^{(i)}, \\ \tilde{\mathbf{v}}_t^{(i)} &= -\nabla \tilde{p}^{(i)}, \\ \tilde{p}^{(i)} &= \tilde{\rho}^{(i)}, \end{aligned} \tag{1.4.13}$$

for i = 0,1 so long as $\epsilon = o(M)$. They can be reduced to a simple wave equation for $\tilde{p}^{(i)}$ or the corresponding acoustic potential $\tilde{\Phi}^{(i)}$.

1.4.2 The Vortical Flow Field

The leading inner solution of (1.4.8) and (1.4.9) subject to the initial condition (1.1.10a) and the far field conditions of (2) and (1.1.12) is the solution of an incompressible viscous vortical flow studied in **Sec.1.2** and **Sec.1.3**. We shall summarize the relevant results derived in **Sec.1.3** on the far field behavior of the vortical field and then study the pressure and density fluctuations and the global dilatation of the inner region.

It was shown in **Sec.1.3** that the far field velocity can be expressed as the gradient of a scalar potential, $\Phi(t, \mathbf{x})$, that is

$$\mathbf{v} = \sum_{n=1,2\cdots} \nabla\Phi^{(n)} \tag{1.4.14}$$

with $\Phi^{(n)} = O(r^{-(n+1)})$. In particular, from (1.3.19) and (1.3.20) we have,

$$\Phi^{(1)}(\mathbf{x}) = \frac{1}{8\pi}\mathbf{E}\cdot[\nabla \frac{1}{r}] \tag{1.4.15a}$$

and

$$\begin{aligned}\Phi^{(2)}(t,\mathbf{x}) &= \frac{1}{8\pi}[\partial_2\partial_3 F_1 + \partial_3\partial_1 F_2 + \partial_1\partial_2 F_3][\frac{1}{r}] \\ &+ \frac{1}{36\pi}[\partial_1{}^2(H_2 - H_3) + \partial_2{}^2(H_3 - H_1) + \partial_3{}^2(H_1 - H_2)][\frac{1}{r}] .\end{aligned} \tag{1.4.15b}$$

We see from (1.4.15a) that $\Phi^{(1)}$ represents a doublet with constant strength and orientation defined by the initial first moment vector $\mathbf{E}/2$, (1.2.12). From (1.4.15b) we see that $\Phi^{(2)}$ represents quadrupoles with strengths given by the second moments of vorticity $\mathbf{F}(t)$ and $\mathbf{H}(t)$ defined by (1.2.14). Only the unsteady terms, i.e. the quadrupoles, can generate far field sound and the corresponding velocity potential is small of order $O(M^3)$ when $r = O(M^{-1})$.

We note that the next order velocity field $M^2\mathbf{v}'$ which obeys the continuity equation (1.4.12) is no longer divergence free due to the density fluctuation, $M^2\rho'_t$, which induces an acoustic source of strength $O(M^3)$. To study the effect of the density fluctuations, we express $\mathbf{v}'$ as the sum of an irrotational term and a divergence-free term, i. e.,

$$\mathbf{v}' = \nabla\phi + \nabla\times\mathbf{A}'. \tag{1.4.16}$$

Equation (1.4.12) becomes,

$$\Delta\phi = -\rho'_t - \nabla\cdot(\rho'\mathbf{v}). \tag{1.4.17}$$

Thus the effect of ρ'_t is contained in the irrotational term, $\nabla\phi$. By repeating what we did for the vector velocity potential $\mathbf{A}$ in **Sec. 1.3** we can show that the far field behavior of the divergence free flow, $\nabla\times\mathbf{A}'$, cannot have a source term and hence the acoustic potential induced by $M^2\mathbf{A}'$ will be at most $O(M^4)$, i. e., at least one order higher than that induced by the unsteady quadrupoles in $\mathbf{A}$. In **Sec. 1.4.4**, we shall show that the irrotational term in (1.4.16) does induce an acoustic source and that its strength is defined by the retarded value of the total density fluctuation which is related to the rate of energy dissipation of the incompressible flow field.

As we explained before, the far field expansion of the vector potential $\mathbf{A}$, (1.2.16–17), remains valid even if the next higher order term for $n = 3$ is included so long as the point $\mathbf{x}$ lies in the inner solution, i. e.,

$$r = o(M^{-1}) \quad \text{but} \quad r \gg 1 , \qquad \text{say} \qquad r = O(M^{-1/4}) , \tag{1.4.18}$$

The far field expansion can therefore be employed to prescribe appropriate boundary data for a numerical solution of the vector Poisson equation (1.1.9) in a finite computational domain $\mathcal{D}$ disregarding effects of compressibility, provided that its size fulfills condition (1.4.18). Thus there is no feed back from the outer acoustic field to the inner flow field. Once a numerical solution of the inner field is obtained we can evaluate the five linear combinations of second moments of vorticity and the rate of energy dissipation. The next step is to relate the rate of energy dissipation to the total pressure and density fluctuations and then to the global dilation of the vortical flow field.

Using the far field behavior of $\mathbf{v}$, we obtain from (1.4.9), $\nabla p' = -\mathbf{v}_t + O(r^{-6})$, and hence the far field behavior of p' is,

$$p'(t,\mathbf{x}) = -\Phi_t^{(2)}(t,\mathbf{x}) + O(r^{-4}) . \tag{1.4.19}$$

The leading term in p' represents quadrupoles and is $O(r^{-3})$. To determine p' in the near field, we have to solve p' from (1.4.9), which reduces to a Poisson equation,

$$\Delta p' = -\nabla \cdot [\mathbf{v} \cdot \nabla \mathbf{v}] = - \sum_{i,j=1,2,3} \partial_i \partial_j [v_i v_j] . \tag{1.4.20}$$

The solution is expressed as the sum of the Poisson integrals of the terms on the right-hand side of (1.4.20),

$$p'(t,\mathbf{x}) = \sum_{i,j=1,2,3} p'_{ij} \quad \text{and} \quad p'_{ij}(t,\mathbf{x}) = \partial_i \partial_j \, q_{ij}(t,\mathbf{x}) , \tag{1.4.21a}$$

where q_{ij} denotes the Poisson integral of the inhomogeneous term $-v_i v_j$,

$$q_{ij}(t,\mathbf{x}) = \frac{1}{4\pi} \int\int\int_{-\infty}^{\infty} \frac{v_i v_j}{|\mathbf{x} - \mathbf{x}'|} d\mathbf{x}' . \tag{1.4.21b}$$

For the later process of matching of the inner and outer solutions we need a particular notion of the volume integral of p'. Since p' is of order $O(r^{-3})$ in the far field, its volume integral over all of space is, in general, divergent. By making use of the far field behavior of p' or rather p'_{ij}, we show that as $R \to \infty$ the limit of the volume integral of p' over a sphere of radius R exists and then define the improper volume integral of p' by the limit, i. e.,

$$< p' > = \lim_{R \to \infty} \int\int\int_S p'(t,\mathbf{x}) \, d\mathbf{x} , \tag{1.4.22}$$

where $\mathcal{S}$ denotes the sphere $|\mathbf{x}| \leq R$.

Let us evaluate one of the volume integrals of the p'_{ij}'s over $\mathcal{S}$, say p'_{11}. It can be reduced to an area integral,

$$\int\int\int_S p'_{11} d\mathbf{x} = \int\int_B dx_2 dx_3 \left[\partial_1 \, q_{11}\right]_{-x_1^*}^{+x_1^*} , \tag{1.4.23}$$

where $\mathcal{B}$ denotes the circular disc $\sigma = \sqrt{x_2^2 + x_3^2} \leq R$ and $x_1^* = \sqrt{R^2 - \sigma^2}$. Note that the upper and lower point, $\mathbf{x}^{\pm} = (\pm x_1^*, x_2, x_3)$, lie on the spherical surface $\partial\mathcal{S}$, i. e., $|\mathbf{x}^{\pm}| = R$. From (1.4.21b), we obtain the leading term of q_{ij} in the far field by means of a power series expansion in $\mathbf{x}'/r$,

$$q_{ij}(t,\mathbf{x}) = \frac{< v_i v_j >}{4\pi r} + O(r^{-2}) \,.$$

Equation (1.4.23) becomes,

$$\int\!\!\int\!\!\int_S p'_{11}\, d\mathbf{x} = -\frac{< v_1 v_1 >}{2\pi} \int\!\!\int_B \frac{\sqrt{R^2 - \sigma^2}\; dx_2 dx_3}{R^3} + O(R^{-1})$$
$$= -\frac{1}{3} < v_1 v_1 > + O(R^{-1}) \,. \tag{1.4.24}$$

By repeating the steps in arriving at (1.4.24) for all of the p'_{ij}'s, we obtain the volume integral of p'

$$\int\!\!\int\!\!\int_S p'\, d\mathbf{x} = -\frac{1}{3} < \mathbf{v}\cdot\mathbf{v} > + O(R^{-1})$$

and (1.4.22) then yields,

$$< p' > = -\frac{1}{3} < |\mathbf{v}|^2 > \,. \tag{1.4.25}$$

Since the rate of decrease of the total kinetic energy is equal to the rate of energy dissipation, (1.2.19), we have

$$< p' >_t = \frac{2}{3} < \Theta > \,. \tag{1.4.26}$$

The density fluctuation, ρ'_t, is defined by the energy equation (1.4.10). In the far field, we get,

$$\rho'_t = p'_t + O(r^{-5}), \tag{1.4.27}$$

which yields the isentropic relationship when $O(r^{-5})$ terms are neglected. As we shall see later, the fluid dilatation due to viscous dissipation induces a source term in the far field whose strength depends only on the volume integral of the density fluctuation. It is obtained by evaluating the volume integrals of both sides of (1.4.10) in the sense of (1.4.22) and by making use of the far field behaviors of $\mathbf{v}$ and p' and then (1.4.26). The result is

$$< \rho' >_t = < p' >_t - [\gamma - 1] < \Theta > = \frac{5 - 3\gamma}{3} < \Theta > \,. \tag{1.4.28}$$

Now we are ready to study the next order velocity field induced by the density fluctuation, i. e., the velocity field associated with the scalar potential ϕ which obeys the Poisson equation (1.4.17) and the far field condition,

$$\nabla\phi = 0 \quad \text{as} \quad r \to \infty. \tag{1.4.29}$$

We call the solution of (1.4.17) and (1.4.29) Problem I. However, to describe the acoustic field we merely need the far field behavior of ϕ and consider a simpler problem, which is to solve (1.4.17) in the far field outside a large sphere S of radius $r = R \gg \ell$. This is called Problem II, for which we replace the inhomogeneous part of (1.4.17) by its far field representation. On account of (1.2.9), (1.4.27) and (1.4.19), we reduce (1.4.17) to

$$\Delta\phi = \Phi_{tt}^{(2)} + O(r^{-4}) , \qquad \text{for} \quad r \geq R . \tag{1.4.30}$$

Solutions are subject to the condition (1.4.29) at infinity and to an appropriate boundary condition on the spherical surface ∂S : $r = R$ consistent with the leading far field solution of Problem I.

We solve Problem II in two steps. First we construct a particular solution $\eta_0(t, \mathbf{x})$ satisfying the above Poission equation without the $O(r^{-4})$ term and the condition (1.4.29). Secondly we derive an appropriate Neumann condition on ∂S from the leading term of the normal velocity of a solution of Problem I. The leading term represents the average mass flux density through ∂S, namely $m/(4\pi R^2)$. Next we construct the homogeneous solution, ξ, of (1.4.30) outside of S so that $\xi + \eta_0$ satisfies the Neumann condition.

Since the total flux of η_0 across ∂S turns out to be zero, it follows that ξ solves the inhomogeneous equation $\Delta\xi = m\delta(\mathbf{x})$ for all $\mathbf{x}$. The source strength m is related to the volume integral of the right-hand side of (1.4.17) over S, that is, to the total density fluctuation. Thus, the contributions, ξ and η_0, of the leading far field potential ϕ reflect the overall fluid dilatation and the quadrupole pressure field, respectively.

By carrying out the aforementioned steps, we obtain the following results:

$$\phi = \xi + \eta_0 + O(r^{-2}) , \qquad r \geq R , \tag{1.4.31}$$

where

$$\eta_0 = -[r^3 \Phi_{tt}^{(2)}]/[6r] \tag{1.4.32}$$

and

$$\xi(t, \mathbf{x}) = -\frac{m(t, R)}{4\pi r} [1 + O(\frac{R}{r})] . \tag{1.4.33}$$

Here $m(t, R)$ is the total flux of ξ through ∂S,

$$m(t, R) = \int\int_{\partial S} \phi_r(t, \mathbf{x}') \, d\Sigma + O(R^{-1}), \tag{1.4.34}$$

where $d\Sigma$ denotes the surface area element on ∂S. The derivation of (1.4.30-34) is given by Ting and Miksis (1990).

Using the divergence theorem, (1.4.17) and (1.4.28), we convert (1.4.34) to

$$m(t, R) = -\int\int\int_S \rho_t'(t, \mathbf{x}) d\mathbf{x} + O(1/R) = \frac{3\gamma - 5}{3} < \Theta > + O(1/R). \tag{1.4.35}$$

From (1.4.31), (1.4.33) and (1.4.35), we get the far field behavior of ϕ. It is

$$\phi(t, \mathbf{x}) = -\frac{\bar{m}(t)}{4\pi r} + \eta_0 + O(\frac{R}{r^2}) \tag{1.4.36}$$

where

$$\bar{m} = \frac{3\gamma - 5}{3} < \Theta > = m(t, R)|_{R\to\infty} . \tag{1.4.37}$$

In deriving (1.4.36-37), we have assumed that $r \gg R \gg \ell = 1$. For example, if $r/\ell = R^2/\ell^2 \gg 1$, then the last term in (1.4.36) is $O(r^{-3/2})$. This is stronger than $O(r^{-1})$, which is all we need; the last term should be much smaller than the first two terms, both of which are $O(r^{-1})$.

Finally, we obtain the far field behavior of the vortical flow, including the second order solution, $O(M^2)$,

$$\mathbf{v} = \nabla\Phi + M^2\nabla\phi + O(M^2 r^{-3}) , \tag{1.4.38a}$$

with

$$\Phi = \Phi^{(1)}(\mathbf{x}) + \Phi^{(2)}(t, \mathbf{x}) + O(r^{-4}) , \tag{1.4.38b}$$

$$M^2\phi = -\frac{M^2\bar{m}(t)}{4\pi r} + M^2\eta_0(t, \mathbf{x}) + O(\frac{M^2}{r^2}) \tag{1.4.38c}$$

and

$$\eta_0 = -r^3\Phi^{(2)}_{tt}/(6r) = O(r^{-1}) . \tag{1.4.38d}$$

In terms of the outer variable $\tilde{r}$, see (1.4.6), the orders of magnitude of the first two terms in Φ and $M^2\phi$ are :

$$\Phi^{(1)} = O(M^2\tilde{r}^{-2}) = O(M^2) , \qquad \Phi^{(2)} = O(M^3)$$

and

$$M^2 r^{-1} = O(M^3) , \qquad M^2\eta_0 = O(M^3) .$$

The first term of Φ representing a doublet is $O(M^2)$ but is independent of t and hence does not contribute to the acoustic field. The second term of Φ and the first two terms of $M^2\phi$ are of the same order and generate an acoustic pressure field of $O(M^3)$. Thus ϵ in (1.4.7) can be equated to M^3. In the next section, we shall match the inner and outer solutions and see that the terms $M^2\Phi^2$ and $M^2\eta_0$ are matched with the first two terms of acoustic quadrupoles while the term $M^2\bar{m}/(4\pi r)$ is matched with the leading term of an acoustic monopole.

1.4.3 The Acoustic Field

To prepare for the matching of the outer solution, governed by the simple wave equation, with the inner solution (1.4.38), we find the behavior of an acoustic monopole and quadrupole for $\tilde{r} \ll 1$.

For an acoustic monopole of strength $M^3\bar{m}(t)$ at the origin, the acoustic potential behaves as

$$\tilde{\phi}(\tau, \tilde{\mathbf{x}}) = -\frac{M^3\bar{m}(\tau)}{4\pi\tilde{r}} = -\frac{M^3\bar{m}(t)}{4\pi\tilde{r}} + \frac{M^3\bar{m}'(t)}{4\pi} + O(M^3\tilde{r}) \quad \text{for} \quad \tilde{r} \ll 1 . \tag{1.4.39}$$

Here $\tau = t - \tilde{r}$ is the retarded time. In terms of the inner variable r, the right-hand side becomes a power series in M. The first term, $O(M^2)$, matches with the potential source in the second order inner solution $M^2\phi$. The second term, $O(M^3)$, has no contribution to the velocity field since it is independent of $\mathbf{x}$.

For a quadrupole at the origin , of strength $M^3 q_{ij}(t)$ and orientation ij, the acoustic potential behaves as

$$\begin{aligned} \frac{M^3}{4\pi}\tilde{\partial}^2_{ij}\left[\frac{q_{ij}(\tau)}{\tilde{r}}\right] &= \frac{M^3 q_{ij}(t)}{4\pi}\tilde{\partial}^2_{ij}\frac{1}{\tilde{r}} \\ &+ \frac{M^3 q''_{ij}(t)}{24\pi}\left[\frac{2\delta_{ij}}{\tilde{r}} - \tilde{r}^2\tilde{\partial}^2_{ij}\frac{1}{\tilde{r}}\right] + O(M^3\tilde{r}^0)\,, \end{aligned} \tag{1.4.40a}$$

where $\tilde{\partial}_i$ stands for partial derivative with respect to $\tilde{x}_i$. In terms of the inner variable r, (1.4.40a) becomes

$$\begin{aligned} \frac{M^3}{4\pi}\tilde{\partial}^2_{ij}\left[\frac{q_{ij}(\tau)}{\tilde{r}}\right] &= \frac{q_{ij}(t)}{4\pi}\partial^2_{ij}\frac{1}{r} \\ &+ \frac{M^2 q''_{ij}(t)}{24\pi}\left[\frac{2\delta_{ij}}{r} - r^2\partial^2_{ij}\frac{1}{r}\right] + O(M^3)\,. \end{aligned} \tag{1.4.40b}$$

The first term on the right-hand side, which is $O(1)$, matches with the potential of the quadrupole of orientation ij in the leading unsteady inner solution $\Phi^{(2)}$. The second term, which is $O(M^2)$, will match with a term in $M^2\phi$ for large r.

We note once more that the first term of Φ in (1.4.38) represents a doublet and its strength is independent of time. Therefore, the corresponding acoustic potential $M^2\tilde{\Phi}^{(1)}(\tilde{\mathbf{x}})$ is an analytic continuation of the inner solution $\Phi^{(1)}(\mathbf{x})$ with $\mathbf{x}$ replaced by $M^{-1}\tilde{\mathbf{x}}$ and it does not generate sound because $\tilde{\Phi}^{(1)}_t = 0$.

To match with the quadrupole of the inner solution in $\Phi^{(2)}(t,\mathbf{x})$, of (1.4.15b), we take note of (1.4.40 a, b) and define the corresponding acoustic potentials by replacing the variable t in the strengths $\mathbf{F}(t)$ and $\mathbf{H}(t)$ by the retarded time τ and the inner variable $\mathbf{x}$ in $\Phi^{(2)}$ by the outer variable $\tilde{\mathbf{x}}/M$, i.e.,

$$\tilde{\Phi}(\tau,\tilde{\mathbf{x}}) = \Phi^{(2)}(\tau,\tilde{\mathbf{x}}/M). \tag{1.4.41a}$$

Using (1.4.40a) again and observing that $q_{11} + q_{22} + q_{33} = 0$, we arrive at the behavior of $\tilde{\Phi}$ for $\tilde{r} \ll 1$,

$$\tilde{\Phi}(\tau,\tilde{\mathbf{x}}) = \tilde{\Phi}(t,\tilde{\mathbf{x}}) - \frac{1}{6}\tilde{r}^2\tilde{\Phi}_{tt}(t,\tilde{\mathbf{x}}) + O(\tilde{r}^0). \tag{1.4.41b}$$

From (1.4.15b), (1.4.41a), and (1.4.38b, c, d), we see that the first term on the right-hand side of the above equation is $O(\tilde{r}^{-3})$ and equal to the first unsteady term $\Phi^{(2)}$ in Φ while the second term is $O(\tilde{r}^{-1})$ and is equal to the second term $M^2\eta_0$ in $M^2\phi$. Since the first term of the acoustic monopole for $\tilde{r} \ll 1$ in (1.4.39) matches with the corresponding term in the inner solution, the first term of $M^2\phi$ in (1.4.38c), we conclude that

- *the leading acoustic pressure is of order $O(M^3)$ and is composed of the contributions from the acoustic quadrupoles $\tilde{\Phi}$ and a monopole $\tilde{\phi}$. It is*

$$\frac{p - P_0}{\rho_0 U^2} \approx M^3\tilde{p}^{(0)} = \tilde{\Phi}_t + \tilde{\phi}_t\,. \tag{1.4.42}$$

The strengths of the acoustic quadrupoles in the first term on the right-hand side are related through (1.4.15b) and (1.4.38a-d) to the retarded values of the five linear combinations of second moments of vorticity, F_1, F_2, F_3, H_1 and H_2, which are defined by (1.2.14). The strength of the monopole $\bar{m}$ in the second term is related to the retarded value of the rate of energy dissipation by (1.4.37). This defines the leading acoustic field in terms of the far field behavior of the viscous vortical flow.

1.4.4 Sound Generation due to Turbulence

It is well known that a turbulent flow is a special class of unsteady flow for which the initial vorticity or velocity is not available or not precisely specified but the flow field has distinct length scales: a length scale ℓ for the mean flow and much smaller multiple scales for the eddies. If the mean flow is unsteady, the initial data for the mean flow, say the mean vorticity distribution in the length scale ℓ, have to be prescribed. Nevertheless, the results obtained above for an unsteady flow remain valid for a turbulent flow. The question is how to obtain the corresponding results for the mean flow alone from those for the resultant flow which is the sum of the mean flow and the turbulent eddies. Let us use bar and check to denote the mean and the turbulent part respectively. For example, we write for the velocity and vorticity,

$$\mathbf{v} = \bar{\mathbf{v}} + \check{\mathbf{v}} \qquad \text{and} \qquad \Omega = \bar{\Omega} + \check{\Omega}. \tag{1.4.43}$$

The volume integral of the pressure fluctuation is equal to that of the mean pressure fluctuation, i. e., $< p'_t > = < \bar{p}'_t >$. The same holds for the density fluctuation. For the rate of energy dissipation, we have

$$< \Theta > = < \Theta_0 > + < \Theta_1 > \tag{1.4.44}$$

with

$$\Theta_0 = \frac{1}{Re} \sum_{j,k=1}^{3} \left(\frac{\partial \bar{v}_j}{\partial x_k} + \frac{\partial \bar{v}_k}{\partial x_j}\right)^2 \qquad \text{and} \qquad \Theta_1 = \frac{1}{Re} \sum_{j,k=1}^{3} \left(\frac{\partial \check{v}_j}{\partial x_k} + \frac{\partial \check{v}_k}{\partial x_j}\right)^2. \tag{1.4.45}$$

The relationships (1.2.14) between the strengths of the quadrupoles and the second moments of vorticity remain valid. Since the latter are linear in Ω, the moments of vorticity are equal to the moments of the mean vorticity. Thus we replace ω_i by $\bar{\omega}_i$ in (1.2.14). The diffusion equation for the mean vorticity is obtainable from the momentum equation for the mean velocity when model equations relating the Reynolds stresses to the velocity gradient are introduced, see for example Schlichting (1978).

The strength of the source $\bar{m}$ is related by (1.4.37) to the rate of energy dissipation in which the effect of turbulent eddies appears directly as the second term on the right-hand side of (1.4.44). An additional turbulence modeling equation is needed to determine the second term, $< \Theta_1 >$. Of course, when the rate of energy dissipation of the eddies and that of the average flow are negligible, the acoustic field induced by a turbulent field will be dominated by acoustic quadrupoles induced by the mean flow.

2. Motion and Decay of Vortex Filaments

In this chapter, we study vortical flows involving two distinct length scales, the normal length ℓ and the short length $\delta^* = \epsilon\ell$ where $\epsilon \ll 1$ is a small parameter. In particular, we consider flow fields composed of a *finite* number of *slender* vortex filaments embedded in a background velocity field of normal length scale ℓ and velocity scale U. The strength Γ of a filament is of the order $U\ell$, its core size is of the order δ^* and hence the vorticity in a filament is of the order $\epsilon^{-2}U/\ell$. For simplicity, we assume here that the background flow is a potential flow. In **Chapter 4** we will discuss the extension of the asymptotic analysis to filaments submerged in a rotational background flow with vorticity of the order U/ℓ.

The characterization of a slender vortex filament, or in short a filament, was given in (4)-(11) in the **Introduction**. We recall that the geometry of a filament is defined by its center line C and by its effective core size. Since vortex lines cannot terminate inside a flow field, a filament will either be of finite length forming a loop or ending on a solid surface or extend to infinity. For a filament of finite length,

$$S(t) = O(\ell), \tag{2.0.1}$$

the parametric representation of the center line, $C(t)$, which is $\mathbf{x} = \mathbf{X}(t,s)\ ,\ 0 \leq s \leq S_0$, has to fulfill the periodicity condition,

$$\mathbf{X}(t, s + S_0) = \mathbf{X}(t,s)\ . \tag{2.0.2}$$

Let $\tilde{s}(t,s)$ denote the arc length, and S(t) the total length of C. They are defined by the equations,

$$\tilde{s}(t,s) = \int_0^s |\mathbf{X}_s(t,s')|\, ds' \qquad \text{and} \qquad S(t) = \tilde{s}(t,S_0)\ . \tag{2.0.3}$$

In case that the parameter s is the initial arc length then we have S_0 equal to the length of the center line at $t = 0$, i. e., if $|\mathbf{X}_s(0,s)| = 1$, then $S_0 = S(0)$.

We now formulate the equations defining the centerline and the effective core size. The line C is understood to be the center line of the large swirling flow in the sense that

1) the velocity component in the plane normal to C at point $\mathbf{X}(t,s)$, the cross-sectional plane $\mathcal{P}$ of the filament, remains of order one, i. e.,

$$|\mathbf{v}(t,\mathbf{X}) \times \hat{\tau}(t,s)|/U = O(1)\ , \qquad \text{as} \quad \epsilon \to 0\ , \tag{2.0.4}$$

where $\hat{\tau}$ denotes the unit tangent vector to C.

2) C is a material line, i.e.,

$$[\dot{\mathbf{X}} - \mathbf{v}(t, \mathbf{X})] \times \hat{\tau} \equiv 0 \,, \tag{2.0.5}$$

where $\dot{\mathbf{X}}$ stands for $\mathbf{X}_t$.

Note that this condition does not require that a point on $\mathcal{C}$ characterized by the parameter s is a material point because the motion of a line is defined by its velocity in its normal plane and is independent of a tangential velocity component. Here we follow the formulation of Callegari and Ting (1978) by setting the tangential component to zero, i. e.,

$$\dot{\mathbf{X}} \cdot \hat{\tau} \equiv 0 \,. \tag{2.0.6}$$

This in turn defines the parameter s for $t > 0$. From (2.0.4)–(2.0.6), we see that the velocity of the filament remains order one, i.e.,

$$\dot{\mathbf{X}}/U = O(1) \qquad \text{for } t > 0. \tag{2.0.7}$$

Note that the order of magnitude has been expressed in powers of ϵ.†

Equation (2.0.6) is a trivial condition for two-dimensional and axisymmetric problems. For a three-dimensional problem, we could require a point $\mathbf{X}(t, s)$ on $\mathcal{C}$ to be a material point, i. e., replace (2.0.5) and (2.0.6) by $\dot{\mathbf{X}} = \mathbf{v}(t, \mathbf{X})$. This condition was used by Ting (1971) for the analysis of a vortex filament with large swirling flow but order one axial velocity so that (2.0.7) is observed, i. e., $\dot{\mathbf{X}}/U$ remains order one. In the analysis of a filament with both large swirling and large axial flow, Callegari and Ting (1978) imposed (2.0.6) in order to preserve condition (2.0.7). Then, (2.0.5) still allows for large axial velocities of order $1/\epsilon$, say.

To define the effective core size, we make use of the result obtained by Ting et al (1966) and (1971) and reported in **Secs.2.2** and **2.3**, that the leading order vorticity distribution in a normal plane $\mathcal{P}$ of $\mathcal{C}$ at $\mathbf{X}$ is axisymmetric with respect to the centerline. Thus we define the effective core size by the size of a contour line on which $|\Omega|$ is a fraction, say $1/e$, of its maximum value in $\mathcal{P}$, i. e.,

$$|\Omega(t, \mathbf{x})| = |\Omega(t, \mathbf{X})| \, [e^{-1} + O(\epsilon)] \qquad \text{for} \quad r = |\mathbf{x} - \mathbf{X}| = \delta(t, s) \quad \text{and} \quad \mathbf{x} \in \mathcal{P} \,. \tag{2.0.8}$$

For filaments submerged in a background potential flow, there are many typical length scales. For example, in addition to the length scale of the background flow field we have typical length scales for the radii of curvature of the center lines of the filaments, the distance between a filament and the boundary of the flow field and the distance between two filaments. We assume that these are all of the order of the normal length scale ℓ, or rather we choose ℓ to be the smallest of all the length scales in the flow away from the core structure.

Because of the rapidly decaying vorticity distribution in the sense of (7), the flow field far away from a vortex filament, i. e., at a distance $r \gg \delta^*$, approaches the potential flow induced by a vortex line along $\mathcal{C}$ with the same strength Γ but zero core radius relative to ℓ as $\epsilon = \delta^*/\ell \to 0$. The potential flow is then given by the Biot–Savart formula (12). With the distance between two filaments being of

† Here a $\ln(1/\epsilon)$ term is considered as $O(\epsilon^0) = O(1)$. In **2.3** we shall show the dependence of the velocity of a filament on $\ln(1/\epsilon)$.

order $O(\ell)$, one can consider all the velocities in the neighborhood of one filament that are induced by the other filaments as additional contributions to background potential flow. Thus, it suffices to analyze only the motion and decay of a single filament submerged in a background potential flow.

The above description of vortex filaments in a three-dimensional flow is applicable to vortices in two dimensions, say in the xy plane, when the vortices are considered as filaments parallel to the z-axis and the flow field is independent of z. For a two-dimensional vortex of zero core size, i. e., in the limit as $\epsilon \to 0$, we call it a vortex point which is equivalent to a vortex line parallel to the z-axis.

In **Sec. 2.1** we review the classical inviscid theories on two and three-dimensional vortical flows. We point out the deficiencies of the inviscid solutions, in particular, the singular behavior of the velocity field near a two-dimensional vortex point or a curved vortex line and the indeterminacy of its motion.

In **Sec.2.1.1** we review the classical theory of a two-dimensional inviscid flow, in which a vortex point, or the center of a small vortex, is assumed to move with the local background velocity. To demonstrate that this well accepted assumption is not true in general, we reproduce in **Sec.2.1.1.1** the solution by Milne-Thomson (1973) of a uniform two-dimensional flow with upstream velocity $U_\infty \hat{\imath}$ around a disc of radius a spinning at a constant angular velocity $\dot{\theta}$. Far away from the disc we see a uniform flow plus the flow field induced by a vortex point of constant strength $\Gamma = 2\pi a \dot{\theta}$ located at the center of the disc. But, the flow field near the disc and the motion of the disc depend on the initial velocity and size of the disc. We examine this solution for the case of a small disc spinning at a high angular velocity, so that $\epsilon = U_\infty a/\Gamma = a/\ell \ll 1$ and $\dot{\theta}\ell/U_\infty = 1/(2\pi\epsilon^2) \gg 1$, where $\ell = \Gamma/U_\infty$. On the normal length scale ℓ and time scale ℓ/U_∞, the leading solution represents a vortex point drifting with the uniform stream. But on the small length scale a and time scale $\epsilon^2\Gamma/U_\infty$, the leading solution shows a high frequency small amplitude oscillation of the disc around a mean trajectory with upstream velocity $U_\infty \hat{\imath}$.

We then connect this solution to that for a free vortex in **Sec.2.1.1.2**. We replace the small spinning disc by fluid rotating at constant angular velocity $\dot{\theta}$ inside an interface which coincides initially, at $t = 0$, with the circular boundary of the disc. Outside of the interface the flow is irrotational. The rotational flow of constant vorticity $2\dot{\theta}$ inside the interface is called a Rankine vortex. The leading solution for the interface is a circle of radius a for $t \geq 0$, and hence the leading order solutions for the velocity fields outside and inside the Rankine vortex coincide with with those for a rigid disc. The correspondence between the motion of the disc and that of the Rankine vortex, the deviation of the interface from the circle and the higher order solutions of the Rankine vortex are presented in **Sec. 2.1.1.2**.

In **Sec.2.1.2**, we consider three-dimensional flows and derive the singular behavior of the velocity field near a vortex line C defined by the Biot-Savart formula (12). Besides the leading singularity of the order $1/r$ associated with a locally straight vortex line tangent to C, there are two additional singular terms. One of them is a logarithmic singularity, while the other one is bounded but multivalued as $r \to 0$. Both these terms are proportional to the curvature of the vortex line and thus account for the essential difference between curved and straight vortex lines.

For the latter, the local average of the velocity over a neighborhood of the vortex much larger than $(\delta^*)^2$ but much smaller than ℓ^2 exists in the limit as $r \to 0$ and is equal to the background velocity. Thus one could redefine the velocity of a vortex point given by the classical two-dimensional inviscid theory as the local average of the velocity field. This new definition is not applicable to a curved vortex line because the corresponding local average in the cross-sectional plane $\mathcal{P}$ of the $\ln r$ term due to the curvature effect does not exist. It is well known that the velocity of a curved vortex line cannot be uniquely defined while the velocity of the center line of a vortex filament with a finite core size is well defined. For example, the velocity of a circular vortex ring with a finite core size was given by Lamb (1932). Because of the curvature effect, the motion of a filament is coupled with the evolution of its core structure. The latter is due to viscous diffusion and stretching of the filament. The stretching effect is present even in an inviscid flow.

Those singularities of an inviscid solution for a vortex line show clearly that the solution is not valid in the neighborhood of the vortex line. In a real fluid, viscous effects, regardless how small, will immediately remove or regularize an inviscid singularity. Hence a vortex line appearing at $t = 0$ will become a vortex filament with an effective core size $\delta(t) > 0$ for $t > 0$. Because of the two length scales, the method of matched asymptotic expansions using ϵ as the small parameter can be employed to study the evolution of the inner core structure including viscous effects while the leading outer solution away from the vortex filament, should be described by the inviscid theory of a vortex line. The idea here is to keep the effects of both viscous diffusion and nonlinear convection in the leading order equations for the inner solution so that they remain valid when either one or both effects are important. To this end, we introduce the standard assumption of "boundary layer" analysis, (13), which says that the Reynolds number based on the outer length scale ℓ is of the order of ϵ^{-2}. We then show that the matched asymptotic solution defines the motion of the vortex filament and its induced flow field. The asymptotic solution is regular everywhere and has a finite total kinetic energy for any nonzero ϵ.

In **Sec.2.2**, we describe in detail the construction and the physical meaning of the matched asymptotic solution for a two-dimensional problem, as given by Ting and Tung (1965). The two-dimensional problem is relatively simple because of the absence of the physical effects of, or rather the mathematical complexity due to, the curvature and torsion of the center line of a filament in space. Yet the essential features of the matched asymptotic analysis are quite similar to those for a general three-dimensional problem. These features are the expansion schemes, the matching procedures and the steps needed to derive both the equations of motion of a vortex center and the equation for the evolution of the core structure.

Since the outer solution is composed of a background potential solution plus the contributions of the vortices, we begin our asymptotic analysis with the assumption that the outer solution depends only on the normal time t and construct the inner solution for a vortex allowing for two time variables, a short time τ in addition to t. We then show that the two-time inner solution becomes an one-time solution in t, provided that the initial data fulfill certain compatibility conditions. The trajectory of the vortex center in the one-time solution agrees with that predicted by the classical theory for a vortex point to the order ϵ.

If the compatibility conditions on the initial data are not satisfied, the two-time solution describes small amplitude but high frequency oscillations about an average trajectory which agrees with that of the one-time solution to the order ϵ. The oscillatory motion turns out to be equivalent to that of a small spinning disc in an inviscid stream. The contributions of the inner solution to the leading and first order outer solutions, including the displacement of the vortex center, are defined by the one-time inner solution while the contributions of the short time fluctuations can appear in the outer solution only as second and higher order corrections. The same correspondence between the one-time inner solution in t and the two-time solution was obtained for axisymmetric problems by Tung and Ting (1966) and Ting (1971).

In **Sec. 2.3** we present the asymptotic analysis of Callegari and Ting (1978) for a filament with large circumferential and axial velocity components in the core. Only the one-time solution in t will be discussed. This allows us to focus our attention on the effects of the filament geometry on the filament motion and the evolution of the core structure. In addition we shall emphasize the physical meaning of the solutions. In particular, we discuss the relationship between the large axial and circumferential velocity components in the core and point out in which part of the analysis we use explicitly the condition (2.0.2) that the filament is of finite length forming a loop. Theories valid for filaments with multiple axial length scales will be discussed in **Sec. 4.1**.

A computational code for the asymptotic solution of Callegari and Ting was developed and numerical results showing the interactions of vortex filaments were presented by Liu, Tavantzis and Ting (1986). Several numerical examples will be reported in **Sec. 2.4**. Here we emphasize the point that in the evaluation of the velocity of the center line of a filament we should keep not only the leading term of order $\ln 1/\epsilon$, but also the order one terms. The reasons are threefold:

1) The logarithm of a large number is not that large. For example, 20 is a typical value for $1/\epsilon$ but $\ln 20 \approx 3$.
2) The velocity given by the $\ln(1/\epsilon)$ term alone is in the binormal direction of the center line $\mathcal{C}$ and thus excludes any elongation of the curve. On the other hand, we know from the analysis of **Sec.2.3** that the stretching of a filament is as important as the viscous diffusion for the evolution of the core structure and hence has a significant influence on the motion of the filament.
3) The $\ln(1/\epsilon)$ term involves only the local curvature and binormal vector of the center line, while the order one terms depend on the core structure, the global geometry of the center line and on the local velocity of the background flow including the velocities induced by all the other filaments.

Thus all the order one terms are needed for an accurate determination of the motion of a filament.

To end **Sec. 2.3**, we apply the analysis of sound generation by a vortical field in **Sec. 1.5** to study the radiation of sound waves by slender vortex filaments. Using the asymptotic solution in **Sec. 2.3**, we show that the leading order strengths of the acoustic quadrupoles induced by a filament depend only on the motion of its center line while the strength of the monopole is proportional to the rate of energy dissipation in the core.

In **Sec.2.4**, we report on several studies of the practical regime of validity of the asymptotic solutions. The answer is obtained by comparing numerical solutions of the N-S equations with the asymptotic solutions. Starting from the same initial data when δ/ℓ is small, Liu, Tavantzis and Ting (1986) carried out the numerical solution and extended the asymptotic solutions until the core size became comparable to ℓ. Comparison of the final results of the two solutions then establishes the practical regime of validity of the asymptotic solution. That is, for a given degree of accuracy, we can specify a practical upper bound for δ/ℓ above which one should switch from the asymptotic solution to numerical solution of the vortical flow problem.

2.1 Inviscid Theories of Vortical Flows and Their Deficiencies

In classical inviscid theories, a slender vortex filament in space is idealized by a vortex line obtained by letting the core size of a filament go to zero while keeping the circulation constant. In a two-dimensional problem, this leads to the model of a vortex point. The inviscid theory of the motion of a curved vortex line has two essential defects. They are (*i*) that the velocity of the fluid on the vortex line is infinite, and that (*ii*) the velocity of the vortex line itself is undefined. We shall review briefly the inviscid theories and point out their defects for two and three-dimensional problems in **Sec. 2.1.1** and **2.1.2** respectively.

2.1.1 Potential Flows around a Spinning Disc and a Rankine Vortex

For a two-dimensional potential flow around a vortex point of strength $\varGamma$ located at (X, Y), the stream function ψ is the sum of the stream function $\varPsi$ for the background potential flow and that for the vortex alone, i.e.,

$$\psi(x, y) = \varPsi(x, y) - \frac{\varGamma}{2\pi} \ln r, \tag{2.1.1}$$

where r is the distance between (x, y) and (X, Y). Here the dependence of the stream functions and the coordinates (X, Y) on time t has been suppressed. The velocity near the point vortex behaves as $\varGamma/(2\pi r)$ and is unbounded as $r \to 0$. The second defect of the inviscid theory is circumvented by the ***assumption*** that the vortex point moves with the local mean velocity which is also the local velocity of the flow field in absence of the vortex. Thus,

$$\dot{X} = \varPsi_y(X, Y) \qquad \text{and} \qquad \dot{Y} = -\varPsi_x(X, Y), \tag{2.1.2}$$

where the dot denotes differentiation with respect to t. Lamb (1932) provided justification for the above assumption on the physical consideration that a vortex point, having zero mass, must move with the local fluid. If not, its velocity $\mathbf{Q}$ relative to the local fluid would produce a finite lift force $\rho|\mathbf{Q}|\Gamma$ acting at the vortex point according to Joukowski's theorem. The lift force could not be balanced by the inertia of the vortex, which is zero, and therefore the relative velocity $\mathbf{Q}$ has to vanish. The above argument is usually referred to as the limit solution of vanishing mass. To illustrate this point and also to explain the physics of a two time solution in its simplest form, we study the motion of a spinning disc in a uniform stream. From the analogy to the motion of a small spinning disc, we get a clear physical understanding of the highly oscillatory motion of a Rankine vortex about its mean trajectory. Later on we shall employ the same analogy to describe the motion of a vortex with a decaying vortical core.

2.1.1.1 Motion of a spinning disc in a uniform stream

Consider the motion of a disc of radius a and uniform density ρ_0 in a two-dimensional incompressible inviscid fluid of density ρ, say in the xy-plane. The disc is spinning at a constant angular velocity $\dot{\theta}$ and is submerged in a uniform stream with velocity U_∞ in the direction of the x-axis, as shown in Fig.2.1.

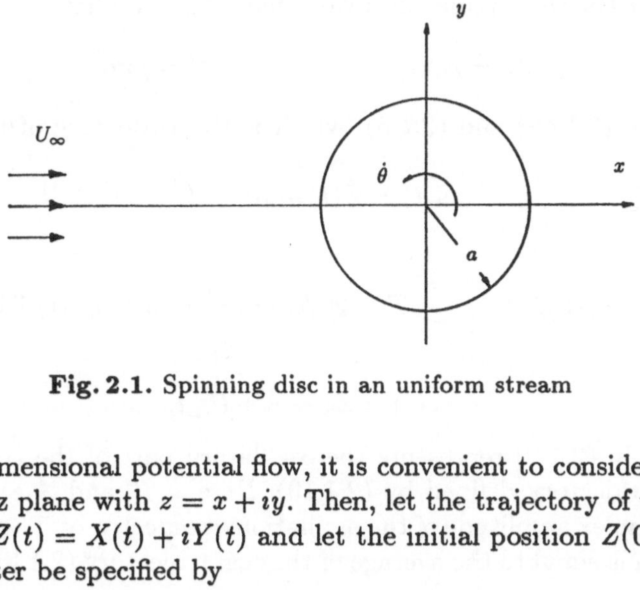

Fig. 2.1. Spinning disc in an uniform stream

For a two-dimensional potential flow, it is convenient to consider the xy plane as the complex z plane with $z = x + iy$. Then, let the trajectory of the disc center be denoted by $Z(t) = X(t) + iY(t)$ and let the initial position $Z(0)$ and velocity $\dot{Z}(0)$ of the center be specified by

$$Z_0 = X_0 + iY_0 \qquad \text{and} \qquad \dot{Z}_0 = \dot{X}_0 + i\dot{Y}_0. \tag{2.1.3}$$

The complex velocity of the uniform stream is then denoted by U_∞. The complex velocity relative to the disc is denoted by

$$Q(t) = U_\infty - \dot{Z}(t) \tag{2.1.4a}$$

and the circulation around the disc by

$$\Gamma = 2\pi a^2 \dot{\theta}. \tag{2.1.4b}$$

The complex potential of the flow field relative to the disc center is the superposition of the potentials of a uniform flow, a point vortex and a doublet located at the vortex center,

$$w(t,z) = Q^*(t)\,(z-Z) + \frac{\Gamma}{2\pi i}\ln\frac{z-Z}{a} + \frac{Q(t)\,a^2}{z-Z}, \tag{2.1.5}$$

where Q^* is the complex conjugate of Q. The flow field matches the normal velocity of the surface because of the contributions from the uniform flow and the doublet, but not the tangential velocity.

In principle, we can compute the force acting on the disc directly from the pressure distribution on the surface. However, there is a more elegant way to derive the equation of motion of the disc from a balance of momentum for a large control area containing the disc, see for example Milne-Thompson (1973). The equation of motion of the center of a disc with mass M is

$$M\,\ddot{Z} = -i\,\rho\Gamma Q - M'\ddot{Z}, \tag{2.1.6}$$

where the first term on the right-hand side represents the Joukowski force, while the second term accounts for the inertia of the surrounding fluid having an equivalent mass M'. For the circular disc with density ρ_0, we have

$$M = \rho_0\pi a^2 \qquad \text{and} \qquad M' = \rho\pi a^2. \tag{2.1.7}$$

The solution of (2.1.4a) and (2.1.6), which is the trajectory of the center of the disc, is

$$Z(t) = \tilde{Z}(t) + Ae^{i2\pi t/T}, \tag{2.1.8}$$

where

$$A = iTQ(0)/(2\pi), \qquad T = 2\pi/\omega = 2\pi^2 a^2(\rho+\rho_0)/(\rho\Gamma) \tag{2.1.9}$$

and

$$\tilde{Z}(t) = Z_0 - A + U_\infty t. \tag{2.1.10}$$

The last term in (2.1.8) represents the oscillatory part of the trajectory relative to the mean trajectory defined by (2.1.10). Here ω, T and A are the frequency, period and complex amplitude of the oscillations, respectively. The mean trajectory (2.1.10), which is equal to the average of the real trajectory (2.1.8) over one period T, i.e.,

$$\tilde{Z}(t) = \frac{1}{T}\int_t^{t+T} Z(t')\,dt', \tag{2.1.11}$$

shows that the disk drifts with the background velocity U_∞. The term $-A$ in (2.1.10) represents the deviation of the initial position from the mean trajectory. If the disc is moving initially at the same velocity as the uniform stream, we have $Q(0) = U_\infty - \dot{Z}_0 = 0$ and hence $A = 0$. Then the oscillatory part of the trajectory (2.1.8) vanishes and the disc drifts along with the uniform stream with zero relative velocity, i.e., $Q(t) = U_\infty - \dot{Z}(t) = 0$.

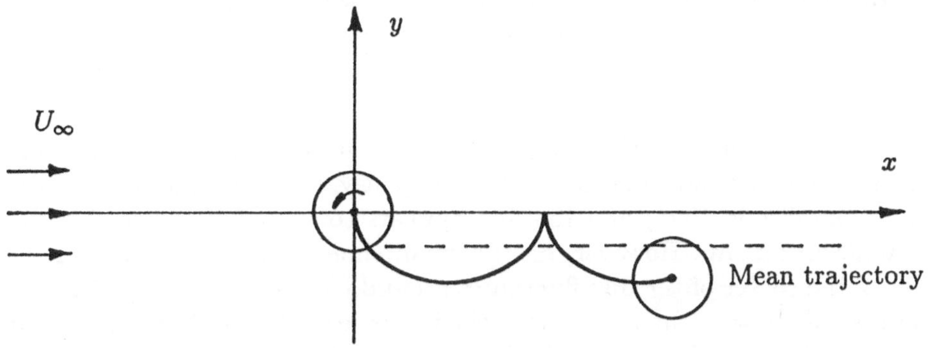

Fig. 2.2. Trajectory of a spinning disc in a uniform stream

Figure 2.2 shows the oscillatory and mean trajectory of a disc that starts with zero velocity from the origin, i.e., $Z_0 = 0$ and $\dot{Z}_0 = 0$. At $t = 0$, the Joukowski force is $\rho U_\infty \Gamma$ acting in the downward direction and the disc is gaining a downward velocity, $\dot{Y} < 0$. This in turn induces a horizontal component of the Joukowski force, increases $\dot{X}$ from zero and thus reduces the horizontal component of relative velocity $U_\infty - \dot{X}$. At $t = T/4$, we have $U_\infty - \dot{X} = 0$ and $-\dot{Y}$ reaches its maximum value. Afterwards, $\dot{X}$ increases beyond U_∞ while $-\dot{Y}$ decreases. At the half period $\dot{Y}$ vanishes, the trajectory reaches a turning point and $\dot{X}$ reaches its maximum value of $2U_\infty$. A similar description of the trajectory holds for the second half period. The wave length, which is the horizontal distance traveled during one period, is $U_\infty T$. The vertical shift $U_\infty T/(2\pi)$ of the mean trajectory given in (2.1.9) is equal to the average of the vertical displacement over one period. This completes the physical description of the oscillatory and mean trajectory of a spinning disc in a uniform stream.

Now we study the limiting case of "vanishing mass" or rather "vanishing radius" with finite Γ and U_∞. In order to set up the analogy for the motion of a Rankine vortex, and later for a vortex with decaying vortical core, we let the density of the disc equal that of the fluid,

$$\rho_0 = \rho = 1 \tag{2.1.12a}$$

and introduce the *normal* length and time scales

$$\ell = \Gamma/U_\infty\ , \qquad \text{and} \qquad \ell/U_\infty = \Gamma/U_\infty^2\ . \tag{2.1.12b}$$

For the limiting case of "vanishing radius", we consider the radius a to be the very small relative to ℓ, i. e.,

$$a = \epsilon\ell = \epsilon\Gamma/U_\infty\ , \qquad \text{with} \qquad \epsilon \ll 1\ . \tag{2.1.13}$$

In order to avoid the introduction of new symbols for dimensionless quantities, we can use ℓ and ℓ/U_∞ as unit length and time with $U_\infty = 1$ and $\Gamma = 1$ and thus render the quantities in (2.1.3) to (2.1.11) dimensionless. In particular, z and t become the *normal* space and time variables. Also, we note

$$T = 4\pi^2\epsilon^2 \qquad \text{and} \qquad A = 2\pi i\, \epsilon^2 Q(0)\,.$$

In the limit as $\epsilon = a \to 0$, the period T, the amplitude A and the initial shift of the mean trajectory vanish as $O(\epsilon^2)$. Consequently, in the normal time scale, the inertia of the disc is $O(\epsilon^2)$ and the disc appears to be moving with the velocity of the uniform stream. However, on the small time scale of the order of ϵ^2, the velocity of the center of the disc fluctuates around U_∞ by order $O(1)$. For example for the case of $\dot{Z}_0 = 0$, the horizontal velocity varies between 0 and $2U_\infty$ and the vertical velocity varies between $\pm U_\infty$ independent of ϵ. This is a standard example of two-time solutions when the initial velocity differs from the required initial value for the one-time solution with "vanishing area".

Now we consider the motion of a small spinning disc in a *non*uniform potential flow with length scale ℓ. For the inner solution describing the flow field near the disc, we let $\bar{r} = r/\epsilon$ and θ denote the polar coordinates with respect to the disc center, (X, Y). The stream function of the background flow on the scale of a is given by its Taylor series at the disc center (X, Y). When the relative coordinates, $(x - X, y - Y)$, are replaced by the inner polar coordinates, the Taylor series becomes,

$$\begin{aligned} \Psi = \Psi^* &+ \epsilon\left[\Psi_x^* \cos\theta + \Psi_y^* \sin\theta\right]\bar{r} \\ &+ \frac{\epsilon^2}{2}\left[\Psi_{xx}^* \cos 2\theta + \Psi_{xy}^* \sin 2\theta\right]\bar{r}^2 + \cdots\,. \end{aligned} \tag{2.1.14}$$

where the superscript * denotes the value at the disc center (X, Y). Using (2.1.14) as the far field behavior of the stream function of the flow field outside the disc, $\bar{r} \geq 1$, we obtain

$$\begin{aligned} \psi = &-\frac{\Gamma}{2\pi}\ln\frac{\epsilon\bar{r}}{a} + \Psi^* + \epsilon\left[\Psi_x^*\bar{r}\cos\theta + \Psi_y^*\sin\theta\right](\bar{r} - \bar{r}^{-1}) \\ &+ \frac{\epsilon^2}{2}\left[\Psi_{xx}^*\cos 2\theta + \Psi_{xy}^*\sin 2\theta\right](\bar{r}^2 - \bar{r}^{-2}) + \cdots\,. \end{aligned} \tag{2.1.15a}$$

In addition to the contribution of the circulation, we have added a doublet and a quadrupole, the $\bar{r}^{-1}$ and the $\bar{r}^{-2}$ terms, to render $\psi = \text{constant}$ on $\bar{r} = 1$. The outer solution of the flow field in the length scale ℓ, i.e., away from the disc, becomes

$$\begin{aligned} \psi = &\Psi - \frac{\Gamma}{2\pi}\ln\frac{r}{\ell} \\ &- \frac{\epsilon^2}{r}\left[\Psi_x^*\cos\theta + \Psi_y^*\sin\theta\right] \\ &- \frac{\epsilon^4}{2r^2}\left[\Psi_{xx}^*\cos 2\theta + \Psi_{xy}^*\sin 2\theta\right] + \cdots\,. \end{aligned} \tag{2.1.15b}$$

The third term on the right-hand side of (2.1.15b), which is a doublet of the order of ϵ^2, represents the contribution of a disc to a uniform background flow. The last term, a quadrupole of order $O(\epsilon^4)$, represents the contribution to the outer flow field due to the interaction of the disc with the background velocity gradients. Similar interaction terms will be found in the outer solution away from a viscous vortex in **Sec.2.2.**

2.1.1.2 Oscillations of a Rankine vortex in a uniform stream

When the disc is replaced by the same fluid rotating as a rigid body with angular velocity $\dot{\theta}$ and *drifting with velocity* $U_\infty = 1$, the flow velocity in the disc area is

$$\mathbf{V}_- = U_\infty + \dot{\theta}(z - Z)i = U_\infty + \frac{i\Gamma(z-Z)}{2\pi a} = O(\epsilon^{-1}) \tag{2.1.16}$$

for $|z - Z| < \epsilon$. The flow is rotational with constant vorticity,

$$\zeta = 2\dot{\theta} = \frac{\Gamma}{\pi a^2} = O(\epsilon^{-2}). \tag{2.1.17}$$

Outside the disc the flow field is that of a potential vortex

$$\mathbf{V}_+ = U_\infty + \frac{i\Gamma(z-Z)}{2\pi|z-Z|^2} \quad \text{for} \quad |z-Z| > \epsilon. \tag{2.1.18}$$

The boundary of the disc, $|z - Z| = \epsilon$, represents an interface ∂D separating the potential flow from the rotational flow. We use the subscripts $-$ and $+$ to denote the limit values inside and outside of the interface respectively. Across the interface the velocity and pressure are continuous in this case. Only the velocity gradient is discontinuous. Equations (2.1.16) and (2.1.17) then represent the inviscid flow field of a Rankine vortex of strength Γ drifting with the uniform stream, i.e., $\dot{Z} = U_\infty$.

When the vortex center has an initial velocity $\dot{Z}_0$ not equal to U_∞, the solution for a spinning disc, (2.1.5), in general is no longer applicable. Because of the doublet term, (2.1.5) yields discontinuities in the tangential velocity and pressure across the boundary of the disc, while an interface is a free boundary across which both the pressure and normal velocity have to be continuous. Since the discontinuities are of the order of $a^2 = \epsilon^2 \ll 1$, we expect the solution for the flow around a spinning disc and the same fluid rotating at constant angular velocity $\dot{\theta}$ inside to be an asymptotic solution for a small free Rankine vortex as $\epsilon \to 0$.

To set up the governing equations for the motion of a free Rankine vortex in general, we introduce the polar coordinates r and θ with respect to the moving center of mass of the vortex. Thus, with $Z = X + iY$, we have

$$z - Z = re^{i\theta}. \tag{2.1.19}$$

The inviscid flow field is composed of a rotational flow of constant vorticity $2\dot{\theta}$ inside an interface ∂D and a potential flow outside of it. Because the fluid is incompressible, the area of D remains equal to $\pi\epsilon^2$ and the center of area coincides with the center of vorticity, (X, Y) where $r = 0$. We seek solutions for which the interface deviates from a circle of radius ϵ by $O(\epsilon^2)$ and express the interface by

$$r = \epsilon\bar{r} = \epsilon + \epsilon^2 R(t, \theta, \epsilon)\ , \tag{2.1.20}$$

where $\bar{r}$ denotes the stretched radial coordinate. The scaled radial displacement R is decomposed in a Fourier series,

$$R = \sum_{n=2,3,\ldots} \{ a_n(\epsilon,t)\cos n\theta + b_n(\epsilon,t)\sin n\theta \} \,. \tag{2.1.21}$$

The symmetric term, $(n = 0)$, is absent because the area of D is conserved. The two first harmonics are absent because the center of area coincides with the moving origin where $\bar{r} = 0$.

The boundary conditions to be imposed at the interface (2.1.20) are the kinematic condition,

$$\epsilon^2 R_t = u_{\pm} - \frac{\epsilon R_\theta}{1+\epsilon R} v_{\pm} \,, \tag{2.1.22}$$

and the balance of pressure,

$$(p_+ - p_-)_\theta + \epsilon\, R_\theta (p_+ - p_-)_{\bar{r}} = 0 \,, \tag{2.1.23}$$

where p stands for the pressure and u and v stand for the radial and circumferential velocity components.

The flow field inside ∂D fulfills the continuity equation,

$$(\bar{r}u_-)_{\bar{r}} + (v_-)_\theta = 0 \,, \tag{2.1.24}$$

and the equation of uniform vorticity

$$[(\bar{r}v_-)_{\bar{r}} - (u_-)_\theta]/\bar{r} = \epsilon^2/\pi \,. \tag{2.1.25}$$

The flow field outside ∂D fulfills the continuity equation (2.1.24) with the subscript $-$ replaced by $+$ and the equation of zero vorticity,

$$(\bar{r}v_+)_{\bar{r}} - (u_+)_\theta = 0 \,. \tag{2.1.26}$$

To relate the jump of pressure gradients in (2.1.23) to the velocity components, we use the momentum equations,

$$\epsilon u_t + uu_{\bar{r}} + vu_\theta/\bar{r} - v^2/\bar{r} = -p_{\bar{r}} - \epsilon U_t \tag{2.1.27}$$

and

$$\epsilon v_t + uv_{\bar{r}} + vv_\theta/\bar{r} + vu/\bar{r} = -p_\theta/\bar{r} - \epsilon V_t \tag{2.1.28}$$

where

$$U = \dot{X}\cos\theta + \dot{Y}\sin\theta \quad \text{and} \quad V = \dot{Y}\cos\theta - \dot{X}\sin\theta \,, \tag{2.1.29}$$

denote the radial and circumferential components of the velocity of the vortex center and U should not be confused with the reference velocity, $U_\infty = 1$, in this subsection. In (2.1.27) and (2.1.28) the subscripts $\pm$ for u, v and p have been suppressed.

In the far field, $\bar{r} \gg 1$, the leading terms should represent the superposition of a uniform and a circulatory flow,

$$u_+ = \cos\theta - U + O(\bar{r}^{-2}) \quad \text{and} \quad v_+ = \frac{\epsilon^{-1}}{2\pi\bar{r}} - \sin\theta - V + O(\bar{r}^{-2}). \qquad (2.1.30)$$

These equations are far field conditions to be imposed on solutions to (2.1.24), (2.1.26). As we shall see later, the higher order terms will include a doublet term due to the scaled area of D. With $\bar{r}^{-2} = \epsilon^2/r^2$, the doublet term will become the term proportional to $\epsilon^2/\bar{r}$ in the stream function for the outer region. From the definition of $\dot{Z}$, we have

$$u_- = 0 \quad \text{and} \quad v_- = 0\,, \qquad \text{at} \quad \bar{r} = 0\,. \qquad (2.1.31)$$

Thus the general formulation of the problem is completed. We now construct the asymptotic solution for $\epsilon \ll 1$. Note that if the solution depends only on the normal time t, all the time derivative terms in the governing equations are of higher order in ϵ. Hence the variable t will appear as a parameter in the asymptotic solution in all powers of ϵ and we are not free to impose any initial data. Should the initial data be incompatible with the one-time asymptotic solution, then it is necessary to keep the time derivative terms in the governing equations by the introduction of a short time variable. Guided by the results for a small spinning disc, (2.1.4), (2.1.8) and (2.1.15), we introduce in addition to the *normal* time variable the *short* time,

$$\tau = t\epsilon^{-2}\,, \qquad (2.1.32a)$$

and replace ∂_t in (2.1.22) and (2.1.27) - (2.1.29) by

$$\partial_t + \epsilon^{-2}\partial_\tau\,. \qquad (2.1.32b)$$

Since the vorticity is $O(\epsilon^{-2})$, the velocity in the near field, $\bar{r} = O(1)$, is $O(\epsilon^{-1})$ and the pressure is $O(\epsilon^{-2})$. This suggests the expansion scheme,

$$f(\epsilon) = \epsilon^{-k}[f^{(0)} + \epsilon f^{(1)} + \dots]\,, \qquad (2.1.33a)$$

where

$$f^{(i)} = f^{(i)}(\bar{r}, \theta, \tau, t) \qquad (2.1.33b)$$

and $k = 1$ when f stands for $u_\pm$ or $v_\pm$ and $k = 2$ when f stands for $p_\pm$. The assumption (2.0.7), that the velocity of the vortex center remain of order $O(1)$, implies that the leading two terms in the series for Z should be independent of τ, i.e.,

$$Z(\tau, t, \epsilon) = Z^{(0)}(t) + \epsilon Z^{(1)}(t) + \epsilon^2 Z^{(2)}(\tau, t) + O(\epsilon^3) \qquad (2.1.34a)$$

and hence

$$\dot{Z}(t) = Z^{(0)}_t + Z^{(2)}_\tau + O(\epsilon)\,. \qquad (2.1.34b)$$

This ansatz allows us to impose any initial velocity, $\dot{Z}^{(0)}(0)$, of order one. Likewise we expand $\dot{X}, \dot{Y}$ (or U, V) and R in regular power series of ϵ, i.e., in the form of (2.1.33) with $k = 0$.

Using the above two-time expansion scheme it is straight-forward to derive the hierarchy of perturbation equations from (2.1.24)–(2.1.26). The leading solutions are

$$u_{\pm}^{(0)} = 0 \,, \quad v_{+}^{(0)} = 1/(2\pi\bar{r}) \quad v_{-}^{(0)} = \bar{r}/(2\pi) \,, \tag{2.1.35}$$

$$p_{+}^{(0)} = -\frac{1}{8\pi^2\bar{r}^2} \,, \qquad p_{-}^{(0)} = \frac{\bar{r}^2 - 2}{8\pi^2} \tag{2.1.36}$$

and the leading order interface is the unit circle, $\bar{r} = 1$. At the next order, the kinematic condition (2.1.22) yields

$$u_{\pm}^{(1)} = R_{\tau}^{(0)} + R_{\theta}^{(0)}/(2\pi) \quad \text{on} \quad \bar{r} = 1 \,. \tag{2.1.36}$$

Corresponding to (2.1.21), we represent $R^{(0)}(t,\theta)$ by

$$R^{(0)} = \Re \sum_{n=2,3,\ldots} c_n(t,\tau)e^{in\theta} \tag{2.1.37a}$$

where

$$c_n = a_n^{(0)} - ib_n^{(0)} \tag{2.1.37b}$$

and $\Re$ stands for "the real part of". The corresponding first order solutions are

$$u_{\pm}^{(1)} = (\phi_{\pm}^{(1)})_{\bar{r}} \,, \quad v_{\pm}^{(1)} = (\phi_{\pm}^{(1)})_{\theta}/\bar{r} \,, \tag{2.1.38}$$

where

$$\phi_{-}^{(1)} = \Re \sum_{n=2,3,\ldots} \{[\frac{(c_n)_\tau}{n} + \frac{ic_n}{2\pi}]\bar{r}^n e^{in\theta}\} \tag{2.1.39a}$$

and

$$\begin{aligned} \phi_{+}^{(1)} =& [(1 - X_t^{(0)} - X_\tau^{(2)})\cos\theta - (Y_t^{(0)} + Y_\tau^{(2)})\sin\theta][\bar{r} + \bar{r}^{-1}] \\ &- \Re \sum_{n=2,3,\ldots} [\frac{(c_n)_\tau}{n} + \frac{ic_n}{2\pi}]\bar{r}^{-n} e^{in\theta} \,. \end{aligned} \tag{2.1.39b}$$

The balance of pressure across the interface, (2.1.23), yields $[p_{+}^{(1)} - p_{-}^{(1)}]_\theta = 0$ and (2.1.28) in turn yields

$$[v_{+}^{(1)} - v_{-}^{(1)}]_\tau + [v_{+}^{(1)} - v_{-}^{(1)}]_\theta/(2\pi) - u_{-}^{(1)}/\pi = 0 \qquad \text{on} \qquad \bar{r} = 1 \,. \tag{2.1.40}$$

Inserting (2.1.38), (2.1.39) into (2.1.40) and equating the coefficients of all harmonics to be zero, we obtain the equations of motion for the vortex center:

$$X_t^{(0)} = 1 \,, \qquad Y_t^{(0)} = 0 \,, \tag{2.1.41a}$$

$$Y_{\tau\tau}^{(2)} - \frac{1}{2\pi} X_\tau^{(2)} = 0 \,, \qquad X_{\tau\tau}^{(2)} + \frac{1}{2\pi} Y_\tau^{(2)} = 0 \tag{2.1.41b}$$

for $n = 1$ and the equations for the variations of higher modes,

$$2\pi(c_n)_{\tau\tau} + i(2n-1)(c_n)_\tau - \frac{n(n-1)}{2\pi} c_n = 0 \,, \tag{2.1.41c}$$

for $n = 2, 3, \ldots$. Equations (2.1.41a) and (2.1.42b) were obtained from the two equations for the first harmonics by the standard two-time method. The τ-averages of those two equations eliminate their dependence on τ and yield those in (2.1.41a), which are also known as the compatibility conditions for the removal of the secular terms in $X^{(2)}$ and $Y^{(2)}$ as $\tau \to \infty$. The solution of (2.1.41a, b) shows that the vortex center is drifting with the background flow in the normal time scale plus a small amplitude high frequency oscillation with period ϵ^2 and an amplitude of the order ϵ^2. Thus the leading order trajectory of the vortex center coincides with that of a corresponding spinning disc with the same Γ, a, U_∞, initial velocity $\dot{Z}_0$ and density, $\rho_0 = \rho$. In addition, the amplitude of the nth mode of oscillation of the interface, governed by (2.1.41c) is given by the general solution,

$$c_n(\tau, t) \doteq C_{n1}(t)e^{-in\tau/(2\pi)} + C_{n2}(t)e^{-i(n-1)\tau/(2\pi)}\ , \quad \text{for} \quad n \geq 2\ . \qquad (2.1.42)$$

The dependence of $X^{(2)}$, $Y^{(2)}$ and the c_n's on the normal time t can only be determined by the compatibility conditions of higher order equations. The initial conditions on the amplitude and rate of change of the nth mode oscillation of the interface define $C_{n1}(0)$ and $C_{n2}(0)$ via (2.1.37):

$$\begin{aligned} C_{n1}(0) + C_{n2}(0) &= a_n^{(0)}(0) - ib_n^{(0)}(0)\ , \\ nC_{n1}(0) + (n-1)C_{n2}(0) &= 2\pi\epsilon[i\dot{a}_n^{(0)}(0) + \dot{b}_n^{(0)}(0)]\ . \end{aligned} \qquad (2.1.43)$$

From (2.1.34), (2.1.35), (2.1.39b) and (2.1.41a, b), we obtain the complex velocity potential in the normal length scale outside of the interface,

$$\begin{aligned} w(z) =& (1 - Z_\tau^{(2)})(z - Z) + \frac{1}{2\pi i}\ln(z - Z) - \frac{\epsilon^2 Z_\tau^{(2)}}{z - Z} \\ & - \sum_{n=2,3,\ldots} \epsilon^{n+2}[\frac{(c_n)_\tau}{n} + \frac{ic_n}{2\pi}](z - Z)^{-n}\ , \end{aligned} \qquad (2.1.44)$$

with the understanding that we have obtained only the leading terms in dZ/dt and in the c_n's. The first two terms on the right-hand side of (2.1.44) represent the flow field of a Rankine vortex drifting with the uniform stream. The third term is the contribution to the flow field when the initial velocity of the vortex center differs from the background velocity. This term represents a doublet, with a strength of the order ϵ^2 oscillating at a high frequency $1/(2\pi\epsilon^2)$. The fourth term represents the flow field induced by the high frequency oscillations of the interface when the initial shape of the interface deviates from the circle $r = \epsilon$ by $O(\epsilon^2)$. This contribution is at most of the order ϵ^4. Since the short time averages of the third and fourth terms are zero, the time average, $\bar{w}(t, z)$ of the velocity potential reduces to the first and second terms which are equivalent to the one time solution (2.1.18).

2.1.2 Potential Flow Induced by a Vortex Line

When the core size of a vortex filament approaches zero while its strength, the circulation $\Gamma > 0$, remains finite, the filament becomes a vortex line C defined by the vector function $\mathbf{X}(s,t)$. Here the parameter s is assumed to increase along the direction of the vorticity vector, $\Gamma\hat{\tau}$, where $\hat{\tau}$ denotes the unit tangent vector. The flow field induced by the vortex line is given by the Biot-Savart (B-S) integral in which the time t can be considered as a parameter. In this subsection we study the singular behavior of the B-S integral with the t-dependence suppressed.

To prepare for the analysis, we recall the Serret-Frenet formulae for C,

$$\begin{aligned} \mathbf{X}_s &= \sigma\hat{\tau} \,, & \hat{\tau}_s &= \sigma\kappa\hat{n} \,, \\ \hat{n}_s &= (T\hat{b} - \kappa\hat{\tau})\sigma \,, & \hat{b}_s &= -T\sigma\hat{n} \end{aligned} \tag{2.1.45}$$

where

$$\sigma = |\mathbf{X}_s| \,.$$

Here $\hat{n}$, $\hat{b}$, κ and T denote, respectively, the unit normal and binormal vectors and the curvature and torsion of C. See for example Struik (1961).

Let us consider a closed vortex line of length S and let the parameter s be the arc length. Then $\mathbf{X}$ is a periodic function of s with period S, i. e.,

$$\mathbf{X}(s+S) = \mathbf{X}(s) \quad \text{and} \quad \sigma \equiv 1 \,. \tag{2.1.46}$$

For a closed vortex line, the induced vector potential given by (1.1.13) exists and becomes

$$\mathbf{A}(\mathbf{x}) = \frac{\Gamma}{4\pi}\int_C \frac{d\mathbf{X}(s)}{|\mathbf{x}-\mathbf{X}|} = \frac{\Gamma}{4\pi}\int_0^S \frac{\mathbf{X}_s\, ds'}{|\mathbf{x}-\mathbf{X}(s')|} \tag{2.1.47}$$

and the induced velocity is given by the well known B-S formula,

$$\mathbf{Q}(\mathbf{x}) = \nabla \times \mathbf{A} = -\frac{\Gamma}{4\pi}\int_C \frac{[\mathbf{x}-\mathbf{X}(s')] \times d\mathbf{X}(s')}{|\mathbf{x}-\mathbf{X}(s')|^3} \,. \tag{2.1.48}$$

A direct derivation of the singular and finite parts of the velocity field near the vortex line from (2.1.48) can be found in Callegari and Ting (1978). Here we shall present a similar analysis for the vector potential and then recover the results for the velocity. To study the local behavior, we introduce a local coordinate system moving with C as shown in Fig. 2.3. For any given point $\mathbf{x}$ close enough to the curve C, we can find a point P, $\mathbf{X}(s)$, on the curve in the neighborhood of $\mathbf{x}$ such that the distance $|\mathbf{x}-\mathbf{X}(s)|$ attains its minimum, denoted by r. Thus we write

$$\mathbf{x} = \mathbf{X}(s) + r\hat{r}(\phi, s) \,, \tag{2.1.49}$$

where $\hat{r} = \hat{n}(s)\cos\phi + \hat{b}(s)\sin\phi$ denote the unit radial vector in the normal plane and (r,ϕ) are the polar coordinates of point $\mathbf{x}$ relative to the point P. The differential of the position vector (at a fixed time t) is given by

$$d\mathbf{x} = \hat{r}\, dr + \hat{\phi}\, r[d\phi + T ds] + \hat{\tau}\,(1 - \kappa r\cos\phi)\, d\phi, \tag{2.1.50}$$

where $\hat{\phi} = \hat{b}\cos\phi - \hat{n}\sin\phi$ denotes the circumferential unit vector. Thus, (2.1.49) relates the Cartesian coordinates to the curvilinear coordinates r, ϕ and s which are not orthogonal because of the term $\hat{\phi} rT\, ds$ in (2.1.50). This term can be removed by replacing ϕ by the new variable

$$\theta = \phi - \theta_0(s) \quad \text{with} \quad d\theta_0 = -T(s)\, ds\ . \tag{2.1.51}$$

Then (r, θ, s) form a set of orthogonal coordinates associated with the orthogonal unit vectors $\hat{r}, \hat{\phi}$ and $\hat{\tau}$, and (2.1.50) becomes

$$d\mathbf{x} = \hat{r}\ dr + \hat{\phi}\ r d\theta + \hat{\tau}\ h_3 ds, \qquad \text{where } h_3 = 1 - \kappa r\cos\phi.$$

From the stretch ratios, in particular h_3, it follows that the coordinate transformation (2.1.49) is one to one so long as

$$0 < r < 1/\kappa. \tag{2.1.52}$$

This condition will be observed because the curvilinear coordinates will be employed only for points close to the curve $\mathcal{C}$, i. e., $r\kappa \ll 1$.†

To extract the singular behavior of $\mathbf{A}(\mathbf{x})$ for points close to $\mathcal{C}$, we can work with the natural curvilinear coordinates r, ϕ and s. Note that the singular behavior of $\mathbf{A}$ is generated by that of the integrand in (2.1.47) and that the integrand becomes singular as $r \to 0$ and $s' \to s$. We therefore introduce the new integration variable $\bar{s} = s' - s$ and find the first few terms of the Taylor series of the integrand in $\bar{s}$ for $r > 0$. These terms become singular as $\bar{s}^2 + r^2 \to 0$, while the remainder approaches a finite limiting value. The integral of the remainder can be evaluated numerically for any $r \geq 0$ while the integral over these first few terms is evaluated analytically for $r > 0$. The analytic result defines the singular part(s) of the B–S integral as $r \to 0$.

The Taylor series expansions of the numerator and denominator of the integrand in (2.1.47) are:

$$\mathbf{X}_s(s') = \hat{\tau}(s') = \hat{\tau}(s) + \bar{s}\kappa\hat{n} + O(\bar{s}^2) \tag{2.1.53}$$

and

$$\frac{1}{|\mathbf{x} - \mathbf{X}(s')|} = \frac{1}{\sqrt{r^2 + \bar{s}^2}}\left[1 + \frac{r\bar{s}^2}{2(r^2 + \bar{s}^2)}\kappa\cos\phi + O\left(\frac{r\bar{s}^3}{r^2 + \bar{s}^2}\right) + O\left(\frac{\bar{s}^4}{r^2 + \bar{s}^2}\right)\right]$$

respectively, where $\kappa, \hat{\tau}$ and $\hat{n}$ are evaluated at s. In this case we have to keep the first two terms in the expansions of the integrand to make sure that the remainder does not contribute to the singular parts of of the velocity, $\nabla \times \mathbf{A}$. Denoting the integrand in (2.1.47) by $\mathbf{F}(r, \phi, s, \bar{s})$ and using the expressions in (2.1.53), we obtain

$$\mathbf{F}(r, \phi, s, \bar{s}) = \frac{\hat{\tau}(s + \bar{s})}{|\mathbf{x} - \mathbf{X}(s + \bar{s})|} = \mathbf{G} + \mathbf{H} \tag{2.1.54a}$$

where

---

† This orthogonal curvilinear coordinate system was introduced by Ting (1971) to express the N-S equations in these coordinates for the analysis of the flow field close to the centerline $\mathcal{C}$ of a slender filament. See also **Sec. 2.3**.

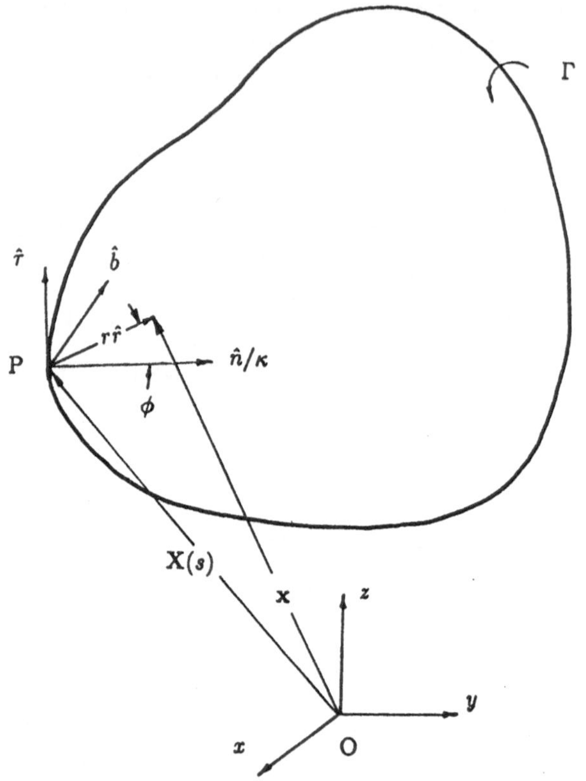

Fig. 2.3. Coordinate system for a vortex line

$$\mathbf{G}(r,\phi,s,\bar{s}) = \hat{\tau}\left[\frac{1}{(r^2+\bar{s}^2)^{\frac{1}{2}}} + \frac{r\bar{s}^2\kappa\cos\phi}{2(r^2+\bar{s}^2)^{\frac{3}{2}}}\right] + \hat{n}\kappa\frac{\bar{s}}{(r^2+\bar{s}^2)^{\frac{1}{2}}} \qquad (2.1.54b)$$

and

$$\mathbf{H}(r,\phi,s,\bar{s}) = \mathbf{F}-\mathbf{G} = O\left(\frac{r\bar{s}^3,\bar{s}^4}{(r^2+\bar{s}^2)^{3/2}}\right) = O(max[r,\bar{s}]). \qquad (2.1.54c)$$

The singular parts of the vector potential are contained in the integral of **G** which can be obtained explicitly as

$$\mathbf{A}^g(\mathbf{x}) = \frac{\Gamma}{4\pi}\int_{-S^-}^{S^+} \mathbf{G}(r,\phi,s,\bar{s})\,d\bar{s} = \mathbf{A}^s + \mathbf{A}^f\,, \qquad (2.1.55)$$

where

$$\mathbf{A}^s = \frac{\Gamma\hat{\tau}}{2\pi}\ln\frac{S}{r} \qquad (2.1.56)$$

denotes the singular part of **A** and

$$\mathbf{A}^f = \frac{\Gamma}{4\pi}\left\{ \hat{\tau}\left[(1+\frac{r\kappa\cos\phi}{2})\ \ln\frac{(S^+ + \sqrt{r^2+(S^+)^2})(S^- + \sqrt{r^2+(S^-)^2})}{S^2} \right.\right.$$
$$\left. - \frac{r\kappa\cos\phi}{2}\left(\ln\frac{r^2}{S^2} + \frac{S^+}{\sqrt{r^2+(S^+)^2}} + \frac{S^-}{\sqrt{r^2+(S^-)^2}}\right)\right]$$
$$\left. + \hat{n}\kappa\left[\sqrt{r^2+(S^+)^2} - \sqrt{r^2+(S^-)^2}\right]\right\} \tag{2.1.57}$$

contributes to the finite part of the integral (2.1.47). Here $-S^-$ and S^+ with

$$S^- > 0\,,\quad S^+ > 0 \qquad \text{and} \qquad S^- + S^+ = S. \tag{2.1.58}$$

are the lower and upper limits of $\bar{s}$ in making one circuit of $\mathcal{C}$. Using (2.1.54a) to (2.1.58), we can express the vector potential as

$$\mathbf{A}(\mathbf{x}) = \mathbf{A}^s + \mathbf{A}^f + \mathbf{A}^h \qquad \text{with} \quad \mathbf{A}^h(\mathbf{x}) = \frac{\Gamma}{4\pi}\int_{-S^-}^{S^+} \mathbf{H}(r,\phi,s,\bar{s})d\bar{s}. \tag{2.1.59}$$

Here $\mathbf{A}^f + \mathbf{A}^h$ represents the finite part of $\mathbf{A}$. Note that the definite integrals for $\mathbf{A}^f$ and $\mathbf{A}^h$ are independent of the choices of S^+ and S^-, so long as they fulfill (2.1.58), because of the periodicity condition (2.1.46). A natural choice is $S^+ = S^- = S/2$.

From the behavior of $\mathbf{A}$ near the vortex line, (2.1.55 – 57), we obtain the singular behavior of the velocity field by computing $\nabla \times \mathbf{A}$. The result is

$$\mathbf{Q}(\mathbf{x}) = \frac{\Gamma}{2\pi r}\hat{\theta} + \frac{\Gamma\kappa}{4\pi}\hat{b}\ln\frac{S}{r} + \frac{\Gamma\kappa}{4\pi}\hat{\theta}\cos\theta + \mathbf{Q}^f\,, \tag{2.1.60}$$

where $\mathbf{Q}^f$ denotes the finite single valued part of the velocity as $r \to 0$. The first singular term on the right-hand side of (2.1.60) represents the circumferential flow around a straight vortex line tangent to the curve $\mathcal{C}$ at $\mathbf{X}(s)$ and the second term represents a singular parallel flow in the direction of the binormal vector $\hat{b}(s)$. The third term points in the circumferential direction and is singular because it has no unique limit as $r \to 0$. Both the second and the third term are proportional to the local curvature κ and are, therefore, absent in the limit of a straight filament. All three terms are vectors in the normal plane of the curve $\mathcal{C}$.

The singular parts of $\mathbf{Q}$ are the same as those derived directly from the B-S formula (2.1.48) by Callegari and Ting (1978). The formula for the finite part Q^f will be presented in Appendix 2.A.

It is evident that the velocity of the vortex line cannot be defined by the velocity of the flow field on the vortex line, $r = 0$, where the velocity is infinite. By intuition one might conjecture that the velocity of the vortex line is defined by the local or circumferential average of the velocity field. This conjecture yields the classical inviscid result for a straight vortex line, with $\kappa \equiv 0$, or for a vortex point in a two-dimensional flow. It does not work for a curved vortex line, $\kappa \neq 0$, because the circumferential average of the second term remains unbounded.

As we shall see in **Sec.2.3**, the velocity of the center line of a slender vortex filament can be defined systematically by the method of matched asymptotics.

The velocity depends on the geometry of the center line and the vortical core structure in addition to the local background velocity. The core structure defines the effective core radius which appears in the binormal velocity as $\Gamma\kappa/(4\pi)$ times $\ln(S/\delta)$ or $\ln(1/\epsilon)+\ln(\epsilon S/\delta)$ and contributes to the finite part $\mathbf{Q}^f$ as $r \to 0$.

To end this subsection, we note that the singular behavior of the vector potential $\mathbf{A}$ was derived for a simple closed vortex line with length $S = O(\ell)$. For an infinitely long vortex line the integral for $\mathbf{A}$ is in general undefined. However, the B-S integral for the velocity is defined because its integrand decays as $|\mathbf{x}-\mathbf{X}(s')|^{-2}$ provided that the length of the segment of C within any finite distance from $\mathbf{x}$ is finite. For example, if the vortex line has the shape of a spiral, it is necessary to assume that the pitch of the spiral is not much smaller than ℓ. This condition is also necessary to insure that the transformation (2.1.49) from the Cartesian coordinates to the curvilinear coordinates in the neighborhood of the vortex line remains one to one. For a vortex line extending to infinity or being very long relative to ℓ, we can construct the vector potential, if needed, in a finite region from the velocity field in order to avoid the difficulty that $\mathbf{A}$ is unbounded at infinity. First we derive the singular and the finite parts of the velocity field using the formulae given by Callegari and Ting (1978). Then, using the equations $\nabla \cdot \mathbf{A} = 0$ and $\nabla \times \mathbf{A} = \mathbf{v}$, we solve for the vector potential $\mathbf{A}$ in the finite region.

2.2 Asymptotic Solutions for Two Dimensional Viscous Vortices

The motion and diffusion of viscous vortices submerged in a potential flow were studied by Ting and Tung (1965) by combining a matched asymptotic analysis in spatial variables with a two-time analysis, in the short and normal time scales. The short time average of the two-time solution was identified as an one-time solution. The physical meaning of the latter and its correspondence with the classical inviscid theory were described. In this section we present the highlights of their analyses.

Consider the N-S equations for an unsteady incompressible flow in the xy plane expressed in terms of the stream function $\psi(t,x,y)$ and vorticity $\zeta(t,x,y)$:

$$\zeta_t + \psi_y\zeta_x - \psi_x\zeta_y = \nu\Delta\zeta \ , \tag{2.2.1}$$

$$\Delta\psi = -\zeta \ . \tag{2.2.2}$$

The initial flow field, say at $t = 0$, is composed of a background potential flow defined by the stream function $\Psi(x,y)$ and the flow field induced by a vorticity distribution $\zeta(0,x,y)$. The vorticity is highly concentrated in a small region, called a vortical core or spot, in which there is a large swirling flow around a point $C(X_0,Y_0)$, called the center of the core.

By a small region we imply that the length scale for the flow field in the vortical core is much smaller than that of the background flow. To characterize this separation of scales, we identify a typical length ℓ and velocity U of the background flow as the reference scales. They will also be referred to as the normal scales. For an incompressible flow, we can always assign the density of the fluid to be unity. Then we choose ℓ, ℓ/U and $\rho\ell^3$ to be the units of length, time and mass respectively, i. e.,

$$\ell = 1 , \qquad U = 1 \qquad \text{and} \qquad \rho = 1 , \tag{2.2.3}$$

so that the symbol for a physical quantity in these units can also be used for the scaled quantity.

To characterize a highly concentrated vorticity distribution in a vortical core, we assume that the vorticity decays rapidly as the distance r to the center C becomes much larger than the effective core size in the sense of (2). The initial effective core size denoted by δ_0 is of the order of a small reference scale, δ^*, which in turn is much smaller than the normal length scale ℓ, i. e., $\epsilon = \delta^*/\ell \ll 1$. The total strength of the vorticity is assumed to be order one, i. e., $\Gamma = O(U\ell)$. Then the scaled vorticity and swirling velocity in the core are $O(\epsilon^{-2})$ and $O(\epsilon^{-1})$ respectively. It is implied that the radial velocity relative to the center C remains of order $O(1)$ so that the path lines inside the core are nearly circular. This completes the description of a *small vortical core* with *highly concentrated* vorticity distribution and a *large* swirling flow.

To account for the diffusion of vorticity while confining it to a small area of the order δ_0^2 on the normal time scale, we introduce the assumption (13) that the Reynolds number of the background flow is large in the sense that

$$Re = U\ell/\nu = 1/\nu = O(\epsilon^{-2}) \qquad \text{or} \qquad \bar{\nu} = \nu/\epsilon^2 = O(1) \; . \tag{2.2.4}$$

With this assumption the asymptotic solution remains valid for any fixed value of $\bar{\nu}$ as $\epsilon \to 0$.

Note that, although we deal with only one viscous vortex, the analysis for a single vortex can be applied to several vortices so long as the distances between them are much greater than their core sizes. Of course, the background flow of one vortex will include the potentials induced by the other vortices and will be unsteady because of the motion of the vortices.

We have assumed that the background potential flow will remain steady. This is not essential because the velocity field is governed by the Laplace equation in which the time t appears merely as a parameter. To compute the unsteady pressure field of the potential flow, i. e., outside the vortical core(s), we assume that the total head, $p+\rho|\mathbf{v}|^2/2$, is uniform initially and hence remains constant for all t. The pressure field is then related to the velocity by the unsteady Bernoulli equation. As we shall see later, the pressure field in a vortical core is defined by the inner velocity field and the matching condition with pressure outside the core.

The evolution of vorticity is governed by (2.2.1). Its left-hand side expresses the transport of vorticity along a particle path line while its right-hand side induces a gradual diffusion of vorticity and a growth of the core size $\delta(t)$. The core size will remain $O(\delta^*)$ on the normal time scale due to the high Reynolds number assumption (2.2.4). From the discussion in **Sec.1.3.4**, we expect that the flow field at an order one distance, $O(\ell)$, away from the vortical core behaves as a potential flow composed of the background flow and the flow induced by a vortex point of strength Γ located at $C(X, Y)$. The question is in what sense and to what degree of accuracy the trajectory of the vortex center C is approximated by the classical inviscid formula (2.1.2). To answer this question we need an inner solution valid in the vortical core. The solution accounts for the diffusion of vorticity and matches with the potential solution (2.1.1) away from the core. Thus we have a typical problem solvable by the method of matched asymptotics.

From our studies of inviscid flows around a spinning disc and a Rankine vortex in **Sec. 2.1.1** and from the above description of a small vortical core centered at point C, we expect that in a neighborhood of C, which is small even relative to δ^*, the circumferential velocity will be $O(1)$ instead of $O(\epsilon^{-1})$. In other words, the fluid or any material point in this small neighborhood should move with order one velocity. We can then introduce the notion of the vortex center $C = (X(t), Y(t))$ which moves as a material point with order one velocity, i. e.,

$$\dot{X}(t) = \psi_y(t, X, Y) = O(1)\ , \quad \dot{Y}(t) = -\psi_x(t, X, Y) = O(1)\ . \tag{2.2.5}$$

This is consistent with (2.0.5) and (2.0.7). The initial position of C is of course (X_0, Y_0), the center of the initial swirling flow.

If the flow field was considered as an initial value problem for the N-S equations formulated in **Sec.1.1**, the velocity field should be defined everywhere including the point C. Note that the initial velocity of the vortex center C is defined by (2.2.5) and can differ from the local background velocity. We now construct the solution of the initial value problem by the method of matched asymptotics using ϵ as the small expansion parameter and expect the velocity of the vortex center to be defined by the asymptotic solution. From the preceding inviscid studies, we see that the flow field has, in addition to the normal time scale, a much smaller time scale of the order ϵ^2 characterizing the high frequency oscillations of the vortical core. The time scale factor ϵ^2 may be assessed from (2.1.9) with density $\rho = 1$ and $a = O(\delta) = O(\epsilon\ell)$.

In the above, we have described many properties of a vortical core based upon physical intuition. It is not clear whether all those properties are consistent with each other and how many of them should be imposed on the solution. In employing the method of matched asymptotics, we introduce ansatzes in the form of expansion schemes for the variables in the inner and outer regions to simulate only two *essential* properties of the vortical flow. The governing equations for the leading and higher order inner and outer solutions will then be obtained systematically. We then verify that the expansion schemes are self consistent and show that the asymptotic solutions have the aforementioned physical or intuitive properties of the flow field, i. e., they are consistent with the two *essential* ones. Usually one states immediately the appropriate expansions for all the variables in order to arrive at the final results as fast as possible. We shall do the same here, but we will single out those few variables for which the expansions are assumed and then show that the expansions for the remaining variables follow from the governing equations.

In the outer region, i. e., at an order one distance away from the vortex center, the flow field is in the normal length and time scales and hence we introduce regular expansions in ϵ, for the stream function and vorticity,

$$\begin{aligned}\psi(t,x,y,\epsilon) &= \psi^{(0)}(t,x,y) + \epsilon\psi^{(1)}(t,x,y) + \cdots \,, \\ \zeta(t,x,y,\epsilon) &= \zeta^{(0)}(t,x,y) + \epsilon\zeta^{(1)}(t,x,y) + \cdots \,.\end{aligned} \tag{2.2.6}$$

In terms of the scaled variables, (2.2.2) remains the same while (2.2.1) becomes

$$\zeta_t + \psi_y\zeta_x - \psi_x\zeta_y = \epsilon^2\bar{\nu}\Delta\zeta \,. \tag{2.2.7}$$

After the substitution of the above series (2.2.6)–(2.2.7) in (2.2.2), making use of (2.2.4) and equating the coefficients of like powers of ϵ, the governing equations for the leading order solution $\psi^{(0)}$ and $\zeta^{(0)}$ are

$$\zeta_t^{(0)} + \psi_y^{(0)}\zeta_x^{(0)} - \psi_x^{(0)}\zeta_y^{(0)} = 0 \qquad \text{and} \qquad \Delta\psi^{(0)} = -\zeta^{(0)} \,. \tag{2.2.8a}$$

They are equivalent to the Euler equations for an inviscid flow. The above expansion scheme permits a background rotational flow of order one vorticity. This problem, which was analyzed by Liu and Ting (1987), will be studied in **Sec. 4.2**. Due to the presence of the vortices, the background vorticity, if nonuniform, will be redistributed and hence the background flow has to be solved simultaneously with the motion of the vortex centers.

Here we present the simpler problem, treated earlier by Ting and Tung (1965), in which the flow field away from the vortical core is assumed to be irrotational at $t = 0$. The leading outer solution, governed by (2.2.8a), remains irrotational , i. e.,

$$\zeta^{(0)} = 0 \,, \quad \Delta\psi^{(0)} = 0 \qquad \text{for } t \geq 0 \,. \tag{2.2.8b}$$

Since the viscous term on the right-hand side of (2.2.7) is two orders higher than the inviscid terms on the left-hand side and the viscous term corresponding to a potential flow is equal to zero, we find that the nth order outer solution is irrotational if the $(n-2)$nd is. It follows by the method of induction that,

$$\zeta^{(n)} = 0 \,, \quad \Delta\psi^{(n)} = 0 \qquad \text{for } n = 0, 1, \ldots \,. \tag{2.2.9}$$

Thus the flow in the outer region is irrotational to all orders in ϵ if it is initially. In particular, the leading outer solution is the sum of a background potential flow and the flow induced by a vortex point in agreement with the classical result (2.1.1), i. e.,

$$\psi(t,x,y) = \Psi(x,y) - \frac{\Gamma}{2\pi}\ln r \,, \tag{2.2.10a}$$

$$\mathbf{v} = \hat{\imath}\psi_y - \hat{\jmath}\psi_x \tag{2.2.10b}$$

and

$$\Psi_x = O(1) \,, \quad \Psi_y = O(1) \,, \tag{2.2.10c}$$

where $r = \sqrt{(x-X)^2 + (y-Y)^2}$ denotes the distance between $\mathbf{x}$ and the center $(X(t), Y(t))$ of the vortical core. Here we have suppressed the superscript (0) on $\psi, \mathbf{v}$ and X, Y and choose the orientation of the coordinate axes to be in the same sense as the circulation, so that we have $\Gamma > 0$. Equation (2.2.10c) restates the fact that the velocity of the background flow is $O(1)$ because its typical velocity has been chosen as the velocity scale in (2.2.3). Note that, although the background flow is steady, the outer stream function ψ is unsteady because of the motion of the vortex center.

The next step is to set up the expansion scheme for the inner solution in the inner spatial variables with two time scales, carry out the inner-outer matching and two-time analyses and identify the meaning of the solution in normal time alone. This step is described in the following subsection **2.2.1**. In the last subsection **2.2.2** we present the leading order solution of the core structure, define the velocity of the vortex center, compare the two-time solution for the trajectory of the vortex center with the one-time solution and then describe the contribution of the inner solution to the higher order outer solution.

2.2.1 The inner solution

Now we study the inner region, in which there is a large swirling flow around the center $C(X,Y)$. The velocity, $\dot{X},\dot{Y}$, of the center is yet to be defined. For the inner region, we replace the coordinates (x,y) by the polar coordinates (r,θ) with respect to the center C and then stretch the radial coordinate r by ϵ. The coordinate transformations are:

$$x = X + \epsilon\bar{r}\cos\theta\,, \quad y = Y + \epsilon\bar{r}\sin\theta \quad \text{and} \quad \bar{r} = r/\epsilon\,. \tag{2.2.11}$$

Let $\tilde{\zeta}$ denote the vorticity in the inner region and $\tilde{\psi}$ the stream function of the velocity field relative to the moving vortex center C. They are functions of two time variables, t and τ, and the stretched polar coordinates $\bar{r},\theta$. Here τ measures on the short time scale $\epsilon^2 l/U$ and is related to the normal time t by

$$\tau = t\epsilon^{-2}\,. \tag{2.2.12}$$

In order to express a function of t, say $f(t)$, as a function $\tilde{f}(\tau,t)$ of two independent variables, t and τ, we assume that $\tilde{f}$ is bounded for all τ and that the average of $\tilde{f}$ over a large interval in τ, called the τ-average, exists. This is the standard ansatz in a two-time analysis. See for example Schneider (1978) or Kevorkian and Cole (1981). The τ-average of $\tilde{f}$, denoted by $\mathcal{M}\tilde{f}$, is defined by

$$\mathcal{M}\tilde{f} = \lim_{T\to\infty}\frac{1}{T}\int_{t\epsilon^{-2}}^{t\epsilon^{-2}+T}\tilde{f}(\tau,t)d\tau\,, \tag{2.2.13}$$

with the understanding that the interval T is large in the short time scale but small in the normal time scale, i. e., $T \ll \epsilon^{-2}$. Note that the τ-average is a function of t only and the averaging operator $\mathcal{M}$ is the identity for a function independent of τ.

With t and τ treated as independent variables, we must replace the partial derivative with respect to t in the original equations with

$$f_t = \frac{\partial\tilde{f}}{\partial t} + \epsilon^{-2}\frac{\partial\tilde{f}}{\partial\tau}\,. \tag{2.2.14}$$

Now the velocity $\tilde{\mathbf{v}}$ is related to the stream function $\tilde{\psi}$ by

$$\tilde{\mathbf{v}} = [\tilde{X}_t + \epsilon^{-2}\tilde{X}_\tau]\,\hat{\imath} + [\tilde{Y}_t + \epsilon^{-2}\tilde{Y}_\tau]\,\hat{\jmath} + \frac{1}{\epsilon}\tilde{\mathbf{V}}\,, \tag{2.2.15a}$$

where

$$\tilde{\mathbf{V}} = \bar{r}^{-1}\tilde{\psi}_\theta\hat{r} - \tilde{\psi}_{\bar{r}}\hat{\theta}\,, \tag{2.2.15b}$$

denotes the velocity relative to the center and $\hat{r}$ and $\hat{\theta}$ represent the unit radial and circumferential vector respectively. The governing equations (2.2.1) and (2.2.2) for the stream function and vorticity as functions of $\tau, t, \bar{r}$ and θ become

$$\tilde{\zeta}_t + \epsilon^{-2}\tilde{\zeta}_\tau + \epsilon^{-2}\tilde{\mathbf{V}}\cdot\bar{\nabla}\tilde{\zeta} = \bar{\nu}\bar{\Delta}\tilde{\zeta}\,, \tag{2.2.16a}$$

$$\epsilon^{-2}\bar{\Delta}\tilde{\psi} = -\tilde{\zeta}\,, \tag{2.2.16b}$$

where $\bar{\nabla}$ and $\bar{\Delta}$ denote the gradient and Laplacian operators in the inner polar coordinates, $\bar{r}, \theta$. The relative velocity $\tilde{\mathbf{V}}$ vanishes at the center C because of (2.2.5), i. e., C is a material point. This leads to two boundary conditions,

$$\tilde{\psi} = 0 \quad \text{and} \quad \tilde{\psi}_{\bar{r}} = 0 \quad \text{at } \bar{r} = 0\,. \tag{2.2.17}$$

The standard matching of the inner and outer solutions from (2.2.15) and (2.2.10) in their overlap region, where $\bar{r} \gg 1$ while $r = \epsilon\bar{r} \ll 1$, yields

$$\frac{d\mathbf{X}}{dt} + \frac{1}{\epsilon}\left[\frac{1}{\bar{r}}\tilde{\psi}_\theta\hat{r} - \psi_{\bar{r}}\hat{\theta}\right] = \hat{\imath}\psi_y - \hat{\jmath}\psi_x = \epsilon^{-1}\frac{\Gamma}{2\pi\bar{r}}\hat{\theta} + (\Psi_y\hat{\imath} - \Psi_x\hat{\jmath}) + O(\epsilon)\,. \tag{2.2.18}$$

The stream function $\tilde{\psi}$ and vorticity $\tilde{\zeta}$ are expanded in power series of ϵ as:

$$\tilde{\psi}(\tau, t, \bar{r}, \theta, \epsilon) = \tilde{\psi}^{(0)}(\tau, t, \bar{r}, \theta) + \epsilon\tilde{\psi}^{(1)}(\tau, t, \bar{r}, \theta) + \cdots\,, \tag{2.2.19a}$$

$$\tilde{\zeta}(\tau, t, \bar{r}, \theta, \epsilon) = \epsilon^{-2}[\tilde{\zeta}^{(0)}(\tau, t, \bar{r}, \theta) + \epsilon\tilde{\zeta}^{(1)}(\tau, t, \bar{r}, \theta) + \cdots]\,. \tag{2.2.19b}$$

Note that, once the expansion (2.2.19a) for $\tilde{\psi}$ is assumed, the expansion (2.2.19b) for $\tilde{\zeta}$ beginning with an $O(\epsilon^{-2})$ term follows from (2.2.16b).

Since the flow field in a vortical core is characterized by a large, $O(\epsilon^{-1})$, swirling flow but order one radial flow, we require

$$\epsilon^{-1}\frac{1}{\bar{r}}\tilde{\psi}_\theta = O(U) \qquad \text{and hence} \qquad \tilde{\psi}^{(0)}_\theta = 0\,. \tag{2.2.20}$$

This is also consistent with the leading order matching condition (2.2.18). Since the outer solution has only a circumferential velocity component of order $O(\epsilon^{-1})$ the leading order matching condition yields

$$\tilde{\psi}^{(0)}_{\bar{r}} \to \Gamma/(2\pi\bar{r}) \qquad \text{as } \bar{r} \to \infty \tag{2.2.21a}$$

and

$$\epsilon^{-2}\tilde{X}_\tau + X_t = O(1)\,, \qquad \epsilon^{-2}\tilde{Y}_\tau + Y_t = O(1)\,. \tag{2.2.21b}$$

Condition (2.2.21b) in turn dictates the following expansions for the coordinates of the vortex center:

$$X(\bar{t},t,\epsilon) = \bar{X}^{(0)}(t) + \epsilon \bar{X}^{(1)}(t) + \epsilon^2 \tilde{X}^{(2)}(\tau,t) + \cdots \, ,$$
$$Y(\bar{t},t,\epsilon) = \bar{Y}^{(0)}(t) + \epsilon \bar{Y}^{(1)}(t) + \epsilon^2 \tilde{Y}^{(2)}(\tau,t) + \cdots \, . \tag{2.2.22}$$

Note that we are using bars and tildes over functions to distinguish a function depending only on the normal time from one that depends on both time variables. The expansion (2.2.22) is consistent with the physical expectation (2.2.5), which says that the center moves at an order one velocity. Thus the expansions for all the variables in the inner region are formulated. We want to point out once more that these expansions follow from only two basic assumptions : The regular expansion (2.2.19a) of the stream function in the inner variables and the assumption (2.2.20) of only a large swirling flow.

Now we carry out the systematic derivations of the governing equations for the leading and higher order solutions in two steps: In step i) we substitute the power series (2.2.19a, b) and (2.2.22) in the differential equations (2.2.16a and b), the auxiliary equations (2.2.15a and b), the boundary conditions (2.2.17) and the matching condition (2.2.18) and in step ii) we equate the coefficients of like powers of ϵ. The initial conditions will be introduced later when we construct the asymptotic solution.

Equation (2.2.16b) yields

$$\bar{\Delta}\tilde{\psi}^{(n)} = -\tilde{\zeta}^{(n)} \, , \quad \text{for } n = 0, 1, \ldots \, . \tag{2.2.23}$$

Likewise, the boundary conditions (2.2.17) yield

$$\tilde{\psi}^{(n)} = 0 \quad \text{and} \quad \tilde{\psi}^{(n)}_{\bar{r}} = 0 \qquad \text{at} \qquad \bar{r} = 0 \, . \tag{2.2.24}$$

From (2.2.23) for $n = 0$ and (2.2.20), we obtain

$$\zeta^{(0)}_{\theta} = 0 \, . \tag{2.2.25a}$$

Using this and (2.2.20), we equate the coefficient of ϵ^{-4} in (2.2.16a) to zero and obtain

$$\tilde{\zeta}^{(0)}_{\tau} = 0 \, . \tag{2.2.25b}$$

Using (2.2.23) for $n = 0$ and (2.2.25a, b) and (2.2.20), we conclude that the leading order vorticity and stream function are independent of τ and θ, i. e.,

$$\tilde{\zeta}^{(0)} = \bar{\zeta}^{(0)}(t,\bar{r}) \tag{2.2.26a}$$
$$\tilde{\psi}^{(0)} = \bar{\psi}^{(0)}(t,\bar{r}) \, . \tag{2.2.26b}$$

The stream function is related to vorticity by solving (2.2.23) and (2.2.24) for $n = 0$,

$$\bar{\psi}^{(0)}(t,\bar{r}) = -\int_0^{\bar{r}} \frac{d\xi}{\xi} \int_0^{\xi} \xi' \zeta^{(0)}(t,\xi') d\xi' = -\int_0^{\xi} \xi' \zeta^{(0)}(t,\xi') \ln \frac{\bar{r}}{\xi'} d\xi' \, . \tag{2.2.26c}$$

By equating the coefficients of ϵ^{-3} and ϵ^{-2} in (2.2.16a) to zero, we obtain respectively,

$$\tilde{\zeta}_{\tau}^{(1)} + \mathcal{L}[\tilde{\psi}^{(1)}, \tilde{\zeta}^{(1)}] = 0 , \tag{2.2.27a}$$

where $\mathcal{L}$ denotes the linear operator,

$$\mathcal{L}[\tilde{\psi}^{(n)}, \tilde{\zeta}^{(n)}] = \frac{1}{\bar{r}} \frac{\partial}{\partial \theta} [\bar{\zeta}_{\bar{r}}^{(0)} \tilde{\psi}^{(n)} - \bar{\psi}_{\bar{r}}^{(0)} \tilde{\zeta}^{(n)}] , \quad n = 1, 2, \ldots \tag{2.2.27b}$$

and

$$\tilde{\zeta}_{\tau}^{(2)} + \mathcal{L}[\tilde{\psi}^{(2)}, \tilde{\zeta}^{(2)}] = \tilde{F}_2(\tau, t, \bar{r}, \theta) + \bar{G}_2(t, \bar{r}) , \tag{2.2.28a}$$

where

$$\tilde{F}_2 = -\frac{1}{\bar{r}} [\tilde{\psi}_{\theta}^{(1)} \tilde{\zeta}_{\bar{r}}^{(1)} - \tilde{\zeta}_{\theta}^{(1)} \tilde{\psi}_{\bar{r}}^{(1)}] , \tag{2.2.28b}$$

$$\bar{G}_2 = -\bar{\zeta}_t^{(0)} + \bar{\nu} \bar{\Delta} \bar{\zeta}^{(0)} . \tag{2.2.28c}$$

By using (2.2.23) for $n = 1, 2$, (2.2.27a) and (2.2.28a) become linear equations for $\tilde{\psi}^{(1)}$ and $\tilde{\psi}^{(2)}$. Equation (2.2.28a) is inhomogeneous. The first inhomogeneous term $\tilde{F}_2$ comes from the nonlinear convective terms of the first order solution while the second term $\bar{G}_2$ represents the normal time derivative and the linear viscous terms for the leading order solution.

The τ-average of (2.2.27a, b) and (2.2.28a, b) become respectively,

$$\mathcal{M}\mathcal{L}[\tilde{\psi}^{(1)}, \tilde{\zeta}^{(1)}] = \frac{1}{\bar{r}} \frac{\partial}{\partial \theta} [\bar{\zeta}_{\bar{r}}^{(0)} (\mathcal{M}\tilde{\psi}^{(1)}) - \bar{\psi}_{\bar{r}}^{(0)} (\mathcal{M}\tilde{\zeta}^{(1)})] = 0 , \tag{2.2.29}$$

$$\begin{aligned} \mathcal{M}\mathcal{L}[\tilde{\psi}^{(2)}, \tilde{\zeta}^{(2)}] &= \frac{1}{\bar{r}} \frac{\partial}{\partial \theta} [\bar{\zeta}_{\bar{r}}^{(0)} (\mathcal{M}\tilde{\psi}^{(2)}) - \bar{\psi}_{\bar{r}}^{(0)} (\mathcal{M}\tilde{\zeta}^{(2)})] \\ &= \mathcal{M}\tilde{F}_2 + \bar{G}_2(t, \bar{r}) . \end{aligned} \tag{2.2.30}$$

The next step is to show that $\tilde{F}_2$ defined by (2.2.28b) can be expressed as a combination of derivatives with respect to θ and τ. By using (2.2.26a, b), (2.2.27a) is rewritten as

$$\tilde{\psi}_{\theta}^{(1)} = \bar{a} \tilde{\zeta}_{\tau}^{(1)} + \bar{b} \tilde{\zeta}_{\theta}^{(1)} , \qquad \text{where } \bar{a}(t, \bar{r}) = -\bar{r}/\bar{\zeta}_{\bar{r}}^{(0)} , \quad \bar{b}(t, \bar{r}) = \bar{\psi}_{\bar{r}}^{(0)}/\bar{\zeta}_{\bar{r}}^{(0)} \tag{2.2.31}$$

which is then used to reduce $\tilde{F}_2$ to

$$\tilde{F}_2(\tau, t, \bar{r}, \theta) = -\frac{1}{\bar{r}} \left[\frac{1}{2} [\bar{a} (\tilde{\zeta}^{(1)})^2]_{\bar{r}\tau} + \frac{1}{2} [\bar{b} (\tilde{\zeta}^{(1)})^2]_{\bar{r}\theta} - [\tilde{\zeta}^{(1)} \tilde{\psi}_{\bar{r}}^{(1)}]_{\theta} \right] \tag{2.2.32a}$$

and the τ-average of $\tilde{F}_2$ becomes

$$\mathcal{M}\tilde{F}_2 = -\frac{1}{\bar{r}} \frac{\partial}{\partial \theta} \mathcal{M} \left[\frac{1}{2} [\bar{b} (\tilde{\zeta}^{(1)})^2]_{\bar{r}} - \tilde{\psi}_{\bar{r}}^{(1)} \tilde{\zeta}^{(1)} \right] . \tag{2.2.32b}$$

Using the above equation, we integrate (2.2.30) with respect to θ from 0 to 2π to obtain the compatibility condition for the second order equation (2.2.30). The condition is $\bar{G}_2 = 0$ or

$$\bar{\zeta}_t^{(0)} = \frac{\bar{\nu}}{\bar{r}} [\bar{r} \bar{\zeta}_{\bar{r}}^{(0)}]_{\bar{r}} . \tag{2.2.33}$$

This in turn serves as the equation to define the leading order vorticity $\bar{\zeta}^{(0)}(t, \bar{r})$. For given initial data $\bar{\zeta}^{(0)}(0, \bar{r})$ the solution of the simple axisymmetric diffusion equation (2.2.3) is (Carslow and Jaeger 1959),

$$\bar{\zeta}^{(0)}(t,\bar{r}) = \frac{1}{2\bar{\nu}t}\int_0^\infty \bar{\zeta}^{(0)}(0,\xi)e^{-(\bar{r}^2+\xi^2)/(4\bar{\nu}t)}\, I_0(\frac{\bar{r}\xi}{2\bar{\nu}t})\,\xi d\xi\ , \tag{2.2.34}$$

where I_0 denotes the modified Bessel function. The corresponding stream function $\bar{\psi}^{(0)}$ is then defined by (2.2.26c) as a weighted integral of $\bar{\zeta}^{(0)}$. These integral representations do not show clearly the behavior of the solution for large t and are too cumbersome to be used in the analysis of the next order equations (2.2.27), in which they appear as coefficients in the linear operator $\mathcal{L}$ on $\tilde{\psi}^{(1)}$ and $\tilde{\zeta}^{(1)}$. In **Sec. 2.2.2**, we construct an alternate solution of (2.2.33) in power series of t^{-1} and discuss the physical meaning of the leading two terms in the series as $t \to \infty$.

The results in (2.2.26a) and (2.2.26b) say that the leading order two-time inner solution is independent of the short time variable τ. The question is whether one obtains the same leading order solution directly from an one-time analysis, in which the solution *is assumed* to depend only on the normal time t from the outset. This question was answered by Ting and Tung (1965). They carried out the one-time analysis and found that the leading order solution is governed by the same simple axisymmetric diffusion equation (2.2.33) with the same initial data and boundary conditions. Thus, we conclude :

- *The leading order solution from the two-time analysis is axisymmetric and independent of the short time variable τ and it is identical with the leading solution from the one-time analysis.*

Now we proceed to analyze the higher order solutions. Since their governing equations are linear, it is convenient to decompose the solutions into symmetric and asymmetric parts, denoted by the subscripts c and a respectively. We write

$$f(\theta) = f_c + f_a(\theta)\ , \qquad \text{with} \qquad f_c = \frac{1}{2\pi}\int_0^{2\pi} f(\theta)d\theta \stackrel{\text{def}}{=} \{f\}\ , \tag{2.2.35}$$

where f stands for $\zeta^{(n)}$ or $\psi^{(n)}$ showing only their dependence on θ. The symmetric part f_c is the circumferential average or the θ-average of f. From here on we use the curly brackets, $\{\ \ \}$, to denote the θ-averaging operator. By using this decomposition for $n = 1$, (2.2.27a) splits into separate equations for the symmetric and asymmetric parts of the first order solution. They are :

$$[\tilde{\zeta}_c^{(1)}]_\tau = 0 \tag{2.2.36a}$$

and

$$[\tilde{\zeta}_a^{(1)}]_\tau + \frac{1}{\bar{r}}[\bar{\zeta}_{\bar{r}}^{(0)}\tilde{\psi}_a^{(1)} - \bar{\psi}_{\bar{r}}^{(0)}\tilde{\zeta}_a^{(1)}]_\theta = 0\ . \tag{2.2.36b}$$

Equation (2.2.36a) says that the symmetric part of the first order vorticity is independent of τ, i. e.,

$$\tilde{\zeta}_c^{(1)}(\tau,t,\bar{r}) = \bar{\zeta}_c^{(1)}(t,\bar{r})\ , \tag{2.2.37a}$$

and from (2.2.23) and (2.2.24) and the far field conditions on $\psi^{(1)}$, we get

$$\tilde{\psi}_c^{(1)}(\tau,t,\bar{r}) = \bar{\psi}_c^{(1)}(t,\bar{r}). \tag{2.2.37b}$$

We conclude at this stage that the symmetric parts of the first order inner solutions, $\bar{\zeta}_c^{(1)}$ and $\bar{\psi}_c^{(1)}$, do not have any fast time variation. Similar to the axisymmetric parts of the leading order solution, their dependence on $\bar{r}$ and t will be determined by compatibility conditions for the third order solutions.

The asymmetric parts, $\tilde{\zeta}_a^{(1)}$ and $\tilde{\psi}_a^{(1)}$, are governed by the linear system (2.2.36b), (2.2.23). With the boundary conditions (2.2.24) on $\tilde{\psi}_a^{(1)}$ and an initial data for $\tilde{\zeta}_a^{(1)}$, the first order asymmetric parts are defined.

Finally, consider the short time averages of the asymmetric parts,

$$\mathcal{M}\tilde{\psi}_a^{(1)}(\tau, t, \bar{r}, \theta) = \bar{\psi}_a^{(1)}(t, \bar{r}, \theta) \tag{2.2.38}$$

and

$$\mathcal{M}\tilde{\zeta}_a^{(1)}(\tau, t, \bar{r}, \theta) = \bar{\zeta}_a^{(1)}(t, \bar{r}, \theta) \,. \tag{2.2.39}$$

They are governed by the quasi-steady equations (2.2.23), (2.2.29) and (2.2.24), in which t appears only as a parameter through the leading order solutions $\zeta^{(0)}(t, \bar{r})$ and $\psi^{(0)}(t, \bar{r})$.

Again we raise the question of whether the symmetric parts of the first order solution, $\bar{\zeta}_c^{(1)}, \bar{\psi}_c^{(1)}$, and the τ-average of the asymmetric parts, $\bar{\zeta}_a^{(1)}, \bar{\psi}_a^{(1)}$, are equivalent to the corresponding symmetric and asymmetric parts of the first order solution obtained directly by the one-time analysis. Again, Ting and Tung (1965) provided the answer. They showed that the asymmetric parts from the one-time analysis obey the same differential equations and boundary conditions as $\bar{\zeta}_a^{(1)}$ and $\bar{\psi}_a^{(1)}$ since the leading order solutions, which appear as coefficients in these equations, are identical. They conclude:

- *The asymmetric first order solution from the one-time analysis is equivalent to the τ-average of the corresponding two-time solution.*

Note that the variable t appears as a parameter in the asymmetric solution from the one-time analysis. Therefore, the solution cannot accept any initial data unless they are compatible with (2.2.23) and (2.2.29). If they are compatible, then the two-time asymmetric solution degenerates to its τ-average, i. e., the one-time solution.

As regards the first order symmetric solutions, the answer to our question is negative, because they are not governed by the same set of equations. The governing equation for $\mathcal{M}\tilde{\zeta}_c^{(1)} = \mathcal{M}\{\tilde{\zeta}^{(1)}\}$, is given by the θ- and τ-average of the third order vorticity equation (2.2.16a). There are nonlinear inhomogeneous terms induced by the averages of the nonlinear convection terms, which involve products of the first order and second order solutions. These terms will be absent in the corresponding equation in the one-time analysis. Those nonlinear inhomogeneous terms of the two-time solutions will be shown explicitly in Appendix A.1. There we present a systematic procedure to derive the governing equations for the symmetric and asymmetric parts of the nth order two-time solution and the equations for their τ-averages.

Now we proceed to derive the equations for the second order solutions. As before, we decompose (2.2.28a) to obtain

$$[\tilde{\zeta}_c^{(2)}]_\tau = \{\tilde{F}_2\}, \tag{2.2.40a}$$

$$[\tilde{\zeta}_a^{(2)}]_\tau + \mathcal{L}[\tilde{\psi}_a^{(2)}, \tilde{\zeta}_a^{(2)}] = \tilde{F}_2 - \{\tilde{F}_2\}, \tag{2.2.40b}$$

where

$$\{\tilde{F}_2\} = \frac{1}{2\bar{r}} \left[\bar{a}\left[(\tilde{\zeta}_c^{(1)})^2 + \{(\tilde{\zeta}_a^{(1)})^2\}\right]\right]_{\bar{r}\tau}. \tag{2.2.40c}$$

Here we made use of (2.2.32a) and (2.2.33). The last equation (2.2.40c) was obtained by taking the θ- and τ- averaging of (2.2.32a).

Since (2.2.40b) involves products of the first order solutions through $\tilde{F}_2$, the τ-average of the second order asymmetric two-time solution is, in general, not equivalent to the corresponding asymmetric solution from the one-time analysis. Only when the first order two-time solution is of a certain special form that renders the τ-average of the products equal to zero, are these two solutions equivalent. This exceptional case will be discussed later in **Sec.2.2.4** where we study the first order two-time solutions.

In the following subsection we present a power series representation of the leading order core structure and explain the physical meaning of the first two terms. In **Sec. 2.2.3**, we obtain the τ-average of the asymmetric first order solutions and then define the velocity of the vortex center in the normal time scale. The two-solutions, their dependence on the core structure and their physical meaning will be presented in **Sec. 2.2.3** and **2.2.4**.

2.2.2 Leading order core structure

We recall that the leading order core structure, $\epsilon^{-2}\bar{\zeta}^{(0)}(t,\bar{r})$ and $\bar{\psi}^{(0)}(t,\bar{r})$, from the two-time analysis is identical to that from the one-time analysis. The solution is related to the initial data by the integral representations (2.2.34) and (2.2.26c). Those integral representations do not show clearly the behavior of the solution for small or large $\bar{r}$ or for large t and are too cumbersome to be used as the coefficients in the governing equations for the higher order solutions. We want to approximate the solution by ***an optimum similarity solution***. We shall introduce a mathematical definition of the optimum solution and explain its physical meaning.

Although we can derive the similarity solution as the leading term of an asymptotic series for large t from the integral representation (2.2.34), we find it more revealing to construct the series solution directly. By making use of the fact that the vorticity decays exponentially in $\bar{r}$, we represent the solution of (2.2.33) in a descending power series of $t + t_0$ where $t_0 > 0$ denotes a constant positive time shift so that the series solution is valid at $t = 0$. An appropriate choice of t_0 will be discussed later. The series solution is constructed in terms of two new variables,

$$\bar{t} = t + t_0 \qquad \text{and} \qquad \bar{\eta}^2 = \bar{r}^2/(4\bar{\nu}\bar{t}). \tag{2.2.41}$$

Equation (2.2.33) is separable and the condition of exponential decay of $\bar{\zeta}^{(0)}$ leads to a discrete spectrum of eigenvalues.† The series solution is

† Note that this is the only occasion where we use explicitly the condition of exponential decay, so that (2) in the **Introduction** holds for all N.

$$\bar{\zeta}^{(0)}(t,\bar{r}) = \sum_{n=0,1\ldots} C_n\, \bar{t}^{-n-1}\, L_n(\bar{\eta}^2) e^{-\bar{\eta}^2} . \tag{2.2.42}$$

Here L_n denotes the nth Laguerre polynomial (Magnus et al 1966), and the coefficients C_n are related to the initial data at $t = 0$ or $\bar{t} = t_0 > 0$ by

$$C_n = 2t_0^{n+1} \int_0^\infty \bar{\zeta}^{(0)}(0, \bar{\eta}\sqrt{4\bar{\nu}t_0}) L_n(\bar{\eta}^2)\, \bar{\eta}\, d\bar{\eta} . \tag{2.2.43}$$

In particular, the first coefficient is $C_0 = \Gamma/(4\pi\bar{\nu})$ and the first term in (2.2.42) is known as the *similarity solution* corresponding to a Lamb vortex created at the instant $t = -t_0$ with zero core radius. It represents the leading term of the vorticity distribution (2.2.42) for large t. We define the optimum time shift t_0^* by the condition that the second term in the series (2.2.39) vanishes, i.e., that $C_1 = 0$. The condition is

$$C_1 = 2(t_0^*)^2 \int_0^\infty \bar{\zeta}^{(0)}(0, \bar{\eta}\sqrt{4\bar{\nu}t_0^*})\, [1 - \bar{\eta}^2]\, \bar{\eta}\, d\bar{\eta} = 0 , \tag{2.2.44a}$$

or

$$2\pi \int_0^\infty \bar{\zeta}(0,\bar{r})\, \bar{r}^3\, d\bar{r} = 4\bar{\nu}t_0^*\Gamma , \tag{2.2.44b}$$

which in turn defines the optimum time shift,

$$t_0^* = \frac{\pi}{2\Gamma\bar{\nu}} \int_0^\infty \bar{\zeta}^{(0)}(0,\bar{r})\, \bar{r}^3\, d\bar{r} . \tag{2.2.44c}$$

The vorticity distribution defined by the first term of the series (2.2.42) with $t_0 = t_0^*$ is called the optimum similarity solution or the optimum Lamb vortex of age t_0^* at the instant $t = 0$ with effective core size $\delta_0 = 4\bar{\nu}t_0^*$. The reasons for this notion are twofold: From the series solution (2.2.42) and condition (2.2.44a), we see the mathematical reason,

- *The optimum similarity solution is the best one term asymptotic solution in the sense that it is the only one that differs from the exact solution by $O(\bar{t}^{-3})$ instead of $O(\bar{t}^{-2})$ as $t \to \infty$, that is,*

$$\bar{\zeta}^{(0)} = \bar{\zeta}^* + O(\bar{t}^{-3}) \quad \text{where} \quad \bar{\zeta}^* = \frac{\Gamma}{4\bar{\nu}(t+t_0^*)} e^{-\bar{r}^2/[4\pi\bar{\nu}(t+t_0^*)]} . \tag{2.2.45}$$

From the fact that the right-hand side of (2.2.44b) is equal to the polar moment of the optimum Lamb vortex, while its left-hand side is the polar moment of the initial vorticity distribution and from the linear diffusion law of the polar moment of vorticity, (1.2.28b), we get the physical justification,

- *the optimum similarity solution and the exact solution (2.2.42) not only have the same total strength Γ but also the same polar moment for all $t \geq 0$.*

Of course, they always have the same first moments which are equal to zero due to symmetry.

Figure 2.4 compares the optimum solution with the exact solution having an initial vorticity distribution in the shape of a top hat, $\zeta(0,r) = \Gamma/(\pi\delta_0^2)H(\delta_0 - r)$,

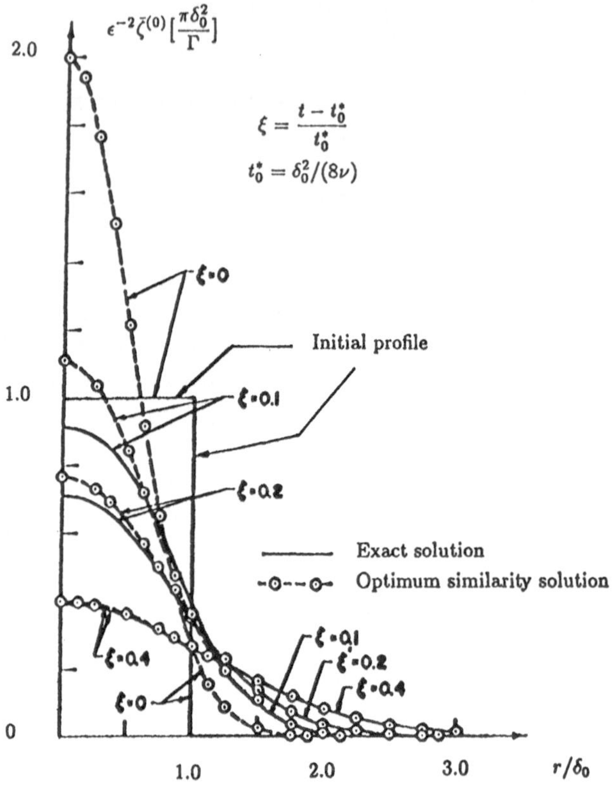

Fig. 2.4. Comparison of the optimum similarity solution and the exact solution. (Ting 1971)

where $H(\cdot)$ stands for the Heaviside function. In this case, the optimum time shift is $t_0^* = \delta_0^2/(8\nu)$. Good agreement between the optimum similarity solution and the exact one is achieved for $\bar{t}/t_0^* \sim 3$. Better agreement would be expected for a more realistic initial vorticity distribution in the shape of a bell, which is a continuous function of $r \geq 0$ decreasing monotonically from its maximum at $r = 0$ to zero. The importance of an optimum similarity solution becomes more apparent when one takes into account that in a real problem, details of the initial data beyond the total strength and an estimate of the effective size are rarely available. The optimum similarity solution could, in such cases, serve as a reasonable model for the core structure for all $t \geq 0$.

The diffusion of the leading vorticity distribution, $\epsilon^{-2}\bar{\zeta}^{(0)}$, which is axisymmetric, is defined by the exact solution (2.2.42) or the optimum similarity solution (2.2.45). The corresponding flow field is the leading circumferential flow, $-\epsilon^{-1}\bar{\psi}_{\bar{r}}^{(0)}$ with

$$-\epsilon^{-1}\bar{\psi}_{\bar{r}}^{(0)}(t,\bar{r}) = \frac{\epsilon^{-1}}{\bar{r}} \int_0^{\bar{r}} \xi\bar{\zeta}^{(0)}(t,\xi)d\xi \ . \tag{2.2.46}$$

As $\bar{r} \to \infty$, it matches with the outer solution of a vortex point as expected,

$$-\epsilon^{-1}\bar{\psi}_{\bar{r}}^{(0)}(t,\bar{r}) \to \frac{\Gamma}{2\pi r} + \cdots \qquad \text{where} \qquad \Gamma = 2\pi \int_0^\infty \bar{\zeta}^{(0)}(t,\bar{r})\,\bar{r}\,d\bar{r} \ . \tag{2.2.47}$$

In particular, the flow field corresponding to the similarity solution is that of a Lamb vortex,

$$-\epsilon^{-1}\psi_{\bar{r}}^{(0)} = \frac{\Gamma}{2\pi r}\left[1 - e^{-r^2/[4\nu(t+t_0^*)]}\right] . \tag{2.2.48}$$

2.2.3 Asymmetric first order solution in normal time scale and the velocity of the vortex

It was concluded in **Sec. 2.2.1** that an asymmetric first order one-time solution in the normal time t can be considered either as the τ-average of an asymmetric first order two-time solution or as the asymmetric solution from the one-time analysis, since these two are equivalent. The asymmetric first order vorticity and stream function are governed by the differential equations (2.2.23 and 29), the boundary conditions (2.2.24) at $\bar{r} = 0$ and the matching conditions (2.2.18) and (2.2.21a, b). In this subsection, we construct the solution of that system of equations, and show that in the normal time scale the leading order velocity of the vortex center is equal to the local background velocity.

By the elimination of $\bar{\zeta}_a^{(1)}$ from (2.2.23 and 29), we obtain a single equation for the asymmetric stream function $\bar{\psi}_a^{(1)}$,

$$[\bar{\psi}_{\bar{r}}^{(0)}\bar{\Delta} + \bar{\zeta}_{\bar{r}}^{(0)}]\,(\bar{\psi}_a^{(1)})_\theta = 0 \ . \tag{2.2.49}$$

It is a linear partial differential equation in $\bar{r}$ and θ while the time t appears as a parameter. If the initial data are inconsistent with (2.2.49) and/or a matching condition, we need a two-time solution in order to fulfill the initial data. This will be discussed in **Sec. 2.2.4**.

Since the solutions have to be periodic in θ, we can express the asymmetric solution in a Fourier series,

$$\bar{\psi}_a^{(1)}(t,\bar{r},\theta) = \sum_{j=1,2,\ldots} [\bar{\psi}_{j1}(t,\bar{r})\cos j\theta + \bar{\psi}_{j2}(t,\bar{r})\sin j\theta] \ . \tag{2.2.50}$$

Substituting (2.2.50) in (2.2.49), we obtain the equations for the Fourier coefficients,

$$\{\frac{\partial^2}{\partial\bar{r}^2} + \frac{1}{\bar{r}}\frac{\partial}{\partial\bar{r}} + [\frac{\bar{\zeta}_{\bar{r}}^{(0)}}{\bar{\psi}_{\bar{r}}^{(0)}} - \frac{j^2}{\bar{r}^2}]\}\,\bar{\psi}_{jk} = 0 \ , \tag{2.2.51}$$

for $k = 1, 2$ and $j = 1, 2 \ldots$.

The boundary conditions (2.2.24) at $\bar{r} = 0$ yield

$$\bar{\psi}_{jk} = 0 \qquad \text{and} \qquad (\bar{\psi}_{jk})_{\bar{r}} = 0 \ . \tag{2.2.52}$$

With the leading order circumferential velocity matched by (2.2.47), the matching condition (2.2.18) becomes

$$\frac{1}{\bar r}\bar\psi^{(1)}_{\theta}\hat r-\bar\psi^{(1)}_{\bar r}\hat\theta \to [-\dot X^{(0)}+\Psi_y(t,X^{(0)},Y^{(0)})]\,\hat\imath-[\dot Y^{(0)}+\Psi_x(t,X^{(0)},Y^{(0)})]\,\hat\jmath\,, \quad (2.2.53)$$

as $\bar r \to \infty$. This in turn yields the conditions on the Fourier coefficients,

$$\bar\psi_{11} \to [\dot Y^{(0)} + \Psi_x(t,X^{(0)},Y^{(0)})]\,\bar r\,, \quad (2.2.54a)$$
$$\bar\psi_{12} \to [-\dot X^{(0)} + \Psi_y(t,X^{(0)},Y^{(0)})]\,\bar r \quad (2.2.54b)$$

and

$$\bar\psi_{jk} \to 0 \qquad \text{for} \quad j=2,3,\dots\,, \quad k=1,2\,, \quad (2.2.54c)$$

as $\bar r \to \infty$. With $\bar\psi^{(0)}_{\bar r\bar r}(t,0) = -\bar\zeta^{(0)}(t,0)/2 \neq 0$, while $\bar\zeta^{(0)}_{\bar r}(t,0) = 0$, the ratio $\bar\zeta^{(0)}_{\bar r}/\bar\psi^{(0)}_{\bar r}$ remains finite as $\bar r \to 0$ and hence (2.2.51) has a regular singular point at $\bar r = 0$. Near this point, the solution $\bar\psi_{jk}$ can be expressed in terms of its two independent solutions as

$$\bar\psi_{jk} \sim c_1(t)\,\bar r^{j} + c_2(t)\,\bar r^{-j}\,. \quad (2.2.55)$$

For any $j \geq 2$, the first condition in (2.2.52) requires that $c_2 = 0$ while the second condition is fulfilled for all c_1. The matching condition (2.2.54c) then requires $c_1 = 0$. Consequently, we have

$$\bar\psi_{jk} \equiv 0 \qquad \text{for} \quad j=2,3,\dots\,, \quad k=1,2\,. \quad (2.2.56)$$

For $j = 1$, the two conditions in (2.2.52) require both $c_1 = 0$ and $c_2 = 0$ and hence

$$\bar\psi_{11} \equiv 0 \qquad \text{and} \qquad \bar\psi_{12} \equiv 0\,. \quad (2.2.57)$$

The matching conditions (2.2.54a and b) then become the equation for the velocity of the vortex center,

$$\hat\imath\,\dot X^{(0)}(t) + \hat\jmath\,\dot Y^{(0)}(t) = \hat\imath\,\Psi_y(X^{(0)},Y^{(0)}) - \hat\jmath\,\Psi_x(X^{(0)},Y^{(0)})\,. \quad (2.2.58)$$

This equation says :

- *In the normal time scale, the leading order velocity of the vortex center is given by the local background velocity.*

The velocity differs from that of the classical inviscid theory by at most $O(\epsilon)$.

From (2.2.56) and (2.2.57), and the equivalence principle restated at the beginning of this subsection, we see that the τ-average of the first order solution has to be symmetric, i. e.,

$$\mathcal{M}\tilde\psi^{(1)}_a = \bar\psi^{(1)}_a = 0, \qquad \text{and} \quad \mathcal{M}\tilde\psi^{(1)}(\tau,t,\bar r,\theta) = \bar\psi^{(1)}_c(t,\bar r)\,. \quad (2.2.59)$$

It follows from (2.2.37) that $\bar\psi^{(1)}_c$ is equal to the symmetric part of the first order two-time solution, whose dependence on t and $\bar r$ is not yet defined. We recall that the governing equation (2.2.33) for the symmetric part of the leading order solution came from the compatibility conditions, i. e., the τ- and θ-averages of the Poisson equation (2.2.23) for $n = 0$ and the second order vorticity diffusion equation (2.2.28a). Similarly, we expect the compatibility conditions for the Poisson equation for $n = 1$ and the third order vorticity diffusion equation to yield the equations for $\bar{\zeta}_c^{(1)}$ and $\bar{\psi}_c^{(1)}$.

The one-time asymmetric solution constructed in this subsection is defined by (2.2.56) – (2.2.58). The solution requires that the initial core structure has to be symmetric to the first order and the initial velocity of the vortex center has to be equal to the local background velocity. If the initial data are inconsistent with these requirements, a two-time asymmetric solution is needed. It is constructed in the following subsection.

2.2.4 Asymmetric first order two-time solution and the oscillatory motion of the vortex center

In **Sec.2.2.1**, we found that the asymmetric first order two-time vorticity and stream function are governed by two linear partial differential equations,(2.2.23) and (2.2.36b), in τ, $\bar{r}$ and θ with the normal time t as a parameter. These two equations can be combined to one equation for the stream function, $\tilde{\psi}_a^{(1)}$. The equation is

$$[\bar{\Delta}\tilde{\psi}_a^{(1)}]_\tau - \frac{1}{\bar{r}}[\bar{\psi}_{\bar{r}}^{(0)}\bar{\Delta} + \bar{\zeta}_{\bar{r}}^{(0)}]\,[\tilde{\psi}_a^{(1)}]_\theta = 0\ . \tag{2.2.60}$$

Since the stream function is periodic in θ, we express the function by its Fourier series in θ,

$$\tilde{\psi}_a^{(1)}(\tau,t,\bar{r},\theta) = \sum_{j=1,2,\ldots} [\tilde{\psi}_{j1}\cos j\theta + \tilde{\psi}_{j2}\sin j\theta]\ , \tag{2.2.61}$$

where the Fourier coefficients $\tilde{\psi}_{jk}$, $j = 1,2,\ldots$, $k = 1,2$ are functions of the variables, τ and $\bar{r}$, and the parameter t. From (2.2.60 and 61), we obtain the equations for the Fourier coefficients,

$$[\bar{\Delta}_j\tilde{\psi}_{j1}]_\tau - \frac{j}{\bar{r}}[\bar{\zeta}_{\bar{r}}^{(0)} + \bar{\psi}_{\bar{r}}^{(0)}\bar{\Delta}_j]\tilde{\psi}_{j2} = 0\ , \tag{2.2.62a}$$

$$[\bar{\Delta}_j\tilde{\psi}_{j2}]_\tau + \frac{j}{\bar{r}}[\bar{\zeta}_{\bar{r}}^{(0)} + \bar{\psi}_{\bar{r}}^{(0)}\bar{\Delta}_j]\tilde{\psi}_{j1} = 0\ , \tag{2.2.62b}$$

with

$$\bar{\Delta}_j = (1/\bar{r})\partial_{\bar{r}}(\bar{r}\partial_{\bar{r}}) - (j^2/\bar{r}^2)\ , \tag{2.2.62c}$$

for $j = 1,2,\ldots$ and $k = 1,2$. For each Fourier coefficient, two homogeneous boundary conditions at $\bar{r} = 0$ are obtained from (2.2.24) and a matching condition with the outer solution is obtained from (2.2.18) as $\bar{r} \to \infty$. Those conditions for the Fourier coefficients are given by (2.2.52) and (2.2.54a, b, c) with the bar accent replaced by the tilde accent and the leading order velocity of the vortex center, $\hat{\imath}\dot{X}^{(0)} + \hat{\jmath}\dot{Y}^{(0)}$ replaced by

$$\hat{\imath}[X_t^{(0)}(t) + X_\tau^{(2)}(t,\tau)] + \hat{\jmath}[Y_t^{(0)}(t) + Y_\tau^{(2)}(t,\tau)]\ . \tag{2.2.63}$$

It was noted in the last subsection that a two-time solution is needed when the initial velocity of the vortex center relative to the local background flow is not equal to zero. Let the initial relative velocity be denoted by $\hat{\imath}\,U_0 + \hat{\jmath}\,V_0 \neq 0$, then the initial condition on the velocity is

$$\hat{\imath}[X_t^{(0)} + X_\tau^{(2)}] + \hat{\jmath}[Y_t^{(0)} + Y_\tau^{(2)}] = \hat{\imath}[\Psi_y(X,Y) + U_0] + \hat{\jmath}[-\Psi_x(X,Y) + V_0] \ , \quad (2.2.64a)$$

at $t = \tau = 0$. In addition we can specify an initial asymmetric first order vorticity profile $Z(\bar{r},\theta)$, i. e.,

$$\tilde{\zeta}_a^{(1)}(0,\bar{r},\zeta) = Z(\bar{r},\theta) \quad (2.2.64b)$$

provided that Z decays rapidly in $\bar{r}$. The initial asymmetric stream function is then defined by (2.2.23).

Since the first order equations are linear, we can express the solution as the sum of two solutions. The first one is a particular solution of the homogeneous equation (2.2.60) fulfilling the homogeneous boundary conditions (2.2.24) at $\bar{r} = 0$ and the inhomogeneous matching condition (2.2.18) which in turn defines the velocity of the vortex center fulfilling the initial condition (2.2.63). The initial vorticity profile of the particular solution is free to be assigned provided that it decays rapidly. The second solution of (2.2.60) fulfills the homogeneous boundary conditions (2.2.52), the homogeneous matching condition (2.2.54) and thus the sum of the two solutions fulfills the initial condition (2.2.64).

Since the matching condition for $j \geq 2$ and the boundary conditions for all j are homogeneous, the Fourier coefficients of the stream function with the trivial initial profile vanish for all $\tau \geq 0$ and $t \geq 0$, i. e.,

$$\tilde{\psi}_{jk}(t,\bar{r}) \equiv 0 \qquad \text{for} \qquad j = 2,3,\dots\ ,\ k = 1,2\ , \qquad \text{if} \qquad Z(\bar{r},\theta) \equiv 0\ . \quad (2.2.65)$$

Now we proceed to construct the particular solution which is independent of the initial profile Z. We consider the particular solution having only the two first harmonics in θ. We try to account for the velocity of the local background flow by using the result (2.2.58) for the one-time solution, i. e., by setting:

$$\dot{X}^{(0)}(t) = \Psi_y(X^{(0)},Y^{(0)}) \quad \text{and} \quad \dot{Y}^{(0)}(t) = -\Psi_x(X^{(0)},Y^{(0)})\ . \quad (2.2.66)$$

We should verify later that this choice does not lead to secular terms in $X^{(2)}, Y^{(2)}$ as $\tau \to \infty$. For the construction of the short time solution, we suppress the t-dependence. The matching condition (2.2.18), or (2.2.54a) and (2.2.54b) with bars replaced by tildes, and (2.2.66) yield

$$\tilde{\psi}_a^{(1)} = \tilde{\psi}_{11}(\tau,\bar{r})\cos\theta + \tilde{\psi}_{12}(\tau,\bar{r})\sin\theta \to V(\tau)\bar{r}\cos\theta - U(\tau)\bar{r}\sin\theta, \quad (2.2.67)$$

as $\bar{r} \to \infty$, with $U(\tau) = X_\tau^{(2)}$, $V(\tau) = Y_\tau^{(2)}$. The initial data for the unknowns $U(\tau)$ and $V(\tau)$ come from the prescribed initial velocity of the vortex center in (2.2.63):

$$U(0) = U_0 = X_\tau^{(2)}(0,0) \qquad \text{and} \qquad V(0) = V_0 = Y_\tau^{(2)}(0,0). \quad (2.2.68)$$

Based on the matching condition (2.2.67), we propose a particular solution of the form,

$$\tilde{\psi}_a^{(1)} = W \sin(\theta - \bar{\omega}\tau - \alpha)\, G(\bar{r}), \tag{2.2.69a}$$

with

$$U(\tau) = W \cos(\bar{\omega}\tau + \alpha), \qquad V(\tau) = W \sin(\bar{\omega}\tau + \alpha), \tag{2.2.69b}$$

where W, α and $\bar{\omega}$ may depend on the parameter t. From the initial data (2.2.68), we require

$$W = W_0 \quad \text{and} \quad \alpha = \alpha_0 \quad \text{at} \quad t = 0\,, \tag{2.2.69c}$$

where

$$W_0 \cos\alpha_0 = U_0 \quad \text{and} \quad W_0 \sin\alpha_0 = V_0\,.$$

The partial differential equations (2.2.62a and b) then reduce to the following ordinary differential equation for $G(\bar{r})$,

$$[\bar{\omega} + \frac{1}{\bar{r}}\bar{\psi}_{\bar{r}}^{(0)}]\bar{\Delta}_1 G + \frac{1}{\bar{r}}\bar{\zeta}_{\bar{r}}^{(0)} G = 0, \tag{2.2.70}$$

the boundary conditions on $\tilde{\psi}^{(1)}$ become,

$$G = 0, \qquad G_{\bar{r}} = 0 \qquad \text{at} \quad \bar{r} = 0, \tag{2.2.71a}$$

and the matching condition (2.2.67) becomes

$$G \to \bar{r} \qquad \text{as} \quad \bar{r} \to \infty. \tag{2.2.71b}$$

Since the vorticity, $\bar{\zeta}^{(0)}$, decays rapidly in $\bar{r}$, the second term in (2.2.70) becomes negligible as $\bar{r} \to \infty$ and the solution G behaves as a first harmonic of a potential solution. Therefore, the asymptotic behavior of $G(t, \bar{r})$ is

$$G(t, \bar{r}) = \bar{r} + \frac{d(t)}{\bar{r}} + o(\bar{r}^{-N}) \tag{2.2.72}$$

for some large N as $\bar{r} \to \infty$. The coefficient of $\bar{r}$ is equal to unity on account of (2.2.71b) while the coefficient $d(t)$ remains to be defined.

Since the differential operator in (2.2.70) is singular at $\bar{r} = 0$, we need to study the behavior of G near $\bar{r} = 0$. From (2.2.33), we expand the leading order vorticity in a series in even powers of $\bar{r}$, i.e., $\bar{\zeta}^{(0)}(t, \bar{r}) = z_0(t) + z_1(t)\bar{r}^2 + \cdots$ and then from (2.2.46), the circumferential velocity by a series in odd powers, $-\bar{\psi}_{\bar{r}}^{(0)} = \frac{1}{2}z_0\bar{r} + \frac{1}{4}z_1\bar{r}^3 + \cdots$. Note that $z_0(t) = \bar{\zeta}^{(0)}(t, 0)$ denotes the vorticity at the vortex center. Now we represent the solution $G(\bar{r})$ by the series,

$$G(\bar{r}) = \bar{r}^{\lambda} \sum_{n=0,1,\ldots} a_n(t)\, \bar{r}^{2n}, \qquad \text{for } \bar{r} \ll 1\,. \tag{2.2.73}$$

By equating the coefficients of the leading term $\bar{r}^{\lambda-2}$ in (2.2.70) we obtain

$$[\bar{\omega} - z_0/2](\lambda^2 - 1)a_0 = 0\,.$$

Because of the boundary conditions (2.2.71a) at $\bar{r} = 0$, we have $\lambda \neq \pm 1$. Hence for a nontrivial solution, we require

$$\bar{\omega} = z_0/2 = \bar{\zeta}^{(0)}(t,0)/2. \tag{2.2.74a}$$

By equating the coefficients of the next term $\bar{r}^\lambda$, we obtain

$$\lambda = 3. \tag{2.2.74b}$$

The coefficients a_n, with $n \geq 1$, in the series are then related to the first coefficient a_0 by recurrence formulae. It is evident that the series representation (2.2.73) fulfills the boundary conditions (2.2.71a), while the unknown coefficient a_0, which can be a function of t, is to be determined by (2.2.71b) or by its asymptotic behavior (2.2.72).

The standard shooting method can be used to determine a_0 for a given zeroth order inner solution, $\bar{\zeta}^{(0)}(t,\bar{r})$, and the numerical result in turn defines the constant $d(t)$ in the asymptotic behavior (2.2.72) of $G(t,\bar{r})$.

For example, if the leading order vorticity is given by the similarity solution (2.2.45), we have

$$\bar{\omega} = \Gamma/(2\pi\bar{\delta}^2) \qquad \text{with} \quad \bar{\delta}^2 = 4\bar{\nu}(t+t_0^*). \tag{2.2.75}$$

By using the similarity variable $\eta = \bar{r}/\bar{\delta}$ and observing (2.2.74a), we set $G(t,\bar{r}) = \bar{\delta}g(\eta)$. Equation (2.2.70) then becomes the following equation for the similarity solution $g(\eta)$,

$$\Big[1 - \frac{1-e^{-\eta^2}}{\eta^2}\Big]\Big[\frac{d^2}{d\eta^2} + \frac{1}{\eta}\frac{d}{d\eta} - \frac{1}{\eta^2}\Big]g - 4e^{-\eta^2}g = 0. \tag{2.2.76}$$

The boundary conditions (2.2.71a and b) become $g \to \eta$ as $\eta \to \infty$ and $g \to a_0^*\eta^3$ as $\eta \to 0$, where $a_0^* = a_0\bar{\delta}^2$. Numerical solutions for $g(\eta)$ by means of a shooting method were explained by Ting and Tung (1965). The correct values for the coefficients $a_0(t)$ in (2.2.73) and $d(t)$ in (2.2.72) are $0.50\,\bar{\delta}^{-2}$ and $1.00\,\bar{\delta}^2$ respectively.

To determine the dependence of the amplitude W and phase α on t, we have to look at the next order equations and find the constraints on W_t and α_t, so as to remove the secular terms as $\tau \to \infty$. Appendix A.2 describes the analysis of the next order solutions including the determination of the ϵ-order correction to the mean trajectory of the vortex center. We mention that Gunzburger (1973) carried the analysis to the third order where the secular terms due to $W(t)$ and $\alpha(t)$ appear for the first time. He found that the amplitude $W(t)$ decays as $1/(t+t_0^*)$ when the leading order vorticity is given by a similarity solution.

To summarize the results obtained in this subsection, we give a qualitative description of the trajectory of the vortex center defined by the two-time solution, in particular, how it differs from that defined by the one-time solution in **Sec. 2.2.3.** For this purpose, it suffices to consider only the initial stage where $\tau = O(1)$ while $t \ll 1$. Thus we consider the first order asymmetric two-time solution (2.2.69a) at $t = 0$.

Recall that the trajectory of the vortex center and its order one velocity are expressed in the form of (2.2.22) and (2.2.65). The trajectory defined by the first

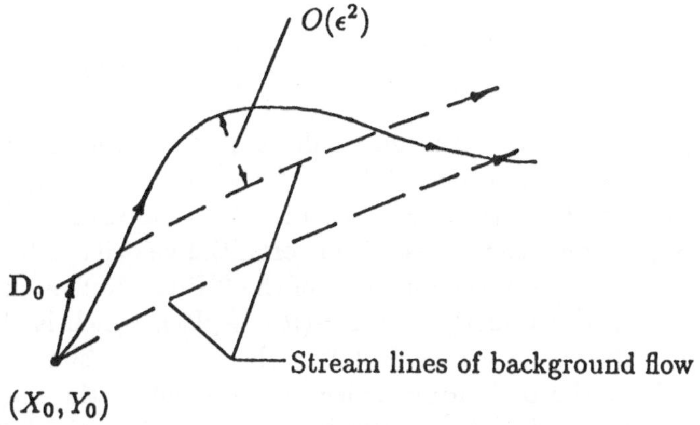

Fig. 2.5. Trajectory of a vortex center with initial velocity different from the local background velocity.

order asymmetric solution in the normal time scale is $\hat{\imath}X^{(0)}(t)+\hat{\jmath}Y^{(0)}(t)$, and the velocity, (2.2.58), is equal to the local velocity of the background potential flow. As shown in Fig. 2.5 the leading order trajectory coincides with the stream line $\mathcal{S}$ of the background flow passing through the initial position (X_0, Y_0). The equation for the stream line $\mathcal{S}$ is $\Psi(x,y) = \Psi(X_0, Y_0)$.

We note from (2.2.22) that the ϵ-order correction to the trajectory has to be in the normal time only and will be defined in **Appendix A.2**. The leading order contribution of the two-time solution which accounts for the initial velocity relative to the background flow, $\hat{\imath}U_0 + \hat{\jmath}V_0$, appears in the second order dispacement of the vortex center, $\epsilon^2[\hat{\imath}X^{(2)} + \hat{\jmath}Y^{(2)}]$. From (2.2.63) and (2.2.67), we see that the second order displacement has an order one velocity in two-time, $\hat{\imath}X^{(2)}_\tau + \hat{\jmath}Y^{(2)}_\tau = \hat{\imath}U + \hat{\jmath}V$. It represents the difference between the velocities defined by the two-time and one-time analyses. From the two-time solution defined by (2.2.69), (2.2.73) and (2.2.74a), we see that the difference is a periodic function of τ with period $\bar{T} = 2\pi/\bar{\omega} = 4\pi/\bar{\zeta}^{(0)}(0, \bar{r} = 0)$. When the dependence of the period on t is taken into account in (2.2.69b) and (2.2.74a), then the velocity difference is in general an almost periodic function with a short period $\epsilon^2 2\pi/\bar{\omega}$ and an order one amplitude, both of which modulate slowly in the normal time scale.

To be more specific, we assume that the leading order vorticity is given by the similarity solution (2.2.45), with its initial core size in the normal length scale equal to $\delta_0 = \epsilon\bar{\delta}_0 = \sqrt{4\nu t_0^*}$. The angular frequency in the short time scale is $\bar{\omega}_0 = \Gamma/(2\pi\bar{\delta}_0^2) = \Gamma/(8\pi\bar{\nu}t_0^*)$. The two-time solution contributes to a second order displacement of the vortex center from the stream line $\mathcal{S}$. During the initial stage the displacement is,

$$\epsilon^2 \int_0^\tau d\tau' \; [\hat{\imath}U(\tau') + \hat{\jmath}V(\tau')] = \mathbf{D}_0 + \frac{\epsilon^2 W_0}{\bar{\omega}_0}[\hat{\imath}\cos(\bar{\omega}_0\tau + \alpha_0 - \frac{\pi}{2}) + \hat{\jmath}\sin(\bar{\omega}_0\tau + \alpha_0 - \frac{\pi}{2})] \quad (2.2.77a)$$

where

$$\mathbf{D}_0 = \frac{\epsilon^2}{\bar{\omega}_0}[-\hat{\imath}V_0 + \hat{\jmath}U_0] = \frac{2\pi\delta_0^2 W_0}{\Gamma}[\hat{\imath}\cos(\alpha_0 + \frac{\pi}{2}) + \hat{\jmath}\sin(\alpha_0 + \frac{\pi}{2})] \, . \qquad (2.2.77b)$$

As shown in Fig.2.5, it represents an oscillation about the mean displacement vector $\mathbf{D}_0$ with a very short period , $T_0 = 2\pi\epsilon^2/\bar{\omega}_0$, and small amplitude , $\epsilon^2 W_0/\bar{\omega}$ in the normal time and length scales. See Fig. 2.5. The magnitude of the mean displacement, $|\mathbf{D}|$, is equal to the magnitude of initial velocity difference W_0 times $T_0/(2\pi)$ and its direction is in the direction of the "Kutta-Joukowski lifting force," obtained by rotating the velocity vector $-(\hat{\imath}U_0 + \hat{\jmath}V_0)$, which is the background velocity relative to the vortex center, by 90° in the sense opposite to the circulation Γ. The amplitude of the oscillating trajectory is equal to the magnitude of the displacement vector, $|\mathbf{D}|$. It is interesting to note that the period of oscillation of the Lamb vortex with core size δ_0 is equal to the period of a Rankine vortex or rotating disk of radius $a = \delta_0$ and has the same circulation.

Finally, we seek the contribution of the oscillatory motion of the vortex center on the far field flow. This comes from the far field behavior (2.2.72) of G as $\bar{r} \to \infty$. The first term on the right-hand side of (2.2.72) matches with the leading order outer solution to account for the difference between the initial velocity of the vortex center and the local background velocity. The second term has to be matched by or induces a doublet of second order strength , $O(\epsilon^2)$, in the outer solution. A doublet in a two-dimensional potential flow represents the displacement effect due to the motion of a rigid body in the fluid. The doublet strength is proportional to the area of the body, which corresponds here to the effective vortical core area of order $O(\epsilon^2)$.

Thus, we conclude :

- *The first order asymmetric two-time solution represents a small amplitude high frequency oscillation of the vortex center around its mean trajectory. The latter is defined by a streamline of the background flow that passes through the initial position of the center plus a second order displacement* $\mathbf{D}_0$*. The amplitude and period of the oscillation are of the order* ϵ^2*.*
- *The high frequency oscillation of the vortex, which accounts for the initial difference between the velocity of the vortex center and local background flow, contributes an* ϵ^2*-order doublet to the outer flow field that accounts for the global effect of the highly oscillatory motion of the vortical core on the outer potential flow*
- *If the initial core structure is symmetric to the first order and the initial velocity of the vortex center happens to be equal to the local background velocity, then the motion and decay of the vortex is given by the one-time solution to the first order, i. e., there is no first order short-time variation and hence its second order contribution, as a doublet, to the outer potential does not materialize.*

We should note that for the first order inner solution, only the local velocity of the outer flow appears in the matching condition. This means that the inner flow field to the first order is in effect submerged in an uniform stream with the local background velocity. Ting and Tung (1965) showed that i) the effect of nonuniformity of the outer flow appeared in the second order matching conditions as inhomogeneous terms proportional to the local velocity gradient and ii) these terms were matched by a second order one-time inner solution which in turn induced a fourth order quadrupole to the outer solution. That is to say:

- *The interaction of the motion of a small vortex with the local velocity gradient of the background flow induces an $O(\epsilon^4)$ quadrupole to the outer potential.*

They also showed that: when the asymmetric first order two-time solution is of the special form (2.2.69), the τ-average of the nonlinear convection terms $\tilde{F}_2$ in the second order vorticity evolution equation (2.2.28a) vanishes and then the equivalence between the first order one-time solution and the τ-average of the first order two-time solution is extended to the second order solutions. This is to say:

- *If $\tilde{\psi}_a^{(1)} = W(t)f(\theta - \omega\tau - \alpha(t))G(\bar{r})$, then the τ-average of the asymmetric second order two-time solution is equivalent to the asymmetric second order solution from the one-time analysis.*

- *The trajectory of the vortex center deviates from the stream line of the background flow passing through the initial position of the vortex center by no more than $O(\epsilon^2)$.*

The studies of the second order one-time and two-time inner solutions will be presented in detail in Appendix A.2.

Now we are ready to study three-dimensional problems. A matched asymptotic analysis in the normal time scale was carried out by Tung and Ting (1967) for the motion of a slender circular vortex ring submerged in an axi-symmetric potential flow. Their analysis includes the evolution of the core structure due to stretching and diffusion. For the special case that the leading vorticity distribution is a similarity solution, the numerical value of a constant in the velocity of the center line of the vortex ring is in error because an incorrect factor of 1/8 was used instead of 1/4 in the similarity solution. This error was corrected by Ting (1971). It was also corrected by Saffman (1970) in an independent study of the circular vortex ring, (see van Dyke 1975 p.248). The corresponding two-time solutions for the axisymmetric problem were presented by Ting (1971) and the meanings of the two-time solutions and correspondence between the one-time solutions and the τ-averages of the two-time solutions were extended from the two-dimensional to the axisymmetric problems. In the same paper, the matched asymptotic analysis in the normal time scale was presented for a slender vortex filament with large circumferential flow. The analysis for the general case of a slender vortex filament with both large axial and circumferential velocities in the vortical core was carried out by Callegari and Ting (1978). The essence of the last reference will be described in the following section.

2.3 Asymptotic Solutions for Vortex Filaments and the Evolution of Their Vortical Cores

Now we study the motion of slender vortex filaments submerged in a background potential flow. The characteristics of a vortex filament, its center line, effective core size and the small parameter ϵ were described at the beginning of this chapter.

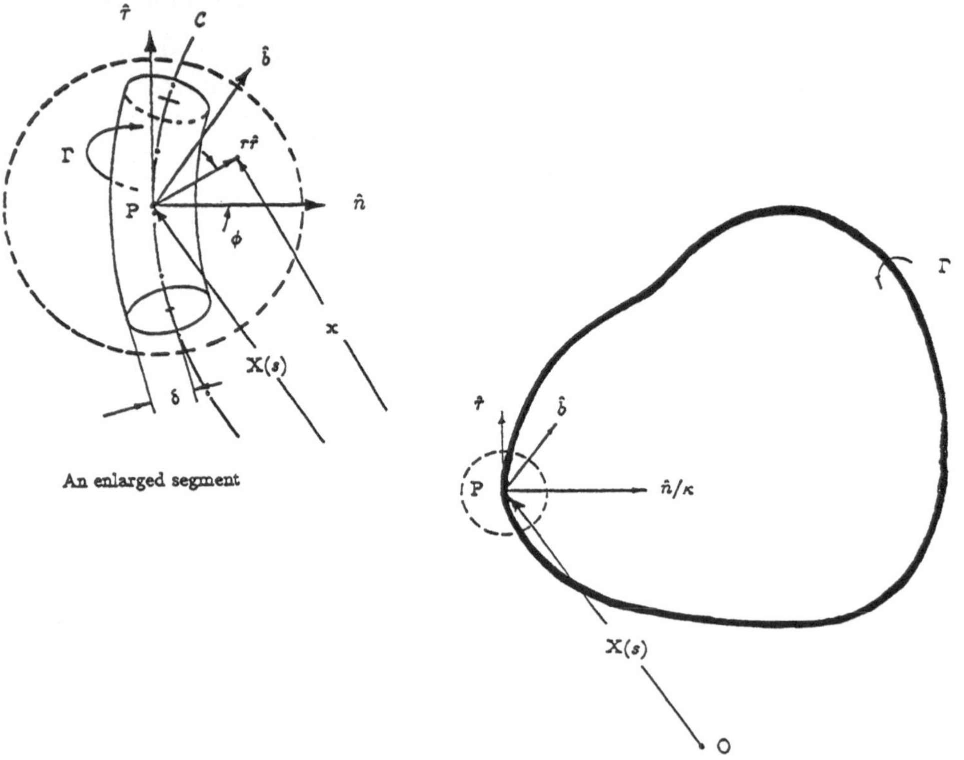

Fig. 2.6. A slender vortex filament and an enlarged segment in the neighborhood of a point P on the centerline C

As sketched in Fig.2.6, the bulk of vorticity of a filament is concentrated in a slender tube-like region with a center line C and an effective core size δ. Also shown in the figure is an enlarged segment of the filament near a point P on C, $\mathbf{x} = \mathbf{X}(t, s)$. By slenderness we mean that $\delta/\ell = O(\epsilon)$, where $\epsilon \ll 1$ denotes a small parameter. The concentration of the vorticity in the filament means that the vorticity decreases rapidly away from C at distances $r \gg \delta$ in the sense of (7). The effective core size measures how fast the vorticity decays as the distance r to C increases relative to δ. For a filament of order one circulation, $\Gamma = O(U\ell)$, the assumption of slenderness implies that there is a large swirling flow around the center line of the order ϵ^{-1} . That is, in the plane normal to C at point P, the fluid has a large circumferential velocity relative to the point P, and an axial (tangential to C) velocity which can also be large of the order of ϵ^{-1}, but the radial velocity component relative to P is *assumed* to be $O(1)$, i. e.,

$$\hat{r} \cdot [\mathbf{v}(t, \mathbf{x}) - \mathbf{v}(t, \mathbf{X})]/U = O(1) . \qquad (2.3.1)$$

Here $\hat{r}$ denotes the unit radial vector in the normal plane and $\mathbf{x} = r\hat{r} + \mathbf{X}$. See Fig. 2.6 and (2.1.49). Note that this condition, which characterizes the core structure, is an additional assumption. It complements the definition of slenderness and the assumptions on the velocity of the centerline. We now restate conditions (2.0.5) and (2.0.6) defining the centerline velocity,

$$[\dot{\mathbf{X}} - \mathbf{v}(t, \mathbf{X})] \times \mathbf{X}_s = 0 \qquad (2.3.2a)$$

and

$$\dot{\mathbf{X}} \cdot \mathbf{X}_s = 0 , \qquad (2.3.2b)$$

where $\dot{\mathbf{X}}$ stands for $\mathbf{X}_t(t, s)$. Condition (2.3.2a) insures that $\mathcal{C}$ is a material line. Following the analysis of Callegari and Ting (1978), we set the tangent component of $\dot{\mathbf{X}}$ to zero, (2.3.2b), to allow for a large axial flow. Under these conditions, the velocity of $\mathcal{C}$ remains order one, i. e.,

$$\dot{\mathbf{X}}/U = O(1) . \qquad (2.3.3)$$

Because of the rapid decay of vorticity away from the center line, the flow field far away from the filament at distances $r/\delta \gg 1$ becomes a potential flow. It resembles that induced by a line vortex along $\mathcal{C}$ with zero core radius, given by the Biot-Savart formula (12). It is evident that the method of matched asymptotics is well suited for the study of such flows. The expansion schemes for the inner core and the outer potential flow and the matching procedure are similar to those in the two-dimensional analysis in **Sec. 2.2**. Again there is the need of a two-time analysis in case that the initial data is incompatible with the one-time solution in the normal time scale. Here we shall assume that the initial core structure and the velocity of the centerline of the filament are compatible with the one-time solution.

The one-time analysis was carried out by Callegari and Ting (1978) for a filament with large swirling and axial flow in its core. They showed that if the leading order swirling flow does not vary along the filament, the same is true for the axial flow and vice versa. Therefore, they consider the restricted but still very complex case where the leading order core structure has no axial variation. Here we begin our analysis with the general case which allows for axial variation of core structure and derive constraints on the axial variations of the leading order axial and circumferential velocity profiles. We then adopt Callegari & Ting's restriction and present highlights of their analysis with emphasis on the physical meaning of the solution.

In order to analyze the flow field near the center line $\mathcal{C}$, Ting (1971) introduced orthogonal curvilinear coordinates attached to $\mathcal{C}$. The same coordinates were employed in **Sec.2.1.2** to study the singular behavior of the inviscid flow field near a vortex line. For the sake of clarity, we restate the definition of the coordinates. Let r denote the minimum distance from a point $\mathbf{x}$ to $\mathcal{C}$. Let $\mathbf{X}(t, s)$ denote the position vector of the point P on $\mathcal{C}$. Then we have

$$\mathbf{x} = \mathbf{X}(t,s) + r\hat{r}, \qquad \text{with} \qquad \hat{r} = \hat{n}\cos\phi + \hat{b}\sin\phi, \tag{2.3.4}$$

where $\hat{r}$ the unit radial vector in the normal plane and ϕ the angle between $\hat{r}$ and $\hat{n}$ (see Fig. 2.3). The unit tangent, normal and binormal vectors $\hat{\tau}$, $\hat{n}$ and $\hat{b}$, are functions of t and s and are related to $\mathbf{X}$ by the Serret-Frenet formulae (2.1.45). Equation (2.3.4) defines the transformation between the Cartesian coordinates x_i $(i = 1,2,3)$, and curvilinear coordinates, r, θ, s. The latter are not orthogonal if the torsion of C is nonzero, i.e., if C is not a planar curve (cf. the discussion after (2.1.49)). To facilitate the transformation of the N-S equations we introduce an orthogonal coordinate system by replacing the angle variable ϕ by

$$\theta = \phi - \theta_0(t,s) \tag{2.3.5a}$$

with

$$\frac{\partial\theta_0}{\partial s} = -\sigma T \ , \tag{2.3.5b}$$

where $\sigma = |\mathbf{X}_s|$ and $T = -\hat{b}_s \cdot \hat{n}\ /\sigma$ denote the linear strain and torsion of C, respectively. Then (r, θ, s) represent radial, circumferential and tangential orthogonal coordinates while $\hat{r}$, $\hat{\theta}$ and $\hat{\tau}$ are the corresponding unit base vectors. The stretch ratios are given by

$$h_1 = 1, \qquad h_2 = r, \qquad h_3 = \sigma[1 - \kappa r\cos(\theta + \theta_0)], \tag{2.3.6}$$

where $\kappa(t,s)$ is the curvature of C. Note that conditions (4) and (5) in the **Introduction** on the core size and curvature imply that for a point near the filament in the inner region, we have $r/\ell = O(\delta/\ell) \ll 1$ or $\kappa r \ll 1$ and hence $h_3 \approx \sigma > 0$. Thus, the transformation (2.3.4) is one to one in the inner region local to $\mathbf{X}(t,s)$.†

The velocity $\mathbf{v}$ in an inertial coordinate system and the associated relative velocity $\mathbf{V}$ in the moving frame are related by

$$\mathbf{v}(t,\mathbf{x}) = \dot{\mathbf{X}}(t,s) + \mathbf{V}(t,r,\theta,s) \ . \tag{2.3.7}$$

We denote the radial, circumferential and axial components of the relative velocity $\mathbf{V}$ by u, v and w, i.e.,

$$\mathbf{V} = u\hat{r} + v\hat{\theta} + w\hat{\tau} \ . \tag{2.3.8}$$

The continuity equation and the N-S equations for the relative velocity $\mathbf{V}$ and pressure p in the curvilinear coordinates are

$$r(w_s + \dot{\mathbf{X}}_s \cdot \hat{\tau}) + (ruh_3)_r + (h_3 v)_\theta = 0 \ , \tag{2.3.9}$$

† There is another condition which is usually implied in the statement that C is a simple curve defined by the parameter s. We shall state this condition in terms of the length scales ℓ and ϵ. We require that the distance between two points P and P' on C associated with different values of s has to be much larger than δ^*, i. e., $|PP'| \gg \delta^*$ if $0 < s < s' \leq S_0$. This condition excludes the situation that the centerline comes close to itself at two different values of s. This situation corresponds to a local self-merging which will be considered in **Sec. 3.1.4**.

$$\ddot{\mathbf{X}} + \frac{1}{h_3}(w - r\hat{r}_t \cdot \hat{\tau})\dot{\mathbf{X}}_s + \frac{d\mathbf{V}}{dt} = -\nabla p + \frac{\nu}{h_3}(\frac{1}{h_3}\dot{\mathbf{X}}_s)_s + \nu\Delta\mathbf{V}. \tag{2.3.10}$$

Note that the velocity of $\mathcal{C}$, $\dot{\mathbf{X}}(t,s)$, which appears in these equations, is defined by (2.3.2a and b) which are equivalent to

$$\dot{\mathbf{X}} \cdot \hat{n} = \mathbf{v} \cdot \hat{n} \,, \qquad \dot{\mathbf{X}} \cdot \hat{b} = \mathbf{v} \cdot \hat{b} \qquad \text{and} \qquad \dot{\mathbf{X}} \cdot \hat{\tau} = 0 \,, \tag{2.3.11}$$

where v stands for $\mathbf{v}(t, \mathbf{X})$. The first two equations in (2.3.11) imply

$$u = 0 \qquad \text{and} \qquad v = 0 \qquad \text{at } r = 0 \,. \tag{2.3.12}$$

Thus (2.3.9), (2.3.10), (2.3.2b) and (2.3.12) form a closed system of differential equations for $\mathbf{V}, p$ and $\mathbf{X}$. In addition to those equations, we have to impose the matching conditions with the outer potential solution and to prescribe the initial values of $\mathbf{V}$, $\mathbf{X}$, and $\dot{\mathbf{X}}$. Similar to the two-dimensional case treated in **Sec. 2.2** and the simple examples in **Sec. 2.1**, we shall see in the next subsection that a one-time expansion scheme for the three-dimensional problem leads to a degeneracy in these equations and, hence, appropriate limitations on the initial data must be imposed.

2.3.1 The Expansion Scheme

To study the solution of (2.3.9)–(2.3.11) in and near the vortical core of a vortex filament, we use the method of matched asymptotic expansions. We introduce the stretched variable,

$$\bar{r} = r/\epsilon \,. \tag{2.3.13}$$

An examination of the expansion of the B-S integral given in (2.1.60) with r replaced by $\epsilon\bar{r}$ indicates that the inner solution can be expanded in a power series in both ϵ and $\ln \epsilon$. In order to simplify the matching of the inner and outer solutions we shall expand the inner solution in powers of ϵ, recognizing that the coefficients in these expansions can be power series in $\ln(\epsilon)$. After matching like powers of ϵ we will then determine the dependence on $\ln \epsilon$. As we shall see later, logarithmic terms will appear in the expansion for $\dot{\mathbf{X}}(t,s)$.

Hence we assume that in the inner region the relative velocity components have expansions in terms of powers of ϵ in the form

$$u(t,\bar{r},\theta,s,\epsilon) = u^{(1)}(t,\bar{r},\theta,s) + \epsilon u^{(2)} + \cdots \,, \tag{2.3.14a}$$

$$v(t,\bar{r},\theta,s,\epsilon) = \epsilon^{-1}v^{(0)}(t,\bar{r},\theta,s) + v^{(1)}(t,\bar{r},\theta,s) + \epsilon v^{(2)} + \cdots \,, \tag{2.3.14b}$$

$$w(t,\bar{r},\theta,s,\epsilon) = \epsilon^{-1}w^{(0)}(t,\bar{r},\theta,s) + w^{(1)}(t,\bar{r},\theta,s) + \epsilon w^{(2)} + \cdots \,. \tag{2.3.14c}$$

We call $u^{(n)}, v^{(n)}$, etc. the nth order solutions and those with $n = 0$ the leading order solutions with the understanding that their dependence on $\ln \epsilon$ has been suppressed. In the above expansions, we allow for both large circumferential and axial components of order ϵ^{-1}. On account of the ansatz (2.3.1), the radial component remains order one.

The radial component of the momentum equation (2.3.10) and the condition that the pressure is $O(1)$ in the outer region require that we choose

$$p(t,\bar{r},\theta,s)=\epsilon^{-2}p^{(0)}(t,\bar{r},\theta,s)+\epsilon^{-1}p^{(1)}(t,\bar{r},\theta,s)+\cdots\,, \tag{2.3.14d}$$

in order to find the nontrivial circumferential velocity $v^{(0)}$. From the definition of vorticity, $\Omega=\nabla\times\mathbf{v}=\nabla\times\mathbf{V}$, we see that the expansion for the vorticity should begin with an order ϵ^{-2} term:

$$\Omega(t,\bar{r},\theta,s)=\epsilon^{-2}\Omega^{(0)}(t,\bar{r},\theta,s)+\epsilon^{-1}\Omega^{(1)}(t,\bar{r},\theta,s)+\Omega^{(2)}+\cdots\,. \tag{2.3.14e}$$

The centerline C of the filament also depends on the parameter ϵ. We assume that it has an expansion in the form

$$\mathbf{X}(t,s,\epsilon)=\mathbf{X}^{(0)}(t,s)+\epsilon\mathbf{X}^{(1)}(t,s)+\cdots\,, \tag{2.3.14f}$$

which begins with an order one term so that the velocity of C remains order one, (2.0.7). The geometric parameters σ, κ and h_3, which are given in terms of the s-derivatives of $\mathbf{X}(t,s,\epsilon)$ by (2.1.45), become

$$h_3=h_3^{(0)}+\epsilon h_3^{(1)}+\cdots=\sigma^{(0)}+\epsilon[\sigma^{(1)}-\sigma^{(0)}\kappa^{(0)}\bar{r}\cos\phi^{(0)}]+\cdots \tag{2.3.15}$$

with $\sigma^{(0)}=|\mathbf{X}_s^{(0)}|$, $\phi^{(0)}=\theta+\theta_0^{(0)}$ and $\theta_0^{(0)}=-\int\sigma^{(0)}T^{(0)}ds$.

By substituting the expansions (2.3.14a, b) into the linear homogeneous boundary conditions (2.3.12) at $\bar{r}=0$ and equating the coefficients of like powers of ϵ, we obtain the boundary conditions for the radial and circumferential velocities,

$$v^{(n)}=0,\qquad u^{(n+1)}=0,\qquad n=0,1,2,\cdots\quad \text{at}\ \bar{r}=0. \tag{2.3.16}$$

In addition, the velocity of the inner flow has to match to that of the outer inviscid flow, which is composed of the velocity $\mathbf{Q}$ induced by the filament itself and a background potential flow $\mathbf{Q}_2$,

$$\begin{aligned}\mathbf{v}&=\dot{\mathbf{X}}+\mathbf{V}=\mathbf{Q}+\mathbf{Q}_2\\&=\frac{1}{\epsilon}\frac{\Gamma}{2\pi\bar{r}}\hat{\theta}+\frac{\Gamma\kappa}{4\pi}\ln\frac{1}{\epsilon\bar{r}}\hat{b}+\frac{\Gamma\kappa}{4\pi}\cos\theta\,\hat{\theta}+\mathbf{Q}_f+\mathbf{Q}_2+\cdots\,,\end{aligned} \tag{2.3.17a}$$

as $\bar{r}\to\infty$. The first three terms on the right-hand side of (2.3.17a) denote the singular parts and $\mathbf{Q}_f$ denotes the finite part of the Biot-Savart integral. See (2.1.48) and (2.1.60). Note that once the leading order outer flow is assumed to be irrotational, the outer flow is irrotational to all orders of ϵ. Thus the matching conditions on the vorticity field yield

$$\Omega^{(n)}\to 0\,,\quad \text{as}\ \bar{r}\to\infty\,,\quad \text{for all}\ n\,.$$

Since a term in the inner solution proportional to $\bar{r}^{-m}$ has to match with an $\epsilon^m r^{-m}$ term in the outer solution, the preceding matching conditions can therefore be replaced by the stronger ones,

$$\Omega^{(n)}\to o(\bar{r}^{-m})\,,\quad \text{as}\ \bar{r}\to\infty\,,\quad \text{for all}\ m\ \text{and}\ n\,. \tag{2.3.17b}$$

The matching conditions for the leading two orders of pressure are

$$p^{(0)} \to 0\ , \quad p^{(1)} \to 0 \qquad \text{as } \bar{r} \to \infty\ . \tag{2.3.17c}$$

Thus we complete the formulation of the expansion scheme and the boundary and matching conditions. In the following subsections we derive the leading and higher order equations systematically by substituting these expansions into the continuity and momentum equations, (2.3.9) and (2.3.10), and equating the coefficients of like powers of ϵ.

2.3.2 The Leading Order Equations

The leading order terms in the continuity and momentum equations are of the order ϵ^{-1} and ϵ^{-3} respectively. Because of the ansatz (2.3.1) or the expansion (2.3.14a), which requires the radial velocity u to remain order one, the coefficient of ϵ^{-1} in the continuity equation yields

$$v_\theta^{(0)} = 0 \qquad \text{or} \qquad v^{(0)} = v^{(0)}(t, \bar{r}, s). \tag{2.3.18}$$

With this result, the coefficients of ϵ^{-3} terms in the tangential, circumferential and radial components of the momentum equation yield, respectively,

$$w_\theta^{(0)} v^{(0)} = 0 \qquad \text{or} \qquad w^{(0)} = w^{(0)}(t, \bar{r}, s)\ , \tag{2.3.19a}$$

$$p_\theta^{(0)} = 0\ . \qquad \text{or} \qquad p^{(0)} = p^{(0)}(t, \bar{r}, s)\ , \tag{2.3.19b}$$

$$\frac{[v^{(0)}]^2}{\bar{r}} = p_{\bar{r}}^{(0)} \qquad \text{or} \qquad p^{(0)} = -\int_{\bar{r}}^{\infty} \frac{[v^{(0)}(t, \bar{r}', s)]^2}{\bar{r}'} d\bar{r}'\ . \tag{2.3.19c}$$

and the leading vorticity components are

$$(\Omega \cdot \hat{r})^{(0)} = 0\ , \tag{2.3.19d}$$

$$(\Omega \cdot \hat{\theta})^{(0)} = -w_{\bar{r}}^{(0)}\ , \tag{2.3.19e}$$

$$\zeta^{(0)}(t, \bar{r}) \equiv (\Omega \cdot \hat{\tau})^{(0)} = \frac{1}{\bar{r}} [\bar{r} v^{(0)}]_{\bar{r}}\ . \tag{2.3.19f}$$

Here ζ denotes the axial component of Ω. It will be easier to solve for $\zeta^{(0)}$ and $w^{(0)}$ instead of $v^{(0)}$ and $w^{(0)}$.

The radial component of the momentum equation (2.3.19c) expresses the balance of the pressure gradient and the centrifugal force due to the large circumferential velocity. In arriving at the equation for $p^{(0)}$, we used the matching condition (2.3.17c). Besides the boundary condition $v^{(0)} = 0$ at $\bar{r} = 0$ from (2.3.16), the matching of the inner velocity with the outer velocity according to (2.3.17a) yields

$$v^{(0)} = \frac{\Gamma}{2\pi \bar{r}} + o(\bar{r}^{-m}), \qquad w^{(0)} = o(\bar{r}^{-m}), \quad \text{for all } m \text{ as } \bar{r} \to \infty\ . \tag{2.3.20}$$

Here we made use of the relations (2.3.19e, f) between the leading order vorticity and velocity and of the matching condition (2.3.17b).

The above equations say that the leading order flow field is axisymmetric, or symmetric, in the normal plane of $\mathcal{C}$. Equations (2.3.14a), (2.3.18) and (2.3.19a) imply that the stream lines are nearly circular helices around C. According to (2.3.19d, e, f) the trajectories of the vorticity field are also nearly circular helices, but in general they differ from the stream lines. If there is no large axial flow, the stream lines become nearly circular helices with small pitch while the vortex lines are nearly parallel to C. This qualitative description is consistent with an intuitive understanding of the flow field in the vortical core of a filament.

Up to now, we have shown that the leading order inner solution is independent of θ. Its dependence on $t, \bar{r}$ and s as well as that of the center line $\mathbf{X}^{(0)}$ on t and s are yet unknown. Similar to the two-dimensional analysis in **Sec. 2.2**, these dependencies will be derived from solvability or compatibility conditions of higher order equations.

We recall the general procedure for the two-dimensional analysis presented in **Sec. 2.2** and **Appendices A.1** and **A.2**. Since the higher order equations, are linear equations, equations for the symmetric part of the nth order solution are separated from those for the asymmetric part by averaging the nth order equations with respect to θ over $[0, 2\pi]$. The θ-averages of the $(n+1)$th order equations, known also as the compatibility conditions, become the governing equations for the symmetric part of the $(n-1)$th order solution. Given the $(n-1)$th order symmetric solution, the asymmetric part of the nth order solution is defined by the nth order equations.

A similar pattern will be observed in a three-dimensional problem. However, additional compatibility conditions on the $(n+1)$st order equations will appear leading to constraints on the dependence of the nth order solutions on s. With these general patterns in mind, we now proceed to the next order analysis.

2.3.3 The First Order Equations and the Asymmetric Solutions

The first order continuity and the three components of the momentum equation are

$$\frac{1}{\bar{r}}[v_\theta^{(1)} + (\bar{r}u^{(1)})_{\bar{r}}] = -\frac{1}{\sigma^{(0)}} w_s^{(0)} - (v\,\kappa \sin\phi)^{(0)}\,, \tag{2.3.21}$$

$$\begin{aligned} u^{(1)} w_{\bar{r}}^{(0)} + \frac{v^{(0)}}{\bar{r}} w_\theta^{(1)} &= -\frac{1}{\sigma^{(0)}}(p_s^{(0)} + w^{(0)} w_s^{(0)}) \\ &\quad - (wv\,\kappa \sin\phi)^{(0)}\,, \end{aligned} \tag{2.3.22a}$$

$$\begin{aligned} u^{(1)} v_{\bar{r}}^{(0)} + \frac{v^{(0)}}{\bar{r}} v_\theta^{(1)} + \frac{v^{(0)} u^{(1)}}{\bar{r}} + \frac{1}{\bar{r}} p_\theta^{(1)} \\ = -\frac{w^{(0)}}{\sigma^{(0)}} v_s^{(0)} + (w^2\,\kappa \sin\phi)^{(0)} \end{aligned} \tag{2.3.22b}$$

and

$$v^{(0)} u_\theta^{(1)} - 2v^{(0)} v^{(1)} + p_{\bar{r}}^{(1)} = -(w^2 \kappa \bar{r} \cos\phi)^{(0)}\,. \tag{2.3.22c}$$

Note that the homogeneous equations of (2.3.21), (2.3.22b) and (2.3.22c) are identical to the equations for $u^{(1)}$, $v^{(1)}$ and $p^{(1)}$ in the two-dimensional case. Here (2.3.22a) is the additional equation needed for $w^{(1)}$. Also (2.3.21) to (2.3.22c) are now inhomogeneous equations because of the axial flow, the s-derivatives and the curvature of C. These three-dimensional effects are absent in the two-dimensional case. For the latter the first order equations are homogeneous and solvable. Now (2.3.21) to (2.3.22c) are solvable only if the compatibility conditions obtained by θ-averaging are satisfied.

2.3.3.1 The Compatibility Conditions

Using (2.2.35), we separate the asymmetric part $f_a(\theta)$ of $f(\theta)$ from its symmetric part f_c. The θ-average of (2.3.21) reads

$$\sigma^{(0)}(\bar{r}u_c^{(1)})_{\bar{r}} + \bar{r}w_s^{(0)} = 0 \,. \tag{2.3.23}$$

The averages of (2.3.22a) and (2.3.22b) yield

$$\sigma^{(0)}w_{\bar{r}}^{(0)}u_c^{(1)} + w^{(0)}w_s^{(0)} = 2\int_{\bar{r}}^{\infty} \frac{v^{(0)}v_s^{(0)}}{\bar{r}'}\,d\bar{r}' \tag{2.3.24a}$$

and

$$\sigma^{(0)}(\bar{r}v^{(0)})_{\bar{r}}u_c^{(1)} + \bar{r}w^{(0)}v_s^{(0)} = 0 \,. \tag{2.3.24b}$$

In arriving at (2.3.24a), we used (2.3.19c) relating $p^{(0)}$ to $v^{(0)}$. The three equations above, which are the compatibility conditions of the first order equations, define the dependence of the leading order core structure, $v^{(0)}$ and $w^{(0)}$, and the symmetric part of the radial flow, $u_c^{(1)}$, on $\bar{r}$ and s while t appears as a parameter.

To simplify these equations and show their physical meaning, we replace the continuity equation (2.3.23) by introducing the stream function $\psi(t,\bar{r},s)$,

$$\psi = \int_0^{\bar{r}} \bar{r}'w^{(0)}(t,\bar{r}',s)\,d\bar{r}' , \tag{2.3.25a}$$

and

$$\sigma^{(0)}\bar{r}u_c^{(1)} = -\psi_s \,. \tag{2.3.25b}$$

Equations (2.3.24a) and (2.3.24b) then become two quasi-steady equations for ψ and $\bar{v}^{(0)}$. They are

$$\psi_{\bar{r}}\psi_{s\bar{r}} - \psi_s\psi_{\bar{r}\bar{r}} + \frac{1}{\bar{r}}\psi_s\psi_{\bar{r}} = 2\bar{r}^2\int_{\bar{r}}^{\infty} \frac{v^{(0)}v_s^{(0)}}{\bar{r}'}\,d\bar{r}' \tag{2.3.26}$$

and

$$(\bar{r}v^{(0)})_{\bar{r}}\psi_s - (\bar{r}v^{(0)})_s\psi_{\bar{r}} = 0 \,. \tag{2.3.27}$$

With ψ related to $w^{(0)}$ by (2.3.25a), (2.3.26) and (2.3.27) become the governing equations for the leading order core structure, $v^{(0)}$ and $w^{(0)}$. The boundary conditions for the quasi-steady equations come from (2.3.16) together with $v_{\bar{r}}^{(0)} \neq 0$ at $\bar{r} = 0$, and the matching condition (2.3.20). They are

$$\bar{r}v^{(0)} \sim \bar{r}^2, \qquad \psi_s \sim \bar{r}^2 \qquad \text{as } \bar{r} \to 0 \tag{2.3.28a}$$

and

$$\bar{r}v^{(0)} = \frac{\Gamma}{2\pi} + o(\bar{r}^{-n}), \qquad \psi_{\bar{r}} = o(\bar{r}^{-n}) \quad \text{for all } n \text{ as } \quad \bar{r} \to \infty. \tag{2.3.28b}$$

Because of assumption (13) of large Reynolds number, $Re = O(\epsilon^{-2})$, the viscous terms will begin to appear only in the second order equations. The equations we obtained in the first order analysis are valid not only for viscous but also for inviscid flows. Thus we conclude the following:

- The leading order core structure, i. e., the circumferential and axial velocities, $\epsilon^{-1}v^{(0)}$ and $\epsilon^{-1}w^{(0)}$, have to fulfill the system of equations, (2.3.26) to (2.3.28b). These constraints on the core structure should be observed for viscous as well as inviscid solutions.
- In particular, (2.3.27) implies a functional relationship between $\bar{r}v^{(0)}$ and ψ,

$$\bar{r}v^{(0)}(t, \bar{r}, s) = \mathcal{G}(t, \psi) \,. \tag{2.3.29}$$

That is , $[\bar{r}v^{(0)}]$ depends implicitly on $\bar{r}$ and s through ψ.

Since $2\pi\psi$ represents the axial mass flux through a circular disc with radius $\bar{r}$, centered on $\mathcal{C}$ and lying in its normal plane, and $2\pi\bar{r}v^{(0)}$ is the circulation around the boundary of the disc, (2.3.29) relates the circulation to the axial mass flux.

Combining (2.3.26) and (2.3.29), we obtain an equation for $\psi(t, \bar{r}, s)$,

$$\psi_{\bar{r}}\psi_{s\bar{r}} - \psi_s\psi_{\bar{r}\bar{r}} + \frac{1}{\bar{r}}\psi_s\psi_{\bar{r}} = 2\bar{r}^2 \int_{\bar{r}}^{\infty} \frac{\mathcal{G}\mathcal{G}_{\psi}\, \psi_s}{(\bar{r}')^3}\, d\bar{r}' \,, \tag{2.3.30}$$

and its boundary conditions are (2.3.28a, b) with $\bar{r}v^{(0)}$ replaced by $\mathcal{G}$.

Since t appears only as a parameter in the system, (2.3.26) to (2.3.28b) or (2.3.29) and (2.3.28a, b), a two-time solution in the normal time t and a short time τ is needed in case that the initial data are incompatible with the system. Following the standard two-time analysis, the dependence of the solution on t will be defined by the compatibility conditions on the second order equations. An in depth study of the quasi-steady system of equations for the one-time solution and the derivation of its evolution equations are not yet available. We note that the system of equations becomes trivial when the leading order core structure does not vary along $\mathcal{C}$, i. e., when it is independent of s. In this case, the dependence of the core structure on t and $\bar{r}$ is defined by the compatibility conditions for the second order equations. From here on we study this degenerated case where the leading order core structure is independent of s.

In the three-dimensional problem studied by Ting (1971), the filament was assumed to have a large swirl but an order one axial velocity, i.e.,

$$w^{(0)} \equiv 0. \tag{2.3.31}$$

With this assumption built into the expansion scheme (2.3.14c), it was concluded from (2.3.23) and (2.3.24a, b) that

$$u_c^{(1)} \equiv 0, \qquad p_s^{(0)} \equiv 0 \tag{2.3.32}$$

and then

$$v_s^{(0)} \equiv 0, \qquad \text{or} \qquad v^{(0)} = v^{(0)}(t, \bar{r}) \,. \tag{2.3.33}$$

Later Callegari and Ting (1978) allowed for both large circumferential and axial velocities in the core. It was observed from the matching condition (2.3.20) that the leading order circumferential velocity for large $\bar{r}$ is independent of s. Motivated by this, the circumferential velocity $v^{(0)}$ in the inner region was *assumed* to be independent of s. Under this assumption, $v_s^{(0)} \equiv 0$, the compatibility condition (2.3.24b) yields $u_c^{(1)} \equiv 0$ and then (2.3.23) yields $w_s^{(0)} \equiv 0$.

From (2.3.24a) and (2.3.25a, b), we see that $w_s^{(0)} \equiv 0$ implies $v_s^{(0)} \equiv 0$. Thus we conclude

$$v_s^{(0)} \equiv 0 \iff w^{(0)} \equiv 0 \,. \tag{2.3.34}$$

In order to gain a physical understanding of this result we shall rederive it by a simple order of magnitude analysis in the spirit of Prandtl's original derivation of the boundary layer theory.

From the order of magnitude of the centrifugal acceleration induced by the large circumferential velocity $v = O(\epsilon^{-1})$ in the radial component of the momentum equation (2.3.10), we see that the pressure gradient p_r in the inner region, where $r = O(\epsilon)$, has to be $O(v^2/r) = O(\epsilon^{-3})$. This is consistent with the scaled equation (2.3.19c). Thus we have

$$\{\, r = O(\epsilon) \quad \text{and} \quad p_r = O(\epsilon^{-3}) \,\} \qquad \Rightarrow \qquad p = O(\epsilon^{-2}). \tag{2.3.35}$$

A nonzero axial gradient of $v^{(0)}$ on the length scale ℓ then induces an axial pressure gradient, $p_s = O(\epsilon^{-2})$, which in turn will force an axial acceleration ww_s. Therefore, both w and w_s have to be $O(\epsilon^{-1})$. Thus we have

$$\begin{aligned} &\text{If} \;\; v_s \neq 0, \;\; v_s = O(v) = O(\epsilon^{-1}), \quad \Rightarrow \quad p_s = O(p) = O(\epsilon^{-2}) \\ &\quad \Rightarrow \quad w\, w_s = O(\epsilon^{-2}) \;\; \text{then} \;\; w_s = O(w) = O(\epsilon^{-1}) \,. \end{aligned} \tag{2.3.36a}$$

Following the same order of magnitude analysis, we find that

$$\begin{aligned} &\text{If} \;\; v_s = O(1), \;\; \text{while} \;\; v = O(\epsilon^{-1}) \quad \Rightarrow \quad p_s = O(\epsilon^{-1}) \\ &\quad \Rightarrow \quad w\, w_s = O(\epsilon^{-1}) \;\; \text{then} \;\; w_s = O(1) \;\; \text{even if} \;\; w = O(\epsilon^{-1}). \end{aligned} \tag{2.3.36b}$$

It says that w_s is order one if v_s is $O(1)$. The converse is also true. Thus we come to the same conclusion (2.3.34) which says

- *The leading order circumferential velocity profile in a cross-sectional plane of the filament remains the same along the filament if the axial velocity profile remains the same and vice versa.*

The regime described by (2.3.36b) becomes interesting for very long filaments of length, $S = O(1/\epsilon)$. Slow variations of the core structure can accumulate along the filament to considerable changes at leading order. This issue will be addressed in **Sec. 4.1**.

In the remainder of this section we shall follow the analysis of Callegari and Ting (1978) in assuming that the leading order circumferential velocity, $\epsilon^{(-1)}v^{(0)}$

is independent of s. It is independent of θ on account of (2.3.18). From (2.3.19) and (2.3.34) we find that $w^{(0)}$ is also independent of θ and s,

$$v^{(0)} = v^{(0)}(t,\bar{r}) \qquad \text{and} \qquad w^{(0)} = w^{(0)}(t,\bar{r}) \,. \tag{2.3.37}$$

It is understood that the one-time solution, which is quasi-steady, is applicable only when the initial data happens to be compatible with the solution. We assume that the initial core structure is compatible with (2.3.37) and has a rapidly decaying vorticity field, (7). We proceed to study the asymmetric solution of the first order equations, (2.3.21) to (2.3.22c).

2.3.3.2 The Asymmetric Solution and the Velocity of the filament

When the leading order core structure is expressed in the form (2.3.37), we have $p_s^{(0)} = 0$ from (2.3.19c) and $u_c^{(1)} = 0$ from (2.3.24b). We replace the first order continuity equation (2.3.21) by introducing a stream function $\psi^{(1)}(t,\bar{r},\theta,s)$ with

$$u^{(1)} = \frac{1}{\bar{r}}\psi_\theta^{(1)}, \qquad \text{and} \qquad v^{(1)} = -\psi_{\bar{r}}^{(1)} + \bar{r}\,(v\,\kappa\cos\phi)^{(0)} \,. \tag{2.3.38}$$

The axial component, (2.3.22a), of the momentum equation becomes an equation for $w_\theta^{(1)}$ uncoupled from the other two equations (2.3.22b and c). We eliminate $p^{(1)}$ from the latter two by cross differentiation and obtain an equation for $\psi_\theta^{(1)}$,

$$v^{(0)}\bar{\Delta}\psi_\theta^{(1)} - \zeta_{\bar{r}}^{(0)}\psi_\theta^{(1)} = -\kappa^{(0)}\sin\phi^{(0)}[(2\bar{r}\zeta^{(0)} + v^{(0)})v^{(0)} + 2\bar{r}w^{(0)}w_{\bar{r}}^{(0)}] \,. \tag{2.3.39}$$

Here

$$\zeta^{(0)}(t,\bar{r}) = \frac{1}{\bar{r}}(\bar{r}v^{(0)})_{\bar{r}} \tag{2.3.40}$$

denotes the leading axial vorticity and $\bar{\Delta}$ the Laplacian in $\bar{r}$ and θ. Equation (2.3.39) is a linear equation for the asymmetric part of $\psi^{(1)}$ because

$$\psi_\theta^{(1)} = (\psi_a^{(1)})_\theta \,. \tag{2.3.41}$$

We expand $\psi_a^{(1)}$ in a Fourier series

$$\psi_a^{(1)} = \sum_{n=1}^{\infty}(\tilde{\psi}_{n1}\cos n\phi^{(0)} + \tilde{\psi}_{n2}\sin n\phi^{(0)}) \tag{2.3.42}$$

and reduce (2.3.39) to a set of ordinary differential equations in $\bar{r}$ for the Fourier coefficients $\tilde{\psi}_{nj}(t,\bar{r},s)$, where t and s appear as parameters. These equations are

$$\left[\frac{\partial^2}{\partial\bar{r}^2} + \frac{1}{\bar{r}}\frac{\partial}{\partial\bar{r}} - \left(\frac{n^2}{\bar{r}^2} + \frac{\zeta_{\bar{r}}^{(0)}}{v^{(0)}}\right)\right]\tilde{\psi}_{nj} = \kappa^{(0)}(t,s)H(t,\bar{r})\delta_{n1}\delta_{j1}, \tag{2.3.43}$$

for $j = 1,2$ and $n = 1,2,\cdots$. Here δ_{nj} is the Kronecker delta and

$$H(t,\bar{r}) = 2\bar{r}\zeta^{(0)} + v^{(0)} + 2\bar{r}w^{(0)}w_{\bar{r}}^{(0)}/v^{(0)}. \tag{2.3.44}$$

The inhomogeneous term in (2.3.43), which is present only for $n = j = 1$, represents three-dimensional effects due to the curvature of the filament. Because of this term, the leading order velocity of the filament will depend on the core structure. This is not the case for a two-dimensional problem for which the inhomogeneous term is absent because $\kappa \equiv 0$.

The condition of zero relative circumferential and axial velocities at $\bar{r} = 0$ requires $\psi_a^{(1)} = 0$ and $(\psi_a^{(1)})_{\bar{r}} = 0$ which in turn yield

$$\tilde{\psi}_{nj} = 0 \quad \text{and} \quad \frac{\partial \tilde{\psi}_{nj}}{\partial \bar{r}} = 0 \quad \text{at} \quad \bar{r} = 0\,, \tag{2.3.45}$$

for $j = 1, 2,\;\; n = 1, 2, \cdots$. The matching condition on the velocity field, (2.3.17a), yields the following far field conditions on the Fourier coefficients of the stream function as $\bar{r} \to \infty$,

$$\begin{aligned} \tilde{\psi}_{11} &\to -\frac{\Gamma\kappa^{(0)}}{4\pi}\bar{r}\ln\frac{1}{\epsilon\bar{r}} - \bar{r}\,(\mathbf{Q}_0 - \dot{\mathbf{X}}^{(0)})\cdot\hat{b}^{(0)}\,, \\ \tilde{\psi}_{12} &\to \bar{r}\,(\mathbf{Q}_0 - \dot{\mathbf{X}}^{(0)})\cdot\hat{n}^{(0)}\,, \\ \tilde{\psi}_{nj} &\to 0 \qquad \text{for } j = 1, 2, \text{ and } n = 2, 3, \cdots\,, \end{aligned} \tag{2.3.46a}$$

where

$$\mathbf{Q}_0 = \lim_{r\to 0}(\mathbf{Q}_f + \mathbf{Q}_2)\,. \tag{2.3.46b}$$

From the matching condition (2.3.20) on $v^{(0)}$ and the definition of $\zeta^{(0)}$ in (2.3.40), we see that $\zeta^{(0)}$ decays rapidly and

$$H(t, \bar{r}) \to \frac{\Gamma}{2\pi\bar{r}} \qquad \text{as } \bar{r} \to \infty\,. \tag{2.3.47}$$

Although the inhomogeneous term and the coefficient $\zeta_{\bar{r}}^{(0)}/v^{(0)}$ in (2.3.43) depend on the leading inner velocity field, which is not yet determined, we do know their behavior for small and large $\bar{r}$ from (2.3.16), (2.3.20) and (2.3.40). Using the behavior and repeating the arguments used in **Sec 2.2.3** for the two-dimensional case, (2.2.51)–(2.2.58), we conclude that

$$\tilde{\psi}_{nj} \equiv 0 \qquad \text{for } j = 1, 2 \text{ and } n = 2, 3, \cdots \tag{2.3.48}$$

because their differential equations, boundary conditions at $\bar{r} = 0$ and far field conditions are homogeneous. For $n = 1$ and $j = 2$, the homogeneous differential equation (2.3.43) and two boundary conditions (2.3.45) at $\bar{r} = 0$ insure that

$$\tilde{\psi}_{12} \equiv 0. \tag{2.3.49}$$

The matching condition for $\tilde{\psi}_{12}$ in (2.3.46a) in turn determines the normal velocity of $\mathcal{C}$,

$$\dot{\mathbf{X}}^{(0)}\cdot\hat{n}^{(0)} = \mathbf{Q}_0\cdot\hat{n}^{(0)}\,. \tag{2.3.50}$$

Now we solve (2.3.43) for $\tilde{\psi}_{11}$. Although $v^{(0)}$ and $w^{(0)}$ are not yet defined, we can express $\tilde{\psi}_{11}$ in terms of these functions by using the fact that $v^{(0)}$ is a solution of the homogeneous equation of (2.3.43) for $n = 1$. The solution of (2.3.43) for $n = j = 1$ fulfilling the boundary conditions (2.3.45) at $\bar{r} = 0$ is

$$\tilde{\psi}_{11}(t,\bar{r},s) = \kappa^{(0)}(t,s)v^{(0)}(t,\bar{r}) \int_0^{\bar{r}} \frac{1}{z[v^{(0)}(t,z)]^2} \Big[\int_0^z \xi v^{(0)}(t,\xi)H(t,\xi)d\xi\Big]\, dz \,. \tag{2.3.51}$$

Note that both t and s are treated as parameters in this expression and that the dependence of $\tilde{\psi}_{11}$ on s comes only from the curvature $\kappa^{(0)}$.

From the behavior of $v^{(0)}$ and $H^{(0)}$ for large $\bar{r}$, we obtain the behavior of $\tilde{\psi}_{11}$, namely

$$\tilde{\psi}_{11} = \bar{r}\kappa^{(0)}(t,s)C^*(t) + \frac{\Gamma\kappa^{(0)}}{4\pi}\,\bar{r}\,\ln\bar{r} + O(\frac{1}{\bar{r}}) \qquad \text{as } \bar{r}\to\infty\,, \tag{2.3.52}$$

where

$$\begin{aligned} C^*(t) =& \frac{\Gamma}{4\pi}\lim_{\bar{r}\to\infty}\Big[\big(\frac{2\pi}{\Gamma}\big)^2 \int_0^{\bar{r}} \xi[v^{(0)}H](t,\xi)d\xi - \ln\bar{r}\,\Big] \\ =& \frac{\Gamma}{4\pi}\Big[\lim_{\bar{r}\to\infty}\big(\frac{4\pi^2}{\Gamma^2}\int_0^{\bar{r}} \xi[v^{(0)}(t,\xi)]^2 d\xi - \ln\bar{r}\big) + \frac{1}{2}\,\Big] \\ &- \frac{2\pi}{\Gamma}\int_0^{\infty} \xi[w^{(0)}(t,\xi)]^2 d\xi \,. \end{aligned} \tag{2.3.53}$$

The matching condition (2.3.46a) then yields the binormal velocity of the filament,

$$\dot{\mathbf{X}}^{(0)}\cdot\hat{b}^{(0)} = \frac{\Gamma\kappa^{(0)}}{4\pi}\ln\frac{1}{\epsilon} + \mathbf{Q}_0\cdot\hat{b}^{(0)} + \kappa^{(0)}C^*. \tag{2.3.54}$$

With $\dot{\mathbf{X}}\cdot\hat{\tau} = 0$, the leading order velocity of the filament, $\dot{\mathbf{X}}^{(0)}$, is specified by its normal and binormal components, (2.3.50) and (2.3.54). Since $\mathbf{Q}_0$ is the sum of the local background velocity and the finite part of the B-S integral defined by (2.1.60), the velocity of the filament depends on the local background velocity as well as on the geometry of the center line $\mathcal{C}$. In addition, the binormal component depends on the leading order core structure, $v^{(0)}$ and $w^{(0)}$, through the term $\kappa^{(0)}C^*$.

Recall that in our expansion of the inner solution in powers of ϵ, terms in $\ln\epsilon$ can appear in the coefficients of the power series. Here we see that a logarithmic term does appear in the leading order binormal velocity. We do not consider this term to be the dominant one in (2.3.54), because in many practical problems $\ln(1/\epsilon)$ is not much larger than unity. For example, with $\epsilon = 0.05$ we have $\ln(1/\epsilon) \approx 3.0$. This observation will be confirmed later by numerical examples which show the effect of the core structure on the motion of the filament via the term $\kappa^{(0)}C^*$ in (2.3.54). However, for a sufficiently large Reynolds number, or a very small ϵ, $\ln(1/\epsilon)$ can be large. Then the dynamics of short wave length distortions of the filament center line becomes important. This new phenomenon has been studied recently by Klein and Majda (1990). An outline of their investigation is presented in **Sec. 4.1**.

In order to evaluate the velocity of the filament, i. e., the right-hand side of (2.3.54), we have to determine the leading order core structure. This will be done in the following sub-section.

2.3.4 The Second Order Equations and the Evolution of the Core Structure

In this section, we derive the compatibility conditions for the second order equations by evaluating their θ-averages. These conditions in turn provide the evolution equations for the leading order core structure, $\epsilon^{-1}v^{(0)}(t,\bar{r})$ and $\epsilon^{-1}w^{(0)}(t,\bar{r})$.

2.3.4.1 The Compatibility Conditions

The second order continuity equation is

$$\begin{aligned}(\bar{r}u^{(2)})_{\bar{r}} + v_\theta^{(2)} + \frac{v^{(0)}}{\sigma^{(0)}}(h_3^{(2)})_\theta = \\ -\frac{\bar{r}}{\sigma^{(0)}}(w_s^{(1)} + \dot{\mathbf{X}}_s^{(0)}\cdot\hat{\tau}^{(0)}) - \frac{\sigma^{(1)}}{\sigma^{(0)}}v_\theta^{(1)} + \bar{r}\kappa^{(0)}(v^{(1)}\cos\phi^{(0)})_\theta \\ -\frac{\sigma^{(1)}}{\sigma^{(0)}}(\bar{r}u^{(1)})_{\bar{r}} + \kappa^{(0)}\cos\phi^{(0)}(\bar{r}^2u^{(1)})_{\bar{r}}\end{aligned} \tag{2.3.55}$$

and its θ-average yields the average radial mass flux $u_c^{(2)}$:

$$u_c^{(2)} = -\frac{1}{\bar{r}\sigma^{(0)}}\int_0^{\bar{r}}[(w_s^{(1)})_c + \dot{\mathbf{X}}_s^{(0)}\cdot\hat{\tau}^{(0)}]\bar{r}'\,d\bar{r}' \ . \tag{2.3.56}$$

Here we used the following results on the first order radial velocity,

$$u_c^{(1)} = 0\ , \quad \bar{r}u_a^{(1)} = \psi_\theta^{(1)} = -\psi_{11}\sin\phi^{(0)} \quad \text{and} \quad (u^{(1)}\cos\phi^{(0)})_c = 0\ . \tag{2.3.57}$$

Carrying out the θ-average of the axial and circumferential components of the second order momentum equations and eliminating $u_c^{(2)}$ by (2.3.56), we arrive at the following two compatibility conditions:

$$\begin{aligned}w_t^{(0)} - \bar{\nu}\frac{1}{\bar{r}}(\bar{r}w_{\bar{r}}^{(0)})_{\bar{r}} = &-\frac{1}{\sigma^{(0)}}(w^{(0)}(w_c^{(1)})_s + (p_c^{(1)})_s) - \frac{w^{(0)}}{\sigma^{(0)}}[\dot{\mathbf{X}}_s^{(0)}\cdot\hat{\tau}^{(0)}] \\ &+\frac{w_{\bar{r}}^{(0)}}{\bar{r}\sigma^{(0)}}\int_0^{\bar{r}}[(w_c^{(1)})_s + \dot{\mathbf{X}}_s^{(0)}\cdot\hat{\tau}^{(0)}]\bar{r}'\,d\bar{r}'\end{aligned} \tag{2.3.58}$$

and

$$\begin{aligned}v_t^{(0)} - \bar{\nu}\left[\frac{1}{\bar{r}}(\bar{r}v_{\bar{r}}^{(0)})_{\bar{r}} - \frac{1}{\bar{r}^2}v^{(0)}\right] = &-\frac{w^{(0)}}{\sigma^{(0)}}(v_c^{(1)})_s \\ &+\frac{(\bar{r}v^{(0)})_{\bar{r}}}{\bar{r}^2\sigma^{(0)}}\int_0^{\bar{r}}[(w_c^{(1)})_s + \dot{\mathbf{X}}_s^{(0)}\cdot\hat{\tau}^{(0)}]\bar{r}'\,d\bar{r}' \ .\end{aligned} \tag{2.3.59}$$

Here $\bar{\nu} = \nu/(U\ell\epsilon^2) = O(1)$ is the scaled diffusion coefficient introduced in (13). The details for the derivation of these two equations were presented in Appendix D of Callegari and Ting (1978).

Since $w^{(0)}$ and $v^{(0)}$ are functions of $\bar{r}$ and t only, the above two equations imply that

$$(p_c^{(1)})_s + 2w^{(0)}(w_c^{(1)})_s - \frac{1}{\bar{r}}\left[w^{(0)}\int_0^{\bar{r}}(w_c^{(1)})_s\,\bar{r}'d\bar{r}'\right]_{\bar{r}} - \frac{1}{2}\bar{r}^3\left(\frac{w^{(0)}}{\bar{r}^2}\right)_{\bar{r}}\bar{s}_{st}^{(0)}$$
$$= -\sigma^{(0)}F_1(t,\bar{r}), \tag{2.3.60a}$$
$$w^{(0)}(v_c^{(1)})_s - \frac{(\bar{r}v^{(0)})_{\bar{r}}}{\bar{r}^2}\int_0^{\bar{r}}(w_c^{(1)})_s\,\bar{r}'d\bar{r}' - \frac{(\bar{r}v^{(0)})_{\bar{r}}}{2}\bar{s}_{st}^{(0)}$$
$$= -\sigma^{(0)}F_2(t,\bar{r}) \tag{2.3.60b}$$

where

$$F_1(t,\bar{r}) = w_t^{(0)} - \bar{\nu}\,\frac{1}{\bar{r}}(\bar{r}w_{\bar{r}}^{(0)})_{\bar{r}} \tag{2.3.61a}$$

and

$$F_2(t,\bar{r}) = v_t^{(0)} - \bar{\nu}\left[\frac{1}{\bar{r}}(\bar{r}v_{\bar{r}}^{(0)})_{\bar{r}} - \frac{v^{(0)}}{\bar{r}^2}\right]. \tag{2.3.61b}$$

Here we used the identity $\dot{\mathbf{X}}_s^{(0)}\cdot\hat{\tau}^{(0)} = [\mathbf{X}_s^{(0)}\cdot\hat{\tau}^{(0)}]_t = \sigma_t^{(0)} = \bar{s}_{st}^{(0)}$, where $\bar{s}^{(0)}(t,s) = \int_0^s \sigma^{(0)}(t,s')ds'$ denotes the leading order arclength at time t.

Following Callegari and Ting (1978), we consider the case where the filament forms a slender torus so that the centerline C is a simple closed curve of length $S(t,\epsilon)$ with initial length $S(0) = O(\ell)$. The case of an infinite vortex filament or one with length much larger than the scaling length, ℓ, will be mentioned in **Sec. 4.1.**

2.3.4.2 Filament in the Form of a Slender Torus

When the centerline C forms a simple closed curve, all physical variables have to be periodic functions of s with period S_0,

$$\mathbf{X}(t,s+S_0,\epsilon) = \mathbf{X}(t,s,\epsilon) \quad \text{and} \quad f(t,\bar{r},\phi,s+S_0,\epsilon) = f(t,\bar{r},\phi,s,\epsilon)\,, \tag{2.3.62}$$

where f stands for p,u,v etc. Using this, we can eliminate the s-derivatives of the first order terms, $w_c^{(1)}$, $v_c^{(1)}$ and $p_c^{(1)}$, in (2.3.60a, b) by integrating with respect to s over the period S_0,

$$w_t^{(0)} - \bar{\nu}\,\frac{1}{\bar{r}}(rw_{\bar{r}}^{(0)})_{\bar{r}} = \frac{1}{2}\bar{r}^3\left(\frac{w^{(0)}}{\bar{r}^2}\right)_{\bar{r}}\frac{\dot{S}^{(0)}}{S^{(0)}}\,, \tag{2.3.63}$$
$$v_t^{(0)} - \bar{\nu}\left[\frac{1}{\bar{r}}(\bar{r}v_{\bar{r}}^{(0)})_{\bar{r}} - \frac{v^{(0)}}{\bar{r}^2}\right] = \frac{1}{2}(\bar{r}v^{(0)})_{\bar{r}}\frac{\dot{S}^{(0)}}{S^{(0)}}\,. \tag{2.3.64}$$

These compatibility conditions in turn are the governing equations for the leading order core structure, $w^{(0)}(t,\bar{r})$ and $v^{(0)}(t,\bar{r})$. The length of the centerline, $S^{(0)}(t)$, which appears in the coefficients on the right-hand side, couples the evolution of the core structure with the equations of motion of the centerline, (2.3.50) and (2.3.54). The terms involving $S^{(0)}(t)$ yield the overall effect of the stretching of the filament while the terms proportional to $\bar{\nu}$ account for the effects of viscous diffusion.

Since it will be easier to work with the axial vorticity $\zeta^{(0)}$ instead of $v^{(0)}$, which are related by (2.3.19f), we convert (2.3.64) to an equation for the axial vorticity by differentiating $\bar{r}$ times (2.3.64) with respect to $\bar{r}$. The result is

$$\zeta_t^{(0)} - \bar{\nu}\,\frac{1}{\bar{r}}(\bar{r}\zeta_{\bar{r}}^{(0)})_{\bar{r}} = \frac{1}{2}(\bar{r}^2\zeta^{(0)})_{\bar{r}}\frac{\dot{S}^{(0)}}{S^{(0)}}\,\frac{1}{\bar{r}}\ . \tag{2.3.65}$$

Equations (2.3.63), (2.3.65), (2.3.50) and (2.3.54) form a closed system of equations for the leading order solutions $\zeta^{(0)}$, $w^{(0)}$ and $\mathbf{X}^{(0)}$ with $v^{(0)}$ and C^* defined by (2,3.19f) and (2.3.53). These equations exhibit the coupling between the inner and outer flows and in particular indicate the effect of the core structure on the motion of the filament. The coupling is given by the integral expressions in $C^*(t)$ in (2.3.53), which depend on both $v^{(0)}(t,\bar{r})$ and $w^{(0)}(t,\bar{r})$.

Before stating the initial and boundary conditions that are compatible with the above system of equations and describing the construction of the solutions, we derive two invariants from (2.3.63) and (2.3.65) and explain their physical meaning including a discussion of the inviscid case.

Noting the rapid decay of $\zeta^{(0)}$ as $\bar{r} \to \infty$, we recover from (2.3.65) the invariance condition,

$$\frac{\partial}{\partial t}\int_0^\infty \bar{r}\zeta^{(0)}(t,\bar{r})\,d\bar{r} = 0\ ,$$

and hence

$$\int_0^\infty \bar{r}\zeta^{(0)}(t,\bar{r})\,d\bar{r} = \frac{\Gamma}{2\pi} = \text{const}\ . \tag{2.3.66}$$

This shows that the leading order solutions satisfy the global constraint of conservation of circulation. Similarly from (2.3.63) we obtain the second invariance condition,

$$m(t)(S^{(0)})^2 = \text{const.} \tag{2.3.67a}$$

where

$$m(t) = 2\pi\int_0^\infty \bar{r}w^{(0)}(t,\bar{r})\,d\bar{r}\ , \tag{2.3.67b}$$

denotes a scaled axial mass flux. In dimensional notation the axial mass flux reads $U\ell^2\epsilon m$.

One can identify the above two invariants as constraints on the average axial velocity and axial vorticity in the filament. We introduce an average cross-sectional area $A(t)$ of the vortex filament such that the conservation of mass yields:

$$AS^{(0)} = \text{const} \tag{2.3.68}$$

and then define the average axial velocity $\bar{w}$ and vorticity $\bar{\zeta}$ as

$$\bar{w} = \frac{2\pi}{A}\int_0^\infty \bar{r}w^{(0)}\,d\bar{r} \qquad \text{and} \qquad \bar{\zeta} = \frac{2\pi}{A}\int_0^\infty \bar{r}\bar{\zeta}^{(0)}\,d\bar{r}\ . \tag{2.3.69}$$

The conservation of circulation (2.3.66) now reads as a constraint on the stretching of average vorticity

$$\bar{\zeta}/S^{(0)} = \text{const} , \tag{2.3.70a}$$

while the invariant (2.3.67a) becomes one on the average axial velocity,

$$\frac{\bar{w}}{A} \sim \bar{w}S^{(0)} = \text{const} . \tag{2.3.70b}$$

It should be noted that these two invariants do not involve the viscosity; therefore they are also valid for an inviscid flow. To be more specific, we can obtain the variation of $w^{(0)}$ and $\zeta^{(0)}$ for this case from (2.3.63) and (2.3.65) by simply dropping the viscous terms. We find that the quantities $w^{(0)}/\bar{r}^2$ and $\zeta^{(0)}\bar{r}^2$ both obey the first order equation

$$\left(\frac{\partial}{\partial t} - \frac{\bar{r}}{2}\frac{\dot{S}^{(0)}}{S^{(0)}}\frac{\partial}{\partial \bar{r}}\right)\chi = 0 .$$

Its general solution is of the form $\chi(\alpha)$ where $\alpha = \bar{r}\sqrt{S^{(0)}(t)/S_0}$. Here s denotes the arc length at $t = 0$ and hence $S^{(0)}(0) = S_0$. With the initial data $w_0(\bar{r})$ and $\zeta_0(\bar{r})$ for the axial velocity and vorticity, the solutions are

$$w^{(0)}(t,\bar{r}) = w_0(\alpha)S_0/S^{(0)}(t) \tag{2.3.71a}$$

and

$$\zeta^{(0)}(t,\bar{r}) = \zeta_0(\alpha)S^{(0)}(t)/S_0 . \tag{2.3.71b}$$

These results are stronger than those given in (2.3.70a,b) for the viscous case. Instead of integral averages, they relate the detailed distributions of $w^{(0)}$ and $\zeta^{(0)}$ with respect to $\bar{r}$ to the initial data through a simple scaling rule by the length of the filament. These results demonstrate once more that it is inconsistent to arbitrarily assign the instantaneous core quantities in an inviscid analysis.

From the above it is straight forward to determine the core evolution in the inviscid case, given the time history $S^{(0)}(t)$ of the filament length. However, even with viscosity the evolution equations (2.3.63) and (2.3.65) are linear partial differential equations for $w^{(0)}(t,\bar{r})$ and $\zeta^{(0)}(t,\bar{r})$, given $S^{(0)}(t)$. Solutions are available in the form of Eigenfunction expansions shown in the next subsection.

2.3.4.3 Evolution of the Core Structure

We begin by rewriting the system of equations for the leading order core structure and the motion of the filament as

$$w_t^{(0)} - \bar{\nu}\,\frac{1}{\bar{r}}(\bar{r}w_{\bar{r}}^{(0)})_{\bar{r}} = \frac{1}{2}\bar{r}^3\left(\frac{w^{(0)}}{\bar{r}^2}\right)_{\bar{r}}\frac{\dot{S}^{(0)}}{S^{(0)}} , \tag{2.3.72a}$$

$$\zeta_t^{(0)} - \bar{\nu}\,\frac{1}{\bar{r}}(\bar{r}\zeta_{\bar{r}}^{(0)})_{\bar{r}} = \frac{1}{2}\,\frac{(\bar{r}^2\zeta^{(0)})_{\bar{r}}}{\bar{r}}\,\frac{\dot{S}^{(0)}}{S^{(0)}} , \tag{2.3.72b}$$

$$\dot{\mathbf{X}}^{(0)} \cdot \hat{\tau}^{(0)} = 0, \tag{2.3.73a}$$

$$\dot{\mathbf{X}}^{(0)} \cdot \hat{n}^{(0)} = \mathbf{Q}_0 \cdot \hat{n}^{(0)} , \tag{2.3.73b}$$

$$\dot{\mathbf{X}}^{(0)} \cdot \hat{b}^{(0)} = \mathbf{Q}_0 \cdot \hat{b}^{(0)} + \frac{\Gamma\kappa^{(0)}}{4\pi}\ln\frac{1}{\epsilon} + \kappa^{(0)}C^*(t,s) , \tag{2.3.73c}$$

where

$$C^*(t) = \frac{\Gamma}{4\pi}[C_v(t) + C_w(t)] \ , \tag{2.3.73d}$$

$$C_v(t) = \lim_{\bar{r}\to\infty}\left(\frac{4\pi^2}{\Gamma^2}\int_0^{\bar{r}} \bar{r}'(v^{(0)})^2\, d\bar{r}' - \ln \bar{r}\right) + \frac{1}{2} \ , \tag{2.3.73e}$$

$$C_w(t) = \frac{-8\pi^2}{\Gamma^2}\int_0^{\infty} \bar{r}'(w^{(0)})^2\, d\bar{r}' \ . \tag{2.3.73f}$$

Here $C_v(t)$ and $C_w(t)$ represent the global contributions of the circumferential and axial velocities in the core to the binormal velocity of the filament. For completeness, we recall the relationship between the circumferential velocity and the axial vorticity,

$$v^{(0)} = \frac{1}{\bar{r}}\int_0^{\bar{r}} z\zeta^{(0)}(t,z)\, dz \tag{2.3.74a}$$

and the definition of the length of the filament,

$$S^{(0)}(t) = \int_0^{S_0} \mid \mathbf{X}_s^{(0)}(t,s) \mid \, ds. \tag{2.3.74b}$$

We now discuss the restrictions on the initial and boundary conditions for the above system. Equations (2.3.72a, b) imply that $w^{(0)}(0,\bar{r}) = w_0(\bar{r})$ and $\zeta^{(0)}(0,\bar{r}) = \zeta_0(\bar{r})$, or equivalently $v^{(0)}(0,\bar{r}) = v_0(\bar{r})$, must be prescribed. We require that w_0 and ζ_0 decay sufficiently rapidly with $\bar{r}$, while v_0 should behave like $\Gamma/(2\pi\bar{r})$ as $\bar{r} \to \infty$.

Next, we observe that (2.3.73a, b, c) contain only first time derivatives of $\mathbf{X}^{(0)}$, while second time derivatives of $\mathbf{X}$ appeared in the full Navier-Stokes equations (2.3.10). Thus, our expansion scheme has introduced a singular perturbation in the time variable. This implies that we cannot prescribe both the initial velocity $\dot{\mathbf{X}}^{(0)}(0,s)$ and the initial position or shape $\mathbf{X}^{(0)}(0,s)$ of $\mathcal{C}$. Instead, we prescribe only the initial position of $\mathcal{C}$ and let the initial velocity be determined by (2.3.73a, b, c) with $v^{(0)}$, $w^{(0)}$ and $\mathbf{X}^{(0)}$ replaced by their initial data.

To treat the general case, where the initial velocity $\dot{\mathbf{X}}^{(0)}(0,s)$ is not compatible with (2.3.73b, c), we have to construct a two-time solution to account for the high frequency fluctuation of the filament velocity. This was done for a vortex in two dimensions by Ting and Tung (1965) and is rederived in **Sec. 2.2.1**. Similar results were derived for a circular vortex ring in axisymmetric flows by Ting (1971). A generalization valid for the three-dimensional case is not yet available. Some aspects of it will be discussed in **Sec.4.1**.

The first pair of equations, (2.3.72a, b), governs the evolution of the core structure $w^{(0)}(t,\bar{r})$ and $\zeta^{(0)}(t,\bar{r})$. The coupling with the motion of the filament appears only in the coefficient $\dot{S}^{(0)}(t)/S^{(0)}(t)$, which represents the overall rate of stretching of the centerline. The second set of equations, (2.3.73a,b,c), defines the centerline velocity, $\dot{\mathbf{X}}^{(0)}(t,s)$. Only two integrals of the core structure over the cross-sectional plane contribute to the filament motion and they appear only in the binormal component of $\dot{\mathbf{X}}^{(0)}$. The relevant integral expressions, $C_v(t)$ and $C_w(t)$, are defined in terms of $v^{(0)}$ and $w^{(0)}$ in (2.3.73e, f).

To construct explicit solutions, we shall decouple these two sets of equations. We solve the system (2.3.72a,b) for the core structure in terms of the initial data with the coefficient $\dot{S}^{(0)}/S^{(0)}$ considered a given function of t. Once this is achieved, we obtain the global contributions of the core structure to the filament motion, $C_v(t)$ and $C_w(t)$, as functionals of this coefficient and of the initial data. In this way, the second system, (2.3.73a, b, c), becomes a closed system of equations for $\mathbf{X}^{(0)}(t,s)$ defining the motion of the filament.

An appropriate transformation of the dependent and independent variables in the first system of equations allows us to eliminate $S^{(0)}(t)$. Afterwards, the function only appears in the transformation rules, but no longer explicitly in the equations themselves. Solutions to this new system are obtained by means of Eigenfunction expansions.

We introduce the new independent variables η and τ_1 by the transformations,

$$\eta = \frac{\bar{r}}{\sqrt{4\bar{\nu}\tau_2(t)}}, \qquad \tau_2 = \tau_1/S^{(0)}(t) \tag{2.3.75a}$$

and

$$\tau_1 = \int_0^t S^{(0)}(t')\,dt' + \tau_{10}. \tag{2.3.75}$$

Here τ_{10} is a positive constant of integration whose choice is at our disposal. A rule for optimizing τ_{10} will be discussed at the end of this subsection.

Next, the axial velocity, $w^{(0)}$, and the axial vorticity, $\zeta^{(0)}$, are replaced, respectively, by

$$W(\eta,\tau_1) = \tau_1 w^{(0)}\, S^{(0)} \qquad \text{and} \qquad Z(\eta,\tau_1) = \tau_1\zeta^{(0)}/S^{(0)}\,. \tag{2.3.76}$$

In terms of these two new variables, equations (2.3.72a, b) both reduce to the common form

$$4\tau_1\chi_{\tau_1} = \frac{1}{\eta}[(\eta\chi_\eta)_\eta + (2\eta^2\chi)_\eta], \tag{2.3.77}$$

where χ stands for W and Z.

Equation (2.3.77) is separable and thus, letting $\chi(\tau_1,\eta) = T(\tau_1)\Psi(\eta)$, we find

$$\tau_1\frac{dT}{d\tau_1} + \lambda T = 0 \qquad \text{or} \qquad T = \tau_1^{-\lambda} \tag{2.3.78a}$$

and

$$\frac{d^2\Psi}{d\eta^2} + \left(\frac{1}{\eta} + 2\eta\right)\frac{d\Psi}{d\eta} + 4(\lambda+1)\Psi = 0\,, \tag{2.3.78b}$$

where λ is the separation constant. Equation (2.3.78b) can be transformed to Laguerre's equation by introducing $\xi = \eta^2$ and $B(\xi) = e^{\xi}\Psi(\sqrt{\xi})$. This leads to

$$\xi\frac{d^2B}{d\xi^2} + (1-\xi)\frac{dB}{d\xi} + \lambda B = 0\,. \tag{2.3.79}$$

Since both $w^{(0)}$ and $\zeta^{(0)}$ are bounded at $\bar{r} = 0$, B has to be bounded at $\xi = 0$. In order to satisfy the matching conditions, (2.3.20), which imply that the axial velocity and vorticity decay *exponentially* in $\bar{r}$, λ must be restricted to be a positive integer. The solutions of (2.3.79) for such values $\lambda = n$, are the Laguerre polynomials (Magnus et al. 1966):

$$B(\xi) = L_n(\xi) = \sum_{m=0}^{n} (-1)^m \binom{n}{m} \frac{\xi^m}{m!} ,$$

and we can represent the functions $\chi = \{W, Z\}$ by a series in $L_n(\xi)\tau_1^{-n}$. Denoting the related expansion coefficients for W and Z by C_n and D_n, respectively, and inverting the transformations in (2.3.76), we obtain $w^{(0)}$ and $\zeta^{(0)}$ in terms of (τ_1, η):

$$w^{(0)} = \frac{1}{S^{(0)}} e^{-\eta^2} \sum_{n=0}^{\infty} C_n L_n(\eta^2) \tau_1^{-(n+1)} \tag{2.3.80a}$$

and

$$\zeta^{(0)} = S^{(0)} e^{-\eta^2} \sum_{n=0}^{\infty} D_n L_n(\eta^2) \tau_1^{-(n+1)} . \tag{2.3.80b}$$

From the relationship between $v^{(0)}$ and $\zeta^{(0)}$, (2.3.19f), we get,

$$v^{(0)} = \frac{2\bar{\nu}}{\bar{r}} \left[D_0 (1 - e^{-\eta^2}) + e^{-\eta^2} \sum_{n=1}^{\infty} \frac{D_n}{\tau_1^n} (L_{n-1}(\eta^2) - L_n(\eta^2)) \right] \tag{2.3.80c}$$

The coefficients C_n and D_n are determined in terms of the initial profiles, $\zeta_0(\bar{r})$ and $w_0(\bar{r})$, the initial length $S^{(0)}(0) = S_0$ and the arbitrary parameter τ_{10} as

$$C_n = 2 S_0 \tau_{10}^{n+1} \int_0^\infty w^{(0)}(0, \eta\sqrt{4\bar{\nu}\tau_{20}}) L_n(\eta^2) \eta \, d\eta , \tag{2.3.81a}$$

$$D_n = \frac{2\tau_{10}^{n+1}}{S_0} \int_0^\infty \zeta^{(0)}(0, \eta\sqrt{4\bar{\nu}\tau_{20}}) L_n(\eta^2) \eta \, d\eta . \tag{2.3.81b}$$

In particular, $C_0 = m_0 S_0^2/(4\pi\bar{\nu})$ and $D_0 = \Gamma/(4\pi\bar{\nu})$. The appearance of the common factor, the exponential function $e^{-\eta^2}$, in (2.3.80a, b) for the axial velocity and vorticity and the definition of η in (2.3.75) suggest the definition of an effective core size for a filament in three dimensions

$$\delta(t) = \epsilon\sqrt{4\bar{\nu}\tau_2} = \sqrt{4\nu\tau_2} . \tag{2.3.82}$$

This is a generization of the definition introduced in **Sec. 2.2.2** for the two-dimensional cases where $t + t_0$ appeared for τ_2.

Note that (2.3.80a, b) express the solutions for the core structure, $w^{(0)}S^{(0)}$ and $\zeta^{(0)}/S^{(0)}$, as functions of two new variables, η and $\tau_1(t)$, with coefficients C_n and D_n depending on the initial data. The dependence of the solutions on the motion of the filament appears implicitly via the new variables, which are related to the physical variables t and $\bar{r}$ and to the length of the filament, $S^{(0)}(t)$, by the transformations (2.3.75a, b). The contribution of the core structure to the velocity of the filament is contained in $C_v(t)$ and $C_w(t)$, defined by (2.3.73e, f). These quantities can be expressed in terms of $S^{(0)}(t)$ when we replace the integration variable $\bar{r}$ in the integrals in (2.3.73e, f) by $\eta/\sqrt{\delta_2(t)}$ and then carry out the integrations in η. Consequently (2.3.73 a–f) form a closed system defining the motion of the filament through its centerline velocity $\mathbf{X}^{(0)}(t,s)$. A list of all relevant equations of the system will be given in Appendix 2A. A computational code solving the system was developed by Liu et al (1986) and employed to study the interaction of filaments. Several numerical examples and their physical meaning will be presented in **Sec.2.4.**

Before doing so, we discuss the determination of an optimal value for the positive constant of integration, τ_{10}, introduced in the definition (2.3.75b) of the new time like variable τ_1. Such an optimum choice of τ_{10} was introduced by Ting (1971) for the case where there is no large axial velocity $w^{(0)} = 0$. The constant τ_{10} is chosen to eliminate the coefficient D_1, in the series representation (2.3.80b) of $\zeta^{(0)}$. In this way, the first term alone becomes an optimum approximation to the series for large t or τ_1 with an error $O(\tau_1^{-3})$ instead of $O(\tau_1^{-2})$. The condition $D_1 = 0$ also insures that the first term yields the correct polar moment, $\langle \bar{r}^2 \zeta^{(0)} \rangle$, for all $t \geq 0$.

When there is also an initial large axial velocity, we cannot expect a single choice of τ_{10} to eliminate both D_1 and C_1 in (2.3.80a, b). This dilemma was noted by Callegari and Ting (1978). It can be resolved when it is realized that the governing equations (2.3.72a and b) for $w^{(0)}$ and $\zeta^{(0)}$ are coupled only indirectly through the stretching of the filament length, $S^{(0)}(t)$. In the transformations of these two equations to (2.3.77) there is no reason to use the same constant τ_{10} in the transformations of independent variables. We assign two different constants τ_{10}^{w} and τ_{10} to set $C_1 = 0$ *and* $D_1 = 0$ for $w^{(0)}$ and $\zeta^{(0)}$, respectively. Noting that $L_1(\xi) = 1 - \xi$, we obtain

$$C_1 = 0\ , \qquad \tau_{10}^{w} = \frac{\pi S^{(0)}}{2\bar{\nu} m(0)} \int_0^\infty w^{(0)}(0,\bar{r})\,\bar{r}^3\, d\bar{r} \tag{2.3.83a}$$

and

$$D_1 = 0\ , \qquad \tau_{10} = \frac{\pi S^{(0)}}{2\bar{\nu}\Gamma} \int_0^\infty \zeta^{(0)}(0,\bar{r})\,\bar{r}^3\, d\bar{r}\ , \tag{2.3.83b}$$

with the mass flux m defined by (2.3.67b). Now the the two series, (2.3.80a, b), are approximated by their first terms accurate to $O(1/t^{-3})$ for large t, provided we use τ_{10}^{w} as the constant in (2.3.80a) and τ_{10} in (2.3.80b).

Velocity of a filament with similar core structure

Equations (2.3.66) and (2.3.67a) imply that a vortex filament has two invariant quantities, the circulation Γ and the axial mass flux times the length square, $m[S^{(0)}]^2$. Let us consider the case that the initial axial velocity and vorticity profiles are consistent with the first terms of (2.3.80a, b) and that they have the same effective core size δ_0. The latter defines the same initial value for τ_1, i. e., $\tau_1(0) = \tau_{10}^{w} = \tau_{10} = S(0)\delta_0^2/(4\nu)$. Then we have $C_n = D_n = 0$ for $n \geq 1$ and the axial velocity and vorticity in the vortical core are given respectively by the two first terms in (2.3.80a,b) for $t \geq 0$. We call them the *similarity solution.* The special core structure, called the *similar* core structure, is a generalization of a rectilinear Lamb vortex with the same initial core size δ_0, circulation Γ but no axial flow to a toroidal filament with circulation Γ and axial mass flux $m(t) = m(0)[S_0/S^{(0)}(t)]^2$. For a Lamb vortex, $t_0 = \delta_0^2/(4\nu) = \tau_{10}/S(0)$, denotes the initial age of the vortex if it was created with zero core size. Assuming that the length of the filament remains order one then τ_{10}/S_0 measures the age of the filament at $t = 0$.

We recall that, for a filament with nonzero initial core size, the first terms in the series (2.3.80a, b) provide the long time behavior of the core structure. In particular, when the optimum choices of τ_{10} for $w^{(0)}$ and $\zeta^{(0)}$ are the same, we call their one term solutions the *optimum similarity solution.* In practical applications and experiments, it is most likely that the circulation, Γ, the axial mass flux, m(0), the effective core size $\delta(0)$ and the initial configuration $\mathbf{X}(0, s)$ are the only initial data available. On the other hand, they are just the initial data needed for the motion of a filament with a similar core structure. For this reason, we assemble below the complete set of equations for the motion and decay of the filament.

Keeping only the first terms in (2.3.80a, b, c), we obtain the similar core structure,

$$w^{(0)} = \frac{C_0}{\tau_1 S^{(0)}} e^{-\eta^2} = \frac{m(t)}{\pi\bar{\delta}^2} e^{-(\bar{r}/\bar{\delta})^2}, \tag{2.3.84a}$$

$$\zeta^{(0)} = \frac{S^{(0)} D_0}{\tau_1} e^{-\eta^2} = \frac{\Gamma}{\pi\bar{\delta}^2} e^{-(\bar{r}/\bar{\delta})^2}, \tag{2.3.84b}$$

$$v^{(0)} = \frac{\Gamma}{2\pi\bar{r}} [1 - e^{-(\bar{r}/\bar{\delta})^2}], \tag{2.3.84c}$$

where

$$\bar{\delta} = \delta/\epsilon = \sqrt{4\bar{\nu}\tau_2}, \qquad \tau_2(t) = \tau_1(t)/S^{(0)}(t) \tag{2.3.84d}$$

and

$$\tau_1 = \int_0^t S^{(0)}(t')dt' + \tau_{10}\,, \qquad m(t) = m(0)\Big[\frac{S_0}{S^{(0)}(t)}\Big]^2\,. \tag{2.3.84e}$$

The global contributions of the core structure to the motion of the filament are

$$C_w = -\frac{2m^2(0)}{\Gamma^2\bar{\delta}^2}\Big[\frac{S_0}{S^{(0)}(t)}\Big]^4 \tag{2.3.85a}$$

and

$$C_v = -\ln\bar{\delta} + \frac{\gamma + 1 - \ln 2}{2} = -\ln\bar{\delta} + 0.442\,, \tag{2.3.85b}$$

where $\gamma = 0.577$ denotes Euler's constant. The three components of the velocity of the centerline of the filament given by (2.3.73a, b, c) become

$$\dot{\mathbf{X}}^{(0)} \cdot \hat{\tau}^{(0)} = 0\,, \tag{2.3.86a}$$

$$\dot{\mathbf{X}}^{(0)} \cdot \hat{n}^{(0)} = \mathbf{Q}_0 \cdot \hat{n}^{(0)} \tag{2.3.86b}$$

and

$$\dot{\mathbf{X}}^{(0)} \cdot \hat{b}^{(0)} = \mathbf{Q}_0 \cdot \hat{b}^{(0)} + \frac{\Gamma\kappa^{(0)}}{4\pi}\Big[\ln\frac{1}{\delta} + 0.442 + C_w\Big]\,. \tag{2.3.86c}$$

Recall that the sign of the parameter s or the direction of the tangent vector of C is chosen so that the circulation, Γ, is positive. From (2.3.38c), we observe that the contribution of the swirling flow to the binormal velocity is always positive and proportional to Γ. The contribution of the axial flow, C_w, is always negative, according to (2.3.85a), regardless whether the axial mass flux is positive or negative. This says,

- *A large axial flow in the core retards the binormal velocity of the filament.*

As mentioned before, $\tau_{10}/S(0)$ measures the initial age of the filament. In case that $\tau_{10} = 0$, the filament is created at $t = 0$ as a vortex line with zero core radius. The contributions, C_v and C_w, to the binormal velocity of the filament are then singular at $t = 0$. From (3.3.85), we see that $C_v \sim \ln t$ is integrable from $t = 0$, while $C_w \sim 1/t$ is not. Therefore, (2.3.86) for the binormal velocity is not valid near the creation of the filament unless the axial velocity is much smaller than the swirling flow, i.e., $w^{(0)} = 0$ or $m(0) = 0$. The reason can be traced to the similarity solutions for the axial and swirling velocities, (2.3.84a) and (2.3.84c). Note that $w^{(0)} \sim \bar{\delta}^2 \sim 1/t$ while $v^{(0)} \sim \bar{\delta} \sim \sqrt{t}$. Therefore, $w^{(0)} \gg v^{(0)}$ as $t \to 0$. To derive a formula for the binormal velocity valid near $t = 0$, we would have to carry out an initial layer analysis with a short time variable and expansions for the velocity components different from (2.3.14a, b, c) so that the axial velocity can be much larger than the swirling velocity. We shall not pursue this analysis here.

From (2.3.73b) (2.3.86), (2.3.74b) and (2.3.75), we see that the motion of the filament with similar core structure is completely determined by the initial configuration of its center line, $\mathbf{X}(0, s)$ and three constants, the circulation Γ, the initial axial flux $m(0)$ and the initial core size $\bar{\delta}(0)$ or its 'initial age' τ_{10}/S_0.

Thus the matched asymptotic analysis for a filament is completed. We have the equations of motion of the centerline of the filament and the inner solution for the core structure. The outer solution for the vector potential $\mathbf{A}$ and the velocity $\mathbf{v}$ induced by the filament are given by the B-S formulae (2.1.47) and (2.1.48).

In the next chapter, we find occasions where a representation for the vector potential $\mathbf{A}$ uniformly valid in and away from the filament is needed. Therefore, we describe here the construction of a uniformly valid composite solution.

The leading order composite solution for the vector potential

The behavior of the outer vector potential near the filament, i. e., in the overlapping region, is described by (2.1.55)–(2.1.59). In particular, the singular part of $\mathbf{A}$ is

$$\mathbf{A}^s = \frac{\Gamma \hat{\tau}}{2\pi} \ln \frac{S}{r} . \tag{2.3.87}$$

We need the inner solution for the vector potential and its behavior in the far field as $\bar{r} \to \infty$, i. e., in the overlapping region, so that we can construct the composite solution.

Since the leading order core structure is independent of θ and s, the inner vector potential can be related to the velocity components by simple quadratures. The results are,

$$\bar{\mathbf{A}} = \hat{\theta}\bar{A}_2 + \hat{r}\bar{A}_3 \ , \tag{2.3.88a}$$

with

$$\bar{A}_2 = \frac{1}{r}\int_0^r \epsilon^{-1} w^{(0)}\, r'\, dr' = \frac{1}{\bar{r}}\int_0^{\bar{r}} w^{(0)}(t,\bar{r}')\, \bar{r}'\, d\bar{r}' \tag{2.3.88b}$$

and

$$\bar{A}_3 = -\int_0^r \epsilon^{-1} v^{(0)}\, dr' = -\int_0^{\bar{r}} v^{(0)}(t,\bar{r}')\, d\bar{r}' \ . \tag{2.3.88c}$$

Here $\bar{\mathbf{A}}$ denotes the leading order solution, which is order one, and it has only circumferential and axial components, $\bar{A}_2$ and $\bar{A}_3$. Since $w^{(0)}$, given by (2.3.80a) decays exponentially in $\bar{r}$, the far field behavior of $\bar{A}_2$ is

$$\bar{A}_2 = \frac{m(t)}{2\pi\bar{r}} + o(\bar{r}^{-m}) \ , \qquad \text{for all } m \ . \tag{2.3.89}$$

Therefore, its contribution to the overlapping region is null. With $v^{(0)}$ given by (2.3.80c), the far field behavior of $\bar{A}_3$ is

$$\bar{A}_3 = \frac{\Gamma}{2\pi}\ \ln\frac{\bar{\delta}}{\bar{r}} + c_3 + o(\bar{r}^{-m}) \ , \qquad \text{for all } m \ , \tag{2.3.90a}$$

where

$$c_3(t) = -\int_0^{\bar{\delta}} v^{(0)}(t,\bar{r}')\, d\bar{r}' - \int_{\bar{\delta}}^{\infty} \Big[v^{(0)}(t,\bar{r}') - \frac{\Gamma}{2\pi\bar{r}'}\Big]\, d\bar{r}' \ . \tag{2.3.90b}$$

Note that the singular part of the outer solution, (2.3.87), is removed or matched by the inner solution. The first two terms on the right-hand side of (2.3.90a) represent the contribution of the inner solution in the overlapping region and should be removed in the composite solution. The composite vector potential is

$$\mathbf{A} = \tilde{\mathbf{A}} + \bar{\mathbf{A}} - \hat{r}\Big[\frac{\Gamma}{2\pi}\ \ln\frac{\bar{\delta}}{\bar{r}} + c_3\Big] \ , \tag{2.3.91}$$

where $\tilde{\mathbf{A}}$ denotes the outer solution defined by the B-S formula (2.1.47).

When the filament has a similar core structure, defined by (2.3.84a, b, c), we obtain from (2.3.88a, b) the formulae for the inner vector potential

$$\bar{A}_2 = \frac{m(t)}{2\pi\bar{r}}\big[1 - e^{-(\bar{r}/\bar{\delta})^2}\big] \tag{2.3.92a}$$

and

$$\bar{A}_3 = -\frac{\Gamma}{4\pi}\int_0^{\eta^2} \frac{1-e^{-\xi}}{\xi} d\xi = -\frac{\Gamma}{4\pi}\Big[E_1(\eta^2) - 2\ln\frac{\bar{\delta}}{\bar{r}} + \gamma\Big] \ , \tag{2.3.92b}$$

where $\eta = \bar{r}/\bar{\delta} = r/\delta$ and E_1 and γ denote the exponential integral and the Euler's constant, respectively. From (2.3.90b), we find that the term $c_3(t)$ in the composite solution (2.3.91) becomes a constant,

$$c_3(t) = -\frac{\gamma\Gamma}{4\pi} = -0.004434\ \Gamma \ . \tag{2.3.93}$$

To end this section, we use the matched asymptotic solution for vortex filament(s) with similar core structure to describe the flow field and then employ the formulae presented in **Sec. 1.4** to evaluate the acoustic field induced by the filament(s). This is done in the following subsection.

2.3.5 Sound Generation by Vortex Filaments

We now collect the formulae presented in **Sec.1.4** for the sound generation induced by a vortical field. The leading order acoustic field is composed of an acoustic monopole (1.4.39) and quadrupoles (1.4.41a).

The strength of the acoustic quadrupoles are related through (1.4.15b) and (1.4.38a-d) to the retarded values of the five linear combinations of second moments, F_1, F_2, F_3, H_1 and H_2, which are defined by (1.2.14). These five moments for a slender filament with center line C : $\mathbf{X}(t, s)$ are

$$F_i(t) = \Gamma \hat{\imath} \cdot \int_C [X_j^2 - X_k^2]\, d\mathbf{X} + O(\epsilon^2) \tag{2.3.94a}$$

and

$$H_i(t) = \Gamma \int_C [2X_jX_k\hat{\imath} - X_kX_i\hat{\jmath} - X_iX_j\hat{k}] \cdot d\mathbf{X} + O(\epsilon^2)\ , \tag{2.3.94b}$$

for $i = 1, 2, 3$ and i, j, k in cyclic order. Also we note the identity, $H_1 + H_2 + H_3 = 0$. Here we used the fact that the core size is order $O(\epsilon\ell)$ and hence the contributions of the second moments relative to C are $O(\epsilon^2)$. Therefore, the strengths of the quadrupoles depend explicitly only on the motion of the filament centerline(s).

From (1.4.39), one observes that the strength of the monopole is related to the retarded value of the strength of the total volume dilatation of the vortical field or the strength of the equivalent source term in the far field representation of the vortical flow. The strength in turn is related to the rate of energy dissipation Θ defined by (1.4.37). For a slender filament with leading order core structure independent of the polar coordinate θ and the axial parameter s, the strength of the monopole is the retarded value of

$$\frac{3\gamma - 5}{3} < \Theta > = \frac{3\gamma - 5}{3Re} < |\Omega|^2 > , \tag{2.3.95a}$$

with

$$< |\Omega|^2 > = 2\pi S^{(0)}(t) \int_0^\infty [(\zeta^{(0)})^2 + (w_{\bar{r}}^{(0)})^2] \bar{r} d\bar{r}\ . \tag{2.3.95b}$$

The strength of the monopole depends only on the core structure and the length of the filament. From (2.3.87a, b) and (2.3.88a, b), we conclude that

- *The motion of the filament(s) creates quadrupole noise while the evolution of the core structure generates monopole noise.*

When the filament has a similar core structure, (2.3.84a, b), (2.3.95b) reduces to

$$< |\Omega|^2 >= \frac{\Gamma^2 S^{(0)}(t)}{\sqrt{2}\bar{\delta}^2(t)} + \frac{m(0)^2 S_0^4}{\sqrt{2}[S^{(0)}]^3}. \tag{2.3.96}$$

We see that the contribution of the core structure to the acoustic monopole is defined by the three parameters, circulation Γ, initial axial flux $m(0)$ and core size $\delta(0) = \bar{\delta}(0)\epsilon$. It was noted before that accurate initial data for the vortical core other than the above three global quantities are very hard to obtain. We could assume that the core structure can be approximated by a similarity solution and then use acoustic measurements of the strength of the monopole to determine the initial core size. The difference between the theoretical predictions and the experimental data on the motion of the filament and the acoustic field would give a measure of the error in the initial axial mass flux and the deviation of the true core structure from the assumed similarity solution.

In the following section, we employ the matched asymptotic solution of filaments to study the interaction of filaments.

2.4 Validity of the Asymptotic Solutions and Their Applications

It is much clearer to present the numerical results and explain their physical meaning for axisymmetric problems than fully three dimensional ones. Therefore, we present in **Sec. 2.4.1** numerical examples simulating the interactions of coaxial vortex rings in axisymmetric flows obtained by Wang (1970) and Gunzburger (1972). We use these examples to demonstrate the importance of the core structure on the motion of the rings. In **Sec 2.4.2**. We quote several numerical examples simulating the interaction of slender toroidal vortex filaments obtained by Liu et al (1986). These examples are chosen to show the complexity of fully three dimensional problems. They show the motion of the filaments leading to merging or intersection and are employed to identify different types of merging problems. The examples also serve the purpose of establishing the practical region of validity of the asymptotic solutions.

2.4.1 Coaxial vortex rings in an axisymmetric flow

We use the axisymmetric examples to show the essential difference between the motion of a filament with curvature and a rectilinear filament or a vortex in two dimensions. We point out once more that in two-dimensional problems the velocity of a vortex center with core size much smaller than the distances to other vortices is defined by the local background velocity independent of its core structure. In contrast, the leading order velocity of a slender vortex filament does depend on its core structure. Furthermore, it is dominated by the logarithm of its effective core size. The core structure changes in time due to viscous diffusion and the stretching of the filament, the latter being an inviscid effect. To demonstrate the relative effect of viscous diffusion and inviscid stretching we quote the numerical investigations of the motion of a slender vortex ring in an axisymmetric flow around a rigid sphere by Wang (1970).

Passage of a vortex ring over a sphere

We consider axisymmetric flows with zero circumferential velocity. As shown in Fig. 2.7, a sphere with radius a is centered at the origin. The upstream velocity $U\hat{k}$ is uniform and parallel to the z-axis which is also the axis of the ring. Here $\sigma = \sqrt{x^2 + y^2}$ denotes the radial coordinate. We consider the case that the initial radius R_0 of the ring at an upstream position is not too small relative to the radius of the sphere, so that the ring can pass over the sphere while its distance to the sphere remains much larger than the core size and the boundary layer thickness along the sphere. The potential flow away from the ring is composed of three parts, an uniform flow around the sphere, the flow induced by the isolated ring with zero core size and that by its mirror image inside the sphere. The last part cancels the normal velocity on the sphere induced by the ring (Lamb 1932).

The image of a vortex ring with centerline radius R lying in the plane $z = Z$, is a coaxial ring inside the sphere. The strength Γ', radius R' and the plane $z = Z'$ of the image are defined by,

$$R' = \frac{a^2 R}{R^2 + Z^2}\,, \quad Z' = \frac{a^2 Z}{R^2 + Z^2} \quad \text{and} \quad \Gamma' = -\Gamma\sqrt{\frac{R}{R'}} = -\Gamma\frac{\sqrt{R^2 + Z^2}}{a} \tag{2.4.1}$$

The stream function of the background flow , i. e., the uniform flow around the sphere plus the flow induced by the mirror image, is

$$\Psi(t, \sigma, z) = \frac{1}{2}U\sigma^2[1 - \frac{a^3}{r^3}] - \frac{\Gamma'}{2\pi}(r'_+ + r'_-)[F(\lambda') - E(\lambda')]\,, \tag{2.4.2}$$

where $r'_\pm = \sqrt{(\sigma \mp R')^2 + (z \mp Z')^2}$, $\lambda' = (r_- - r'_+)/(r'_- + r'_+)$ and F and E are the complete elliptic integrals of the first and second kind respectively.

The vortex ring is assumed to have a similar core structure with initial core size δ_0. Because of symmetry, the self induced velocity of the ring given by the asymptotic theory has only a binormal component, pointing in z-direction. Evaluating the finite part of the B-S integral for the circular centerline, we obtain

$$\dot{Z}_1(t) = \frac{\Gamma}{4\pi R}[\ln\frac{8R}{\delta} - 0.558] \tag{2.4.3}$$

where

$$\delta^2(t) = \frac{4\nu\tau_1(t)}{S(t)} = \frac{4\nu\int_0^t R(t')\,dt'}{R(t)} + \delta_0^2\,. \tag{2.4.4}$$

The velocity of the ring submerged in the uniform stream over a sphere is equal to the local velocity of the background flow (2.4.2) plus the self induced velocity (2.4.3). The result is

$$\hat{\sigma}\dot{R} + \hat{k}\dot{Z} = \frac{1}{\sigma}\left(\hat{\sigma}\Psi_z(t, R, Z) - \hat{k}\Psi_\sigma(t, R, Z)\right) + \hat{k}\dot{Z}_1(t)\,. \tag{2.4.5}$$

In the numerical examples quoted here, we have $\Gamma = Ua/2$ and Re $= \Gamma/\nu = 3.125 \times 10^4$. At $t = 0$, the centerline of the ring lies in the upstream station $z = Z(0) = -10a$ and its radius is $R(0) = a/2$.

The variation of the core size, δ versus t is shown in Fig. 2.8 for two different initial core sizes, $\delta_0/a = 0.01$ and $\delta_0/a = 0.001$. In the inviscid theory, the core size decreases to compensate for the stretching of the ring as it passes over the sphere and returns to its original size afterwards. In the viscous theory the core size tends to grow all the time due to viscous decay and hence retards the motion. Only when the ring is passing over the sphere, does the inviscid stretching effect overcome the viscous effects. Figure 2.8 shows that the viscous effects are pronounced for the case with a much smaller initial core size especially near the initial stage. The core size grows more than ten fold from $\delta_0 = 0.001a$ before the ring begins its passage over the sphere.

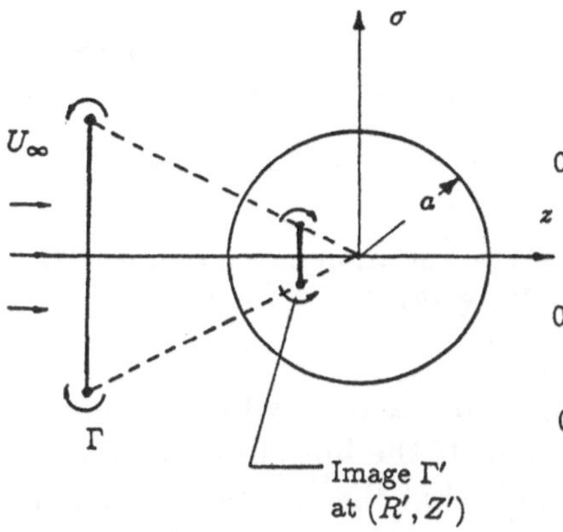

Fig. 2.7. A vortex ring moving towards a sphere and the "image" of the ring

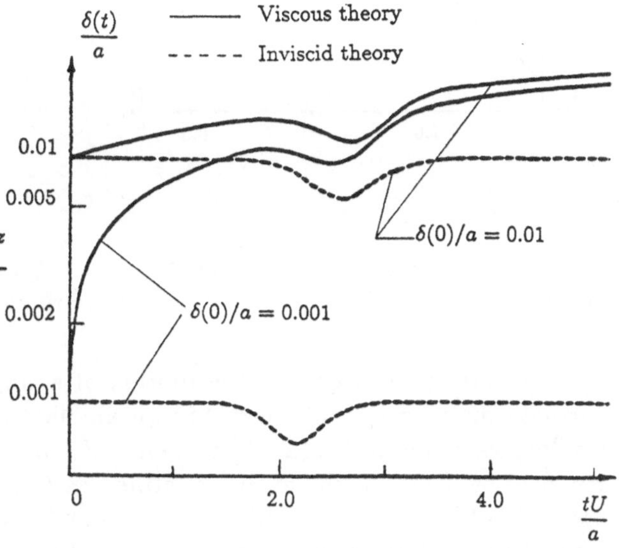

Fig. 2.8. Variation of the core size (Wang 1970)

The trajectories of the vortex ring, Z versus t, and the variation of the ring radius, R versus t, are shown in Fig. 2.9 and Fig. 2.10, respectively. From both figures we see that the enlargement of the ring radius begins when the ring is passing over the sphere and that the forward motion of the ring is faster for the inviscid theory. When the ring is passing over the sphere, the viscous effect is enhanced since the core size becomes smaller due to the stretching of the ring. Again we see from Figs. 2.9 and 2.10 that the viscous effect of retarding the forward motion is larger for the ring with a smaller initial core. To acquire a feeling of the relative order of magnitude of the viscous effect and that of the initial core structure, we consider the case of a slender ring with $\delta_0/R(0) = 10^{-2}$. The dominant term inside the square bracket in (2.4.3) becomes $\ln(8R/\delta) = \ln 800 \sim 6.68$. A doubling of the core size due to diffusion will make a change of $-\ln 2 \sim -0.693$ which is already 10% of the dominant term and is comparable to the constant -0.558 in (2.4.3).

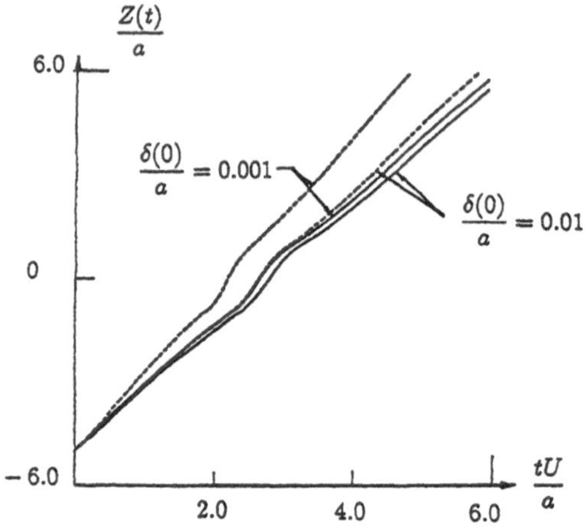

Fig. 2.9. Forward motion of the ring (Wang 1970)

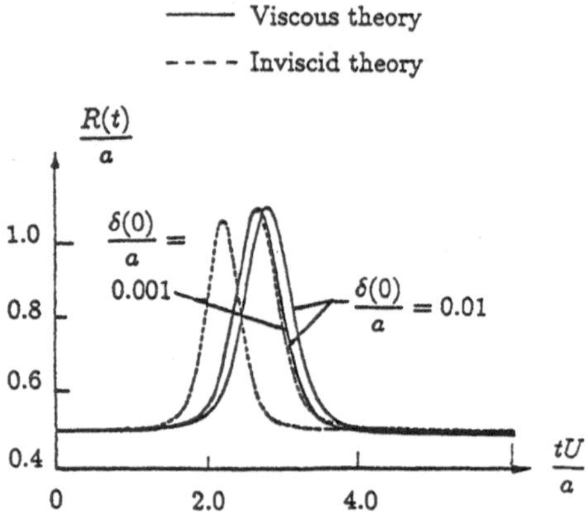

Fig. 2.10. Variation of the radius of the ring (Wang 1970)

Note that (2.4.3) defines the motion of a single vortex ring with a similarity vorticity distribution submerged in an ambient fluid. If the initial profile is nonsimilar, we can still use (2.4.3) to define $\dot{Z}_1$ provided that the constant -0.558 is replaced by $K(t)$ which can be identified as $C_v(t) - \ln(8R/\ell)$ in (2.3.73c). It was pointed out in **Sec. 2.3.4** that we can define the optimum initial core size δ_0^* of an equivalent similarity solution by matching the polar moment of the initial vorticity distribution with that of a similarity distribution. Then we have

$$K^*(t) = -0.558 + O(t^{-2}) \tag{2.4.6}$$

and the differences between a nonsimilar and an optimum similar core structure on the forward motion of vortex rings will be small. To confirm this expectation, we quote numerical examples shown in Gunzburger (1972).

Interaction of two coaxial vortex rings with nonsimilar core structure

The initial geometry of the two rings is shown in Fig. 2.11. The initial vorticity distribution is assumed to be that of a Rankine vortex, namely

$$\zeta_i(t_0, r) = \varpi_i \, H(\delta_{0i} - r) \ , \quad i = 1, 2 \ , \tag{2.4.7}$$

where $\varpi_i = \Gamma_i/(2\pi\delta_{0i}^2)$ and H represents the Heaviside function. The equations of motion of the vortex rings for nonsimilar core structure were integrated numerically for the case, $\Gamma_1 = I_2 = 1$ and $R_e = 10^6$. The initial configuration at $t = t_0$ is

$$R_1 = R_2 = Z_2 \ , \quad \delta_1 = \delta_2 = R_1/100 \ . \tag{2.4.8}$$

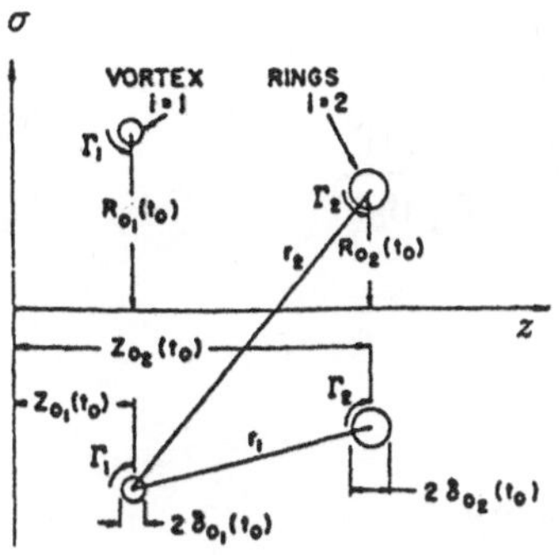

Fig. 2.11. Initial configuration of a pair of vortex rings

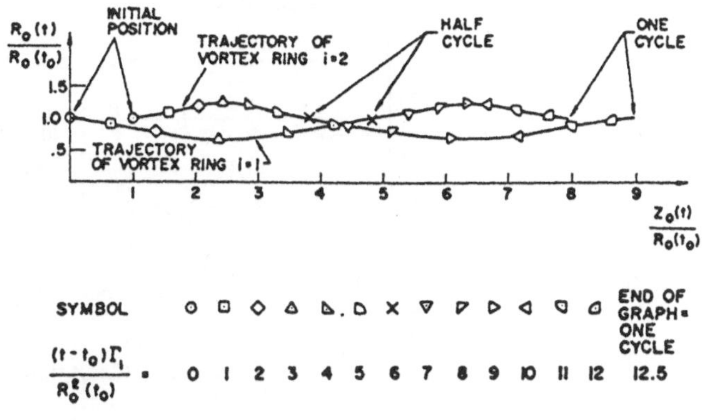

Fig. 2.12. Trajectories of the vortex rings (Gunzburger 1972)

The resulting motions of the rings are presented in Fig. 2.12. As t increases, the ring in the rear decreases in size, $\dot{R}_1(0) < 0$, the ring in the front increases in size, $\dot{R}_2(0) > 0$ and the first ring is gaining up on the second ring. When the ring that was initially behind the other moves ahead, the roles are reversed. The rings therefore take turns overtaking each other and going through each other. This behavior is qualitatively similar to that of the inviscid theory, Lamb (1932). For the inviscid theory, the core size changes due to stretching only. At the end of each periodic relative motion the cores return to their initial sizes. With viscous diffusion, the motion of the rings is aperiodic. The changes of core size during the first period are shown in Fig. 2.13.

The results displayed in Figs. 2.12 and 2.13 were obtained for the non-similar solution by using enough terms in the series for $\bar{\zeta}^{(0)}$ to insure an accuracy of 0.01% in $R_i(t)$, $Z_i(t)$ and $\delta_i(t)$ for $i = 1, 2$. Results were also obtained using the optimum similarity solution, and these results were within 0.5% of the "exact" (nonsimilarity) solution shown in Figs. 2.12 and 2.13. These results confirm our expectations that:

- *Viscous effects cannot be ignored in the motion of vortex rings whereas the difference between the non-similar solution and the corresponding optimum similarity solution is small.*

We now study the interaction of slender filaments without axisymmetry.

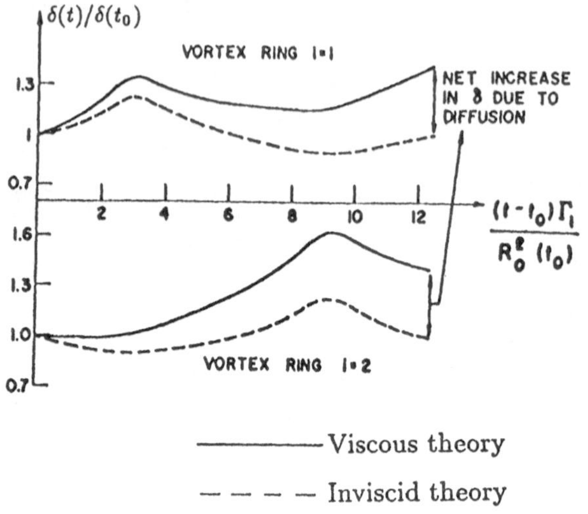

Fig. 2.13. Core sizes vs. time for one cycle (Gunzburger 1972)

2.4.2 Interactions of Vortex Filaments

A computational code for the equations of motion of a slender vortex filament was developed by Liu et al (1986). These equations were obtained in matched asymptotic analysis by Callegari and Ting (1978) and rederived in **Sec. 2.3**. They are summarized in Appendix 2.A. For a filament with similar core structure, the required initial data are: the shape of the centerline, $\mathbf{X}(0,s)$, the circulation Γ, the axial mass flux $m(0)$ and the core size δ_0. If the core structure is nonsimilar, then the initial profiles of the axial vorticity and velocity have to be specified. Since the effect of core structure, similar or nonsimilar, was demonstrated by examples of axisymmetric problems above, we shall quote here numerical examples of filaments with similar core structure to demonstrate the following effects not accounted for in the axisymmetric examples:

1) The effects of the axial mass flux, $m(0)$ and viscosity on the motion of a filament, given its initial shape, $\mathbf{X}(0,s)$, circulation, Γ, and core size, δ_0.
2) The effect of the initial orientations of two filaments relative to each other on their motion and interaction leading to merging or intersection and the onset of different types of merging problems.

It was pointed out repeatedly that the asymptotic analysis was formulated under the conditions that the effective core size $\delta(t)$ for a filament is much smaller than the other length scales in the flow field and that each filament is in the form of a slender torus of finite length. For flows induced by free vortex filaments in an unbounded domain, those length scales are: (1) the minimum radius of curvature $R_{\min}$ of a centerline $\mathcal{C}$, (2) the minimum distance $d_{\min}$ between two "distinct" points on $\mathcal{C}$, and (3) the minimum distance d_{ij} between two centerlines $\mathcal{C}_i$ and $\mathcal{C}_j$. The definition of a small core size implies the following conditions on the three length ratios:

$$(R_{\min}/\delta)_i \geq k_1 \gg 1 \ , \quad \text{for} \quad i = 1, \ldots, N \ , \tag{2.4.9}$$

$$(d_{\min}/2\delta)_i \geq k_2 \gg 1 \ , \quad \text{for} \quad i = 1, \ldots, N \ , \tag{2.4.10}$$

$$d_{ij}/(\delta_i + \delta_j) \geq k_3 \gg 1 \ , \quad \text{for} \quad i,j = 1, \ldots, N \ , \quad i \neq j \ . \tag{2.4.11}$$

On account of (2.4.9), we define two "distinct" points of $\mathcal{C}$ by the condition that the arc length between them along $\mathcal{C}$ is finite, say, larger than πR_{min}. These three conditions (2.4.9), (2.4.10) and (2.4.11) exclude self-merging, self-interaction, and intersection of two filaments, respectively. When there is a background potential flow, such as the flow around a body, additional conditions should be imposed so that the vortical core will not overlap with the boundary layer along the body surface.

The onset of merging is defined as the instant when the equality sign holds in any one of (2.4.9), (2.4.10) and (2.4.11). In **Chapter 3**, we will use these criterions to classify the types of global and local mergings. Now we have to assign the values of k_i , $i = 1, 2, 3$ and would like to assign only moderate values noting that the vorticity in the merged region is of the order of $\exp(-k_i^2)$. This formal extension of the asymptotic theory to moderate values of the k_i has been verified by comparing the asymptotic solutions with numerical solutions of the N-S equations. Let t_0 and t_1 denote the times where the length ratios in (2.4.9) – (2.4.11) predicted by the asymptotic solutions are greater or equal to $2\,k_i$ and k_i, respectively. Then the computations were initialized with the asymptotic solution at $t = t_0$ and continued to $t = t_1$. The accuracy of the extended asymptotic solutions at $t = t_1$ was established by comparison with the computational results. These verifications were carried out by Liu and Ting (1982) for the axisymmetric problems and by Chamberlain and Weston (1984) and Chamberlain and Liu (1985) for the fully three dimensional problems. All these computations showed that the difference between the asymptotic solution and the numerical solution remains sufficiently small say less than 2%, if we choose $k_1 = 2$ and $k_2 = k_3 = 1.5$. These values are used in the following examples to define the practical region of validity of our asymptotic solutions.

Now we quote examples to show the effect of "large" initial mass flux in the vortical core and viscous diffusion on the motion of a filament.

Effect of axial mass flux and Reynolds number on the motion of a filament

From the scaling introduced in **Sec .2.3**, a large axial velocity in the vortical structure is $O(\epsilon^{-1})$ while the cross-sectional area is $O(\epsilon^2)$, therefore, the "large" axial mass flux in the scales of ℓ and U is $\epsilon\, m(t)$.

The circulation of the filament used in the examples is $\Gamma = 5$. The initial shape $\mathbf{X}(0, s)$ of the centerline is an ellipse lying in the xy plane with the major axis equal to 2 along the x-axis and the minor axis equal to 1.5 along the y-axis. The initial core structure is similar with core size $\delta_0 = \sqrt{4\nu}$.

For the reference case, we choose $m(0) = 0$, i. e., no initial axial flux, and $\nu = 0.0045$ and hence $\delta_0 = 0.134$ and $Re_0^{-1/2} = \sqrt{\nu/\Gamma} = 0.03$. Shown in Fig. 2.14 are the side and top views of the centerline $\mathbf{X}(t, s)$ of the filament at various instants from $t = 0$ to the final instant $t = 21.02$. The latter marks the commencement of the local self merging as the core size becomes comparable to the minimum radius of curvature of the filament according to criterion (2.4.9).

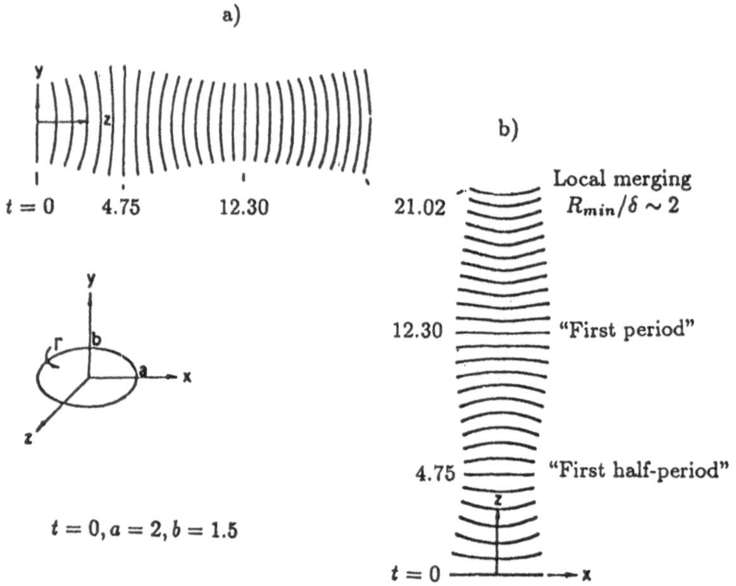

Fig. 2.14. Motion of the centerline of an initially elliptical vortex filament. a) side view, b) top view. (Liu, Tavantzis and Ting 1986)

From Fig. 2.14 we see that the centerline ceases to be planar for $t > 0$ due to the variation of the radius of curvature along its arc length. The centerline becomes almost planar again at $t = 4.75$, which we identify as "the first half-period", with its shape nearly the same as the initial one but with an interchange of the major and minor axes. At $t = 12.30$, "the first period", the centerline becomes almost planar again with its shape nearly the same as the initial one. Due to the decay of the core structure, the motion becomes more aperiodic as the core size increases with time.

Let $x^*(t)$ and $y^*(t)$ denote the maxima of the x- and y-coordinates of the centerline, $\mathbf{X}(t, s)$. Then the envelopes in the top view and the side view of $\mathbf{X}(t, s)$ in Fig. 2.14 show the variations of $x^*(t)$ and $y^*(t)$, respectively.

The motion of the centerline for a filament with axial mass flux and a different Reynolds number looks qualitatively similar to the reference case. Therefore, we use the first half period of the motion of the centerline as a quantitative measure of its difference from the reference case. Figure 2.15 illustrates the dependence of the "first period" of x_1^* on the Reynolds number $Re_0 = \Gamma/\nu$ and the initial axial mass flux in the core. We see from Fig. 2.15 that the movement of the filament slows down for larger viscosity, ν, or larger $1/Re$ because of the larger core size. The axial flow in the core tends to slow down the movement as expected from

(2.3.85a) and (2.3.86c). The effect of the axial mass flux is more pronounced for larger ν/Γ.

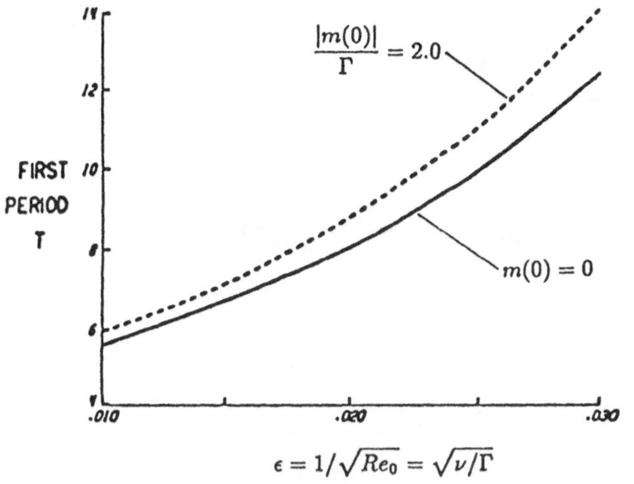

Fig. 2.15. Effect of the axial flow on the first period of motion (Liu, Tavantzis and Ting 1986)

Now we show examples of the interaction of two filaments leading to different types of merging problems.

Interaction of two filaments

Before our systematic study of the interaction of two filaments with different initial orientations, we show an example in which two filaments tend to touch each other with the vorticity vectors in the adjacent segments of the filaments in opposite directions. Thus in the scale of core size, these two segments appear to be merging as two rectilinear filaments of opposite strength. We expect that the subsequent local merging of the filament segments will result in the mutual cancellation of the vorticity. This example is of special interest because the subsequent merging and cancellation of vorticity in the overlapping segments of the filaments and their reconnection to a single filament were investigated experimentally by Oshima et al (1988) and by numerical solution of N-S equations by Ishii et al (1989). Both investigations will be reported in **Sec. 3.3**.

The initial configuration of two vortex rings is shown in Fig. 2.16. They simulate two vortex rings created by two coplanar orifices facing the same direction at $t = 0$. They have the same strength, $\Gamma_1 = \Gamma_2 = 5$ and their centerlines are circles of radius $R_0 = 1.5$ lying in the xy plane centered at $(\pm 2.5, 0, 0)$. Both core structures are similar profiles with the same core size $\delta_0 = \sqrt{4\nu}$ and have no large axial flow, $m(0) = 0$. With $\nu = 0.0005$, we have $\delta_0 = 0.0447$ and $Re_0 = \Gamma_1/\nu = 10,000$. The initial minimum distance between the rings is 2, $d_{12}(0) = 2 = 44.7\delta_0$. As time progresses, the centerline of the rings become nonplanar and tilted toward each other. The shape of a ring deviates from a circle and the deviation is more pronounced for the segments close to each other, where the effect of interaction is maximal. Shown in Fig. 2.16 are the three views of the centerlines at different instants from $t = 0$ to $t_f = 7.04$. At t_f, we have $\delta = 1.25$ and $d_{12} = 0.36 < 3\delta$. The two filaments touch each other according to the criterion (2.4.11). Thereafter, local merging takes place and the asymptotic solution is no longer applicable. The subsequent merging and cancellation of vorticity in the overlapping segments of the filaments and their reconnection to a single filament will be discussed in **Sec. 3.3.**

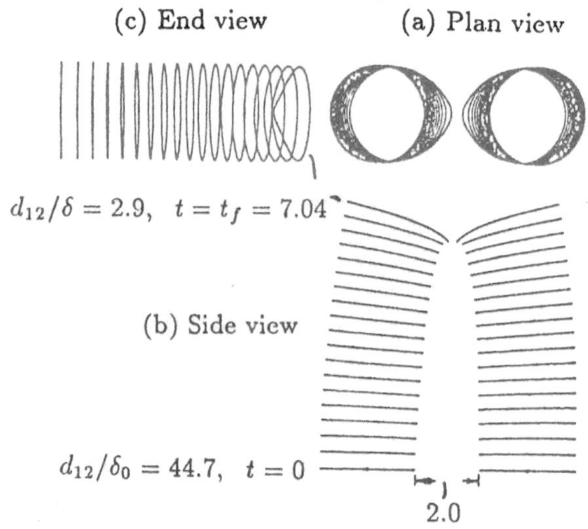

Fig. 2.16. Interaction of two filaments of the same strength, $\Gamma_1 = \Gamma_2 = 5$ with $\nu = 0.0005$, $\delta_0 = \sqrt{0.0002}$ and ring radius $R_0 = 1.5$ centered at $(\mp 2.50, 0, 0)$. (Liu, Tavantzis and Ting 1986)

Now we present three numerical examples simulating different types of interactions of two vortex filaments. These examples differ from each other in one parameter which is the ratio of the initial distance between the two filaments and their size. By varying this parameter, we demonstrate how we can use the matched asymptotic solutions to study the interaction of filaments leading to the onset of different types of merging.

In these three examples, the two filaments have opposite strengths, $\Gamma_1 = -\Gamma_2 = 5$. We choose $\nu = 0.0005$ and hence $Re_0 = 10,000$. The initial core structures of both filaments are similar with core size $\delta_0 = \sqrt{0.002}$ and there is no axial mass flux ($m_0 = 0$). Their centerlines are initially circles of the same radius, $R_0 = 1.5$. These two circles lie in the two parallel planes, $z = \mp 0.25$, and are centered at $(\mp x_0, 0, \mp 0.25)$ respectively. The initial data in figures 2.17 - 2.19 differ only in one parameter, the ratio between the radius R_0 and the distance $2x_0$ between the centers of the two circles. The projections of these two circles onto the xy plane overlap each other for $|x| < R_0 - x_0$. We define

$$\lambda^* = (R_0 - x_0)/R_0 \tag{2.4.12}$$

as the initial overlap parameter and call the filament on the left side of the yz plane the first filament. We use the polar angle θ for each circle as the parameter s for the centerline. Due to the symmetry with respect to the yz plane, it suffices to consider $x_0 \geq 0$, i. e., $\lambda^* \leq 1$. There is no initial overlap if $\lambda^* < 0$. The choice of the signs of Γ_1 and Γ_2 are such that the two rings will begin to move towards each other.

When $\lambda^* = 1$, we have the head-on collision of two circular vortex rings. This is an axisymmetric problem. The asymptotic solution provides the solution of the problem until the criterion (2.4.11) is met, i.e., the distance between these two center lines is equal to 1.5 times the sum of their core sizes. Thereafter, the cores of the rings merge or overlap and finite difference solutions of N-S equations will be employed in **Sec. 3.3.1** to simulate the subsequent merging process resulting in the cancellation of vorticity.

When $\lambda^* < 1$, the problem is no longer axisymmetric. For $x_0/R_0 \ll 1$, or $\lambda^* \to 1^-$, we have a nearly head-on collision problem and the merging will take place along the entire length of the rings. That means we have a global merging problem.

When $x_0/R_0 \gg 1$ or $-\lambda^* \gg 1$, these two rings are far apart laterally and the effect of interaction is weak. They will move in opposite directions, nearly parallel to the z-axis as isolated rings and pass each other when they cross over the xy plane. Eventually, the core of each ring will grow due to diffusion and become comparable to the ring radius. Then we have a global self merging and the structure of a slender ring disappears.

In between these two limiting cases, merging or intersection of the rings can take place locally, so that only small segments of the filaments will merge or overlap each other. Because of symmetry, the merging takes place when the filaments are crossing over the xy plane. In this intermediate range, we choose three values of λ^*, namely, 0.75, 0.133 and 0.0833. The motion or interaction of the filaments corresponding to these three values of λ^* are shown in Figs. 2.17, 2.18 and 2.19 respectively. These three figures illustrate three different types of filament interaction leading to 1) intersection or merging in two local regions, 2) touching or merging in one local region and 3) passing through the xy plane without intersection or touching with local self-merging appearing afterwards, respectively.

Shown in Fig. 2.17 is the motion of the centerlines of the two filaments with $\lambda^* = 0.75$. They move toward each other, i. e., towards the xy plane, with the "non-overlapping" part of the centerline remaining nearly planar while the "overlapping" part bends backward. As $t_f = 0.2217$, the core size δ is 0.04581, and the two filaments intersect in two local regions near the yz plane at $\theta^* = \pm 75^o$, where there distance is down to $d_{12} = 0.1372 \leq 3\delta$ and the criterion (2.4.11) is met. The projection of a centerline onto the xy plane deviates from a circle, and the deviation is more pronounced near θ^*, where the interaction is the strongest and local merging is about to take place.

Shown in Fig. 2.18 are the motion of the two centerlines with $x_0 = \ell_c = 1.30$ or $\lambda^* = 0.133$. Here ℓ_c is a critical distance found through numerical experiments. For $x_0 < \ell_c$, the filaments merge, while for $x_0 > \ell_c$ they manage to pass by each

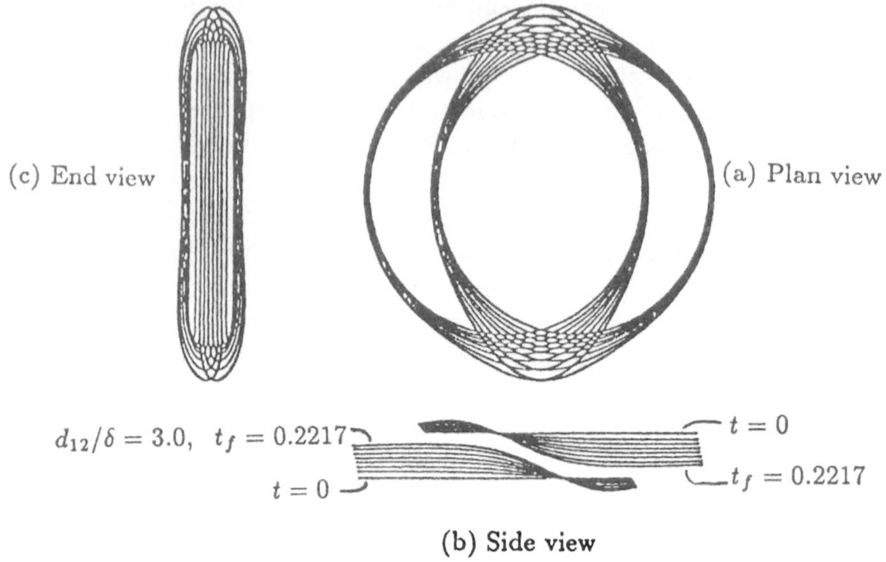

Fig. 2.17. Interaction of two filaments of opposite strength leading to intersection or merging in two local regions (Liu, Tavantzis and Ting 1986)

other without any noticeable merging or overlapping of the cores. Qualitatively, the deformation of the centerlines in Fig. 2.18 prior to merging resembles that in Fig. 2.17 except that the overlapping portion is smaller but bends back more sharply. At $t_f = 0.2324$ the two filaments intersect at one local region near $\theta^* = 0$, where the criterion (2.4.11) is met. They are in effect touching each other. The incipient stage of merging differs from that in Fig. 2.17. Now the vorticity vectors in the adjacent segments of the two filaments are nearly in the same direction and hence we expect the vorticity field in the merging region to intensify during the subsequent merging stage. This is completely different from the local merging region appearing in Fig. 2.16 where the vorticity vectors in the two segments are in the opposite direction and cancellation of vorticity is expected.

We note that the time t_f of intersection decreases as x_0 increases from zero to ℓ_c or the initial overlap parameter decreases from 1 to 0.133. Also the critical value x_0/R_0 depends on the scaled initial core size, δ_0/R_0 and the viscous effect, Re_0. If the initial core size is not too small or Re is not too large, it is possible that the critical value x_c/R_0 exceeds 1, i. e., $\lambda^* < 0$.

For $x_0 > \ell_c$ we expect the two filaments to pass each other without intersection although their projections on the xy plane in this case do overlap initially. Shown in Fig. 2.19 is the motion of the two centerlines with $x_0 = 1.375$ or $\lambda^* = 0.0833$. We see from Fig. 2.19 that the two centerlines move toward each other with the nonoverlapping portion retaining nearly the same circular shape while the small overlapping region bends backward sharply at almost a right angle to the xy plane. When $t^* \sim 0.1973$, the portions of the centerlines which were initially in the nonoverlapping region are now nearly coplanar in the xy plane while the small portions initially in the overlapping region bend backward sufficiently to allow the mutual passage of the two filaments. For $t > t^*$, the two filaments are moving away from each other, i. e., from the xy plane, and the effect of interaction is reversed. The portion that previously bent backward sharply begins to stretch forward laterally until the criterion (2.4.9) is met and local *self-merging* takes place at $t_f = 0.3193$.

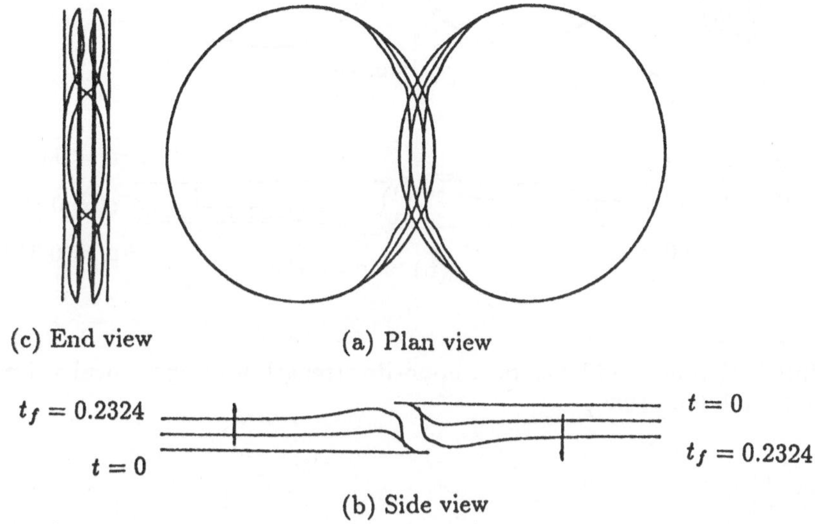

Fig. 2.18. Interaction of two filaments of opposite strength leading merging or touching in one local region (Liu, Tavantzis and Ting 1986)

As x_0 increases beyond 1.375, the mutual interaction between filaments becomes weaker and we are reaching the regime of weaker interaction mentioned above.

From the examples discussed in this section, we conclude that the asymptotic solution can be extended to its practical region of validity to identify the type of merging which is going to take place and be employed to generate the initial data for the numerical solution of the N-S equations to study the merging process. Numerical simulations of the merging process will be presented in the following chapter.

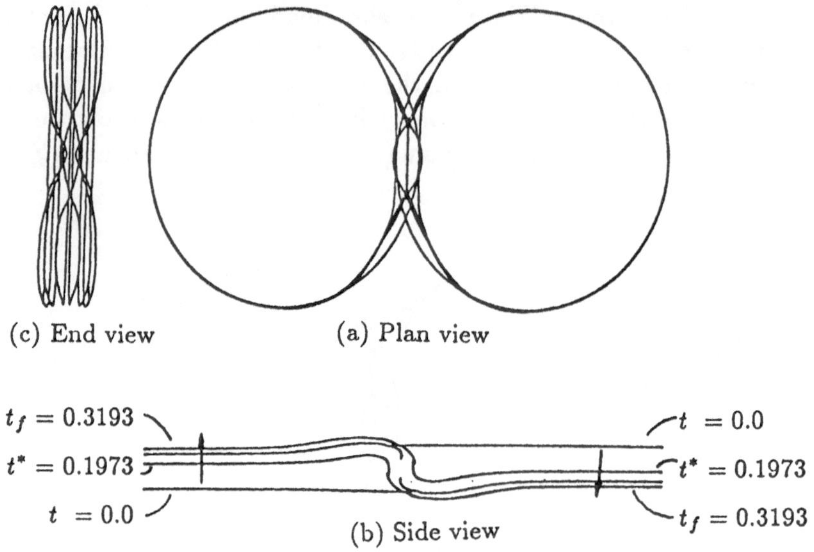

Fig. 2.19. Interaction of two filaments of opposite strength leading to local self merging (Liu, Tavantzis and Ting 1986)

3. Numerical Solutions of Viscous Vortical Flows

In the first chapter, we set up the initial value problem for a vorticity field in which the diffusion length scale is of the order of the typical scale of the flow field. In this case, viscosity affects the flow globally. In the second chapter, we showed the importance of the viscous diffusion for the evolution of the core structure and for the motion of well separated slender vortices or filaments. In this chapter, we emphasize that the viscous diffusion of vorticity play a dominant role in the merging and/or cancellation of vortices. We show that such processes, even if taking place only in local region(s), change the topology of the vorticity field. Consequently, viscous diffusion has, in this way, an indirect but important influence on the global flow structure. To study the merging of vortices, we have to resort to numerical solution of the N-S equations. In this chapter, we explain several numerical schemes and the results of numerical simulations of viscous vortical flows.

The numerical algorithms to be presented have to be restricted to a finite computational domain, $\mathcal{D}$. The boundary conditions imposed on $\partial\mathcal{D}$ should be consistent with the far field behavior of the original problem in $\mathbb{R}^d$. Using higher order boundary approximations based on the far field expansions in **Chapter 1**, we can reduce the size of $\mathcal{D}$ without sacrificing the accuracy of the numerical solution. By comparing the errors introduced through the numerical scheme in the interior of $\mathcal{D}$ and those from the approximate boundary conditions on $\partial\mathcal{D}$, we determine an optimal size of $\mathcal{D}$, that is, a minimum number of grid points for a given degree of accuracy. In this way, we arrive at an efficient numerical scheme.

We explain the basic concepts for the choice of the computational domain and the formulation of boundary conditions tailored for problems with several specific types of initial data. From the structure of the initial data, we expect either global merging of filaments or local merging of small segments of filaments. The types are classified according to the relative order of magnitude of the three length scales inherent in the initial data. They are the size of the vortical field $L = O(\ell)$, the decay length $\ell_d = O(\delta^*)$ and the size of the merging region ℓ_m.

Krause, Liu and Ting (1985) introduce the following classifications and identify the required size, D, of the computational domain, $\mathcal{D}$:

I. When $\ell_m = O(\ell)$, we have *a global merging.*
   (a) When $\ell_d = O(\ell)$, we need $D \gg \ell$ say $D = 4\ell$,
   (b) When $\ell_d \ll \ell$, we merely need $D - L \gg \ell_d$.

II. When $\ell_m \ll \ell$, we have a *local merging.*
   We merely need to satisfy $D \gg \ell_m$, while D may be much smaller than ℓ.

Examples of local merging are often found in three dimensional problems when the core size of a filament becomes comparable to its local radius of curvature, or when the center lines of two slender filaments intersect at a finite number of points. A local merging region generally contains only small segment(s) of the filament(s).

The essential ideas in the choice of the size of $\mathcal{D}$ and the formulation of the appropriate boundary conditions for global merging problems of type Ia and Ib are explained in **Sec.3.1.2** and **3.1.3** respectively. In **Sec.3.1.4** we describe in detail the numerical scheme for the simulation of local merging processes and show how numerical and asymptotic methods can compliment each other in describing the entire flow field.

According to the above classifications, any merging of two dimensional vortices is a global merging problem of type Ia or Ib. Related numerical examples are presented in **Sec.3.2.1**.

In **Sec.3.2.2**, we present yet a different numerical method for two dimensional merging problems. The key idea for the scheme is best explained by the example of a pair of two-dimensional vortices. By applying Poincaré's relationships on the conservation of the total strength and first moments of vorticity and using the linear growth of its polar moment, (**Sec. 1.3.4**), rules for the merging of two vortices with viscous cores to a single Lamb vortex with an optimum core structure are formulated. Continuing along the same line of thought, an approximate numerical method was proposed by Ting (1986) and Ting and Liu (1986). The vorticity distribution is expressed as a sum of Lamb vortices with finite effective core sizes which may overlap each other. The N-S equations are then satisfied approximately by using a minimum principle to govern the motion of the vortex centers. The diffusion of vorticity is included through the individual spreading of the Lamb vortices. The results of numerical examples given in **Sec.3.2.3** are in good agreement with the corresponding finite difference solutions in **Sec.3.2.1**.

Examples for the global merging of axisymmetric vortex rings and vortex filaments are presented in **Sec.3.3.1** and **3.3.2** respectively.

3.1 Classification of Merging Problems and Efficient Numerical Schemes

In this section we explain our classification of merging problems, elaborate on the meaning of an efficient numerical scheme and then outline a numerical scheme for each type of merging.

3.1.1 Classification of merging problems

From the structure of initial data that exhibit the merging of vortices or the commencement of their merging, we can identify three characteristic length scales. They are the size of the vortical field L, the size of the merging region(s) ℓ_m and a vorticity decay length ℓ_d. The size L is usually taken to be the normal length scale ℓ of the flow field introduced in the preceding two chapters. The decay length ℓ_d depends on the Reynolds number Re of the flow field and is of the order $L/\sqrt{Re}$.

Depending on the order of magnitude of ℓ_d relative to L, the types of merging are classified as type I, denoting a global merging with $\ell_m = O(L)$ and type II, which is a local merging with $\ell_m << L$. The global merging can be further subdivided into two types: type I(a) when $\ell_d = O(L)$ and type I(b) when $\ell_d << L$.

In summary, we identify the characteristic lengths

$$\begin{aligned} L &: \text{ overall extend of the vorticity distribution,} \\ \ell_m &: \text{ size of the merging region,} \\ \ell_d &: \text{ vorticity decay length,} \end{aligned}$$

and then classify the different types of merging by

$$\begin{aligned} &\text{type I(a)} : \ell_m/L = O(1), \quad \ell_d/L = O(1), \\ &\text{type I(b)} : \ell_m/L = O(1), \quad \ell_d/L \ll 1, \\ &\text{type II} \quad : \ell_m/L \ll 1. \end{aligned}$$

An example of global merging of type I(a) is shown in Fig. 3.1(a). This situation might arise when the core size, δ, of a vortex filament ceases to be small relative to its radius of curvature so that the criterion (2.4.9) is violated. During this process, we have $\delta = O(L)$ and as the merging progresses the filament structure disappears as indicated in the figure. In a numerical simulation, the vortical field can be effectively confined inside a domain $\mathcal{D}$ with size $D >> L = O(\ell_m)$ for a finite time interval provided that the domain is allowed to move with the center of vorticity.

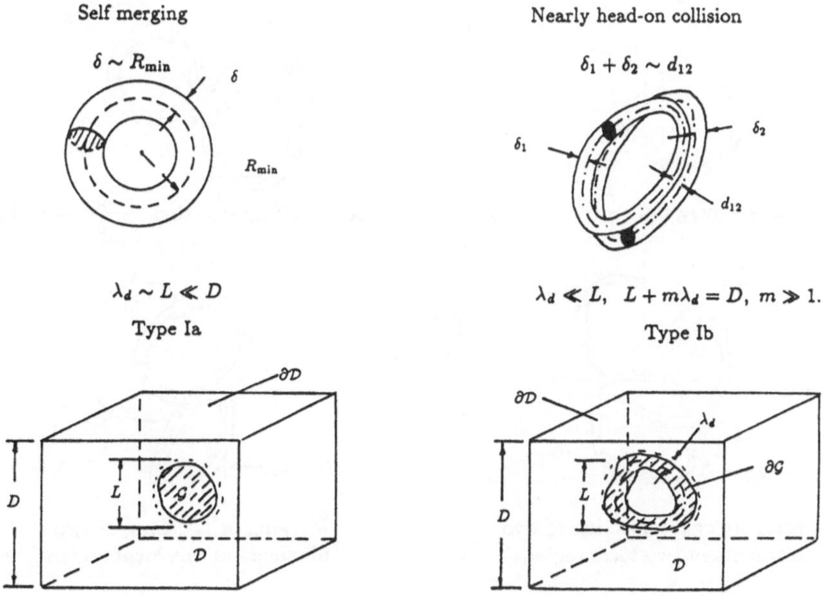

Fig. 3.1. Global merging of Filament(s)

An example of global merging of type I(b) is shown in Fig. 3.1(b). Two filaments merge along their entire length almost simultaneously, while the core sizes remain much smaller than the minimum radius of curvature of the filaments. The criterion

(2.4.11) is violated while (2.4.9) and (2.4.10) remain valid. During the merging, the vorticity distribution of both filaments can be effectively confined in a slender toroidal region or in a thin rectangular domain with a radius or thickness much larger than the core size but still much smaller than the size of the filament(s). These two types of global mergings can be described as follows:

For type I(a), the initial vorticity distribution is assumed to decay exponentially on a length scale on the order of the size, L, of the vorticity field. Therefore, the far field behavior of the vorticity distribution obeys

$$|\Omega(t,\mathbf{x})| = O(e^{-r^2/L^2}) \quad \text{as } r \to \infty \quad \text{for } t > 0 \;, \tag{3.1.1}$$

where $r = |\mathbf{x}|$. The dominant part of Ω is contained in a sphere of radius $L = O(\ell)$ and the size, D, of the computational domain has to be much larger than L, i.e., $D >> L$ (see Fig. 3.1a).

For type I(b), the initial vorticity distribution is effectively contained in a domain $\mathcal{G}$ of size L, but it is assumed to decay exponentially outside of $\mathcal{G}$ with a much smaller decay length $\ell_d \ll L$. Therefore, the vorticity behaves outside of $\mathcal{G}$ as

$$|\Omega| = O(e^{-h^2/\ell_d^2}) \;, \quad \text{with } \ell_d << L \;, \tag{3.1.2}$$

where h denotes the distance normal to the boundary $\partial\mathcal{G}$ of $\mathcal{G}$ (see Fig. 3.1b). The computational domain may then be of size $D = L + m\ell_d$ with $m >> 1$ but $m\ell_d = O(L)$.

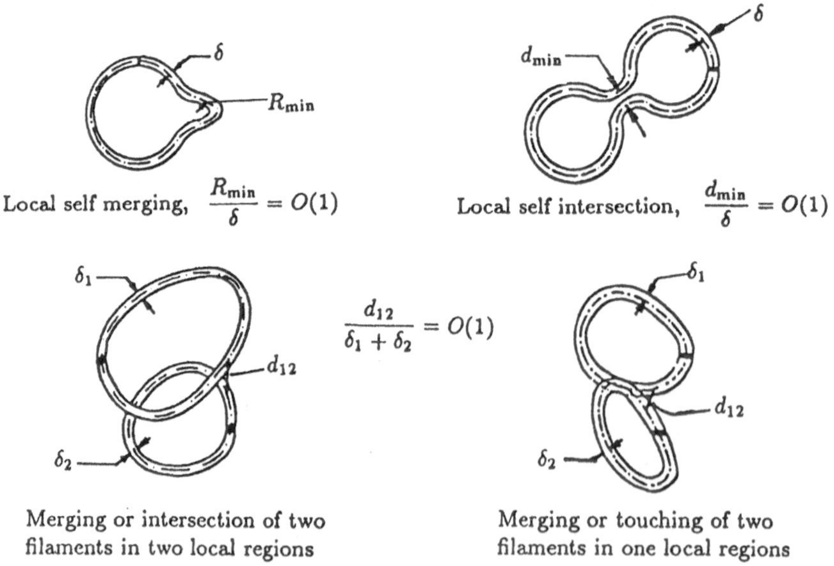

Fig. 3.2. Local merging and/or interaction of filament(s)

Examples of local mergings of type II are shown in Fig. 3.2. They demonstrate two cases of local self-merging of a filament in violation of criterion (2.4.9) and (2.4.10) and two cases of the intersection of two filaments in local region(s) in violation of (2.4.11). The size of a local merging region is of the order of the core size and is much smaller than the size of the filament(s). Thus $\ell_d, \ell_m = O(\delta) << L$. Away from the local merging regions the slender filament structures persist and it is clear that it would be very inefficient to study the merging process by numerical solution of the N-S equations on a domain containing the entire vortical field.

Before describing efficient numerical schemes tailored for specific types of merging problems we shall first identify different sources of errors in the numerical solutions and then describe the essence of an efficient scheme specifically for problems of type Ia. The corresponding notions for problems of type Ib and II will then become apparent.

3.1.1.1 Source of Numerical Errors

For a finite-difference scheme, the errors e_t and e_s in the temporal and spatial derivatives are usually assessed in terms of the nondimensional time step Δ_t/t_{ref} and of the scaled grid size $\Delta_s/\ell_{\text{ref}}$, where t_{ref} and ℓ_{ref} are characteristic time and length scales for significant local changes of the vorticity (see for example Richtmyer and Morton (1967)). Estimates for the global discretization errors can be expressed in the form

$$e_t = O\big([\Delta_t/t_{\text{ref}}]^{p_t}\big) \ll 1 \qquad \text{and} \qquad e_s = O\big([\Delta_s/\ell_{\text{ref}}]^{p_s}\big) \ll 1 \;, \tag{3.1.3}$$

where $p_s + 1$ and $p_t + 1$ are integers specifying the order of local approximations of the difference scheme. In our discussion the relevant reference length is the decay length ℓ_d and for initial data of the type I(a) it coincides with the overall extension, L, of the vorticity distribution. The relevant reference time for type I(a) is L/U where U denotes an average velocity in the initial flow field.

For a given degree of accuracy, say e_a, we require

$$e_s \sim e_t \sim e_a \;. \tag{3.1.4}$$

This, in turn, defines the grid size Δ_s and the time step Δ_t through (3.1.3) for a given set of p_t and p_s, i. e., for a particular discretization.

Since the numerical solution can be carried out only for a finite computational domain, $\mathcal{D}$, other sources of error are introduced due to the replacement of the far field conditions by approximate conditions applied on the boundary, $\partial\mathcal{D}$. Consider first the condition of exponential decay of $|\Omega|$ in (3.1.1). If this condition is replaced by

$$\Omega = 0 \qquad \text{on} \quad \partial\mathcal{D} \;, \tag{3.1.5}$$

the error is exponentially small, namely

$$e_\Omega = O(e^{-D^2/L^2}) \;. \tag{3.1.6}$$

On the other hand, we have shown in **Sec.1.2** that the velocity and vector potential $\mathbf{A}$ decrease only in descending powers of L/D. Thus, if we impose

$$\mathbf{A} = 0 \qquad \text{on} \quad \partial\mathcal{D} \;, \tag{3.1.7a}$$

the error is

$$e_A = O(L^2/D^2) \ , \tag{3.1.7b}$$

(because the zeroth moments of vorticity vanish for all t.) Therefore, the error in this set of boundary data is dominated by e_A.

By using the far field expansion procedure described in **Sec.1.2**, (1.2.15)–(1.2.17), we can reduce the error to

$$e_A = O((L/D)^{m+2}) \ , \tag{3.1.8}$$

where m denotes the highest moments of vorticity retained in the far field expansion. Such an improvement of the boundary data for **A** provides a considerable reduction of the computational efforts needed to achieve a given degree of accuracy. To clarify this aspect, we determine the number of grid points, N, needed to formally keep the errors in the boundary data and the errors of the discrete approximation errors in $\mathcal{D}$ at the order e_a. With

$$N = O(D^3/\Delta_s^3) \ , \tag{3.1.9}$$

equations (3.1.3) and (3.1.8) imply

$$N = O(e_a^{-\mu}) \qquad \text{where} \qquad \mu = \frac{3}{p_s} + \frac{3}{m+2} \ . \tag{3.1.10}$$

Thus, an increase in the accuracy of the boundary data has the same potential for reducing the numerical effort as an increase in the order of the discretization, provided that the constraints (3.1.4) on the errors are observed.

To give a concrete illustration of the effect of a more accurate boundary condition on **A**, we consider p_s fixed and show the savings in the number of grid points, N, when m is increased from 0 to 2 by retaining up to second moments of vorticity in the far field expansion. Let μ_j denote the exponent μ in (3.1.10) with $m = j$, $j = 0, 2$ and N_j denote the corresponding number of grid points needed. With the exponent μ_0 reduced to $\mu_2 = \mu_0 - 3/4$, the number of grid points is reduced by a factor $e_a^{-3/4}$. For example, even for a moderate value of $e_a = 1/256$, we have $N_2 = N_0/64$ leading to a considerable savings in the computational efforts.

In arriving at (3.1.10), we required that the error in the boundary data is of the order of the error in the discretization, i. e., $e_A \sim e_s \sim e_a$. In principle, we could allow a larger error, e_A, relative to e_s since the errors induced in the boundary conditions have a weaker influence on the global accuracy. (The reason is that the boundaries are a manifold of dimension one less than that of the computational domain.) However, we prefer the stronger condition on e_A because the decisions on the number N and the computational domain $\mathcal{D}$ are based on the estimates for the characteristic lengths in the structure of the initial data. As time proceeds, the size of the vorticity field L increases accompanied by an increase in the error of the boundary data while that of the discrete approximation diminishes. Since we do not want to change the computational domain frequently, we choose a value for e_A smaller than necessary, that is, a computational domain larger than necessary at the beginning. This practical consideration was reported by Chamberlein et al. 1984. In a series of numerical simulations using different order boundary conditions, they observed spurious generation of vorticity at the boundaries after a finite time T for lower order approximations. This spurious generation can be delayed to a much later time by increasing the size of $\mathcal{D}$ at a few time steps before T or by using higher order boundary conditions to begin with.

3.1.1.2 Meaning of an Efficient Numerical Scheme

For a scheme to be efficient, the number of operations for the evaluation of the approximate boundary data for **A** should be at most of the same order as that for the solution of the finite-difference equations at each time step. Since we are interested in time accurate approximations for transient phenomena (we are not seeking numerical schemes with large Δ_t), the vorticity field can be advanced in time by solving an explicit finite difference approximation of the vorticity diffusion equation (1.1.3). The number of operations per time step is then of order N for most such methods. Therefore, the overall number of operations for each time step is dominated by that for the solution of the Poisson equation for the vector potential **A**. Using a fast Poisson solver, this number is of order $N \ln N$, (Swarztrauber and Sweet (1979)). Since $\ln N$ may be considered as a finite number relative to N, our objective is to increase the accuracy of the boundary data so as to reduce the size of $\mathcal{D}$ and hence the number of grid points, while keeping the number of operations in the computation of the boundary data at order $O(N)$.

It was noted by Ting (1983) that, although boundary data obtained by a direct evaluation of the Poisson integral will reduce the error e_A to that of e_Ω in (3.1.1), the number of computational steps for a straight-forward numerical integration is $O(N^{5/3})$. Hence, this procedure is not acceptable and a more efficient scheme had to be devised. Such a method and related accuracy estimates will be the topics of **Sec. 3.1.3**.

In principle, we can reduce the error e_A in (3.1.7) to $O([L/R]^{m+2})$ by including additional terms in the series (1.1.14)–(1.1.17) up to the mth moment of vorticity and reduce accordingly the size of $\mathcal{D}$. In general, the number of moments that must be evaluated at each time step is $4\Sigma j - 7 = 2m(m+1) - 7$ for $m \geq 2$, (see the integral invariants discussed in **Sec. 1.2**). The number of computational steps for the evaluation of the boundary values of **A** then becomes $O(2m^2 N)$. Furthermore, the error induced by omitting the vorticity outside of $\mathcal{D}$ in an approximate evaluation of an mth moment of vorticity is $O([D/L]^m\, e^{-D^2/L^2})$, in which the significance of the factor $[D/L]^m$ increases as m increases. These two observations imply a practical upper bound on m for a given D/L. The upper bound increases as D/L increases.

If we increase the accuracy requirement by choosing a smaller e_a while retaining the same number of grid points, and the same discrete approximation, we have to reduce the grid size Δ_s and the time step Δ_t accordingly. Hence, we have to shrink the computational domain. If $\mathcal{D}$ and R are the domain and its size for the required accuracy, and $\mathcal{D}_1$ and R_1 are those for $e_{a1} < e_a$, we have

$$\mathcal{D}_1 \subset \mathcal{D} \qquad \text{and} \qquad R_1 < R\,.$$

To reduce also the error e_A in the approximate boundary conditions to the order e_{a1}, we have to increase the number of terms in the far field expansion, (1.1.14). However, in this case, the error of the numerical evaluation of an mth moment due to the neglect of the vorticity outside of $\mathcal{D}_1$ can be larger than e_{a1}.

This difficulty can be overcome by using a larger domain, say $\mathcal{D}_2$, only in updating the vorticity distribution and for the numerical evaluation of the mth moments. The boundary condition for the vorticity, $\Omega = 0$, is then imposed on $\partial\mathcal{D}_2$, while approximate boundary conditions for the vector potential are still imposed on $\partial\mathcal{D}_1$. Without lowering our requirements of efficiency, we are allowed to use a larger domain, and hence a larger number of grid points, N_2, for the vorticity calculation, because the number of steps in updating the vorticity is of order N_2. So long as $N_2 = O(N_1 \ln(N_1))$, the number of computational steps per time step remains order $N_1 \ln(N_1)$. This "*two-domain method*" will be described in detail in **Sec. 3.3.1**.

For initial data of the type Ib, we use the fact that the vorticity decays on a short length scale $\ell_d \ll L$. This allows us to reduce the size of the computational domain $\mathcal{D}$ from a size much larger than L to a size $D = L + O(k\lambda)$ with $k >> 1$, and $k\lambda$ being at most of order L. In this case, the vorticity on the boundary is small, of order $\exp(-k^2)$, but the velocity is order one because $D/L = O(1)$. Hence, it is *incorrect*, not just inaccurate, to impose $\mathbf{A} = 0$ on $\partial\mathcal{D}$. It is imperative that approximate boundary data be derived for $\mathbf{A}$ such that the error e_A is reduced to say $O(k^{-4})$. This procedure will be explained in **Sec. 3.1.3.**

For a local merging problem of type II, as shown in Fig. 3.2, the computational domain may be much smaller than the size of the vortical field. Now it is not only incorrect to impose $\mathbf{A} = 0$ on the boundary $\partial\mathcal{D}$, but also to impose $\Omega = 0$. For this problem we have to develop a scheme which combines the asymptotic solution outside $\mathcal{D}$ with the numerical solution in $\mathcal{D}$. For the latter, we need to construct approximate boundary data for both $\mathbf{A}$ and Ω and assess their errors. This scheme will be outlined in **Sec. 3.1.4.**

3.1.2 Approximate Boundary Data for Merging of Type I(a)

For this type, the initial vorticity field, Υ, is assumed to decay exponentially in r with a decay length scale $\ell_d = O(\ell)$ and the dominant part of Υ is contained in a sphere $\mathcal{G}$ of radius $L = O(\ell)$ centered at the origin. The flow field has only one length scale and the far field behavior of Ω is defined by (3.1.1). The computational domain can be a cube of size D, also centered at the origin with $D >> L$ (see Fig. 1a).

In order to derive an approximate boundary condition for the vector potential, $\boldsymbol{A}$, with accuracy higher than that of $\boldsymbol{A} = 0$, we repeat the procedure presented in **Sec. 1.2**. The vector potential satisfying the vector Poisson equation (1.1.9) and the far field condition, (1.1.12) is expressed as a Poisson integral

$$\mathbf{A}(t, \mathbf{x}) = \frac{1}{4\pi} \int\int\int_{\mathbb{R}^3} \frac{\Omega(t, \mathbf{x}')}{|\mathbf{x} - \mathbf{x}'|} d\mathbf{x}' \,. \tag{3.1.11}$$

The far field behavior of $\mathbf{A}$, in terms of powers of r^{-1} is

$$\mathbf{A}(t,\mathbf{x}) = \sum_{n=0}^{m} \mathbf{A}^{(n)}(t,\mathbf{x}) + O(r^{-m-2}) \,. \tag{3.1.12}$$

The $\mathbf{A}^{(n)}$ were expressed in terms of moments of vorticity in **Sec.1.1**, (1.1.14 - 17). For the first three terms in (3.1.12), we have

$$\mathbf{A}^{(0)} \equiv 0 \,, \tag{3.1.13}$$

$$4\pi r^3 \, \mathbf{A}^{(1)} = \sum_{j=1}^{3} \langle x'_j \Omega' \rangle x_j \tag{3.1.14}$$

and

$$4\pi r^5 \, \mathbf{A}^{(2)} = \sum_{j=1}^{3} \sum_{k=1}^{3} \langle x'_j x'_k \Omega' \rangle \, \frac{1}{2}(3x_j x_k - r^2 \delta_{jk}) \,. \tag{3.1.15}$$

Formulae (3.1.13 - 15) were derived in **Sec.1.2**, (1.2.16 - 17), by using the integral invariants of Truesdell and Moreau. In the numerical scheme, we use $\langle f \rangle$ to denote the volume integral of f over $\mathcal{D}$ instead of $\mathbb{R}^3$. The error introduced in this approximation for an mth moment is $O([D/L]^m \, e^{-D^2/\ell^2})$.

By inserting (3.1.13 - 15) in (3.1.12), we obtain the second order ($m = 2$) boundary data for $\boldsymbol{A}$ with error

$$e_A = O(\ell^4/D^4) \,. \tag{3.1.16}$$

We assume that the size of $\mathcal{D}$ is so chosen that the error in the vector potential, e_A, and the errors in the approximations of the mth moments are within the required accuracy e_a. So far, the integral invariants have been employed to simplify the evaluations of the higher order boundary data of $\mathbf{A}$. Those invariants are also useful for a posteriori error checking. For that purpose, there are additional integral identities, for example, the global decay law for $< \mathbf{A} \cdot \Omega >$, presented in **Sec.1.2.2** (1.2.18). It is

$$\langle \mathbf{A} \cdot \Omega \rangle = \langle \mathbf{A} \cdot \Omega \rangle_{t=0} - 2\nu \int_0^t \langle \Omega \cdot \Omega \rangle \, dt \,. \tag{3.1.17}$$

After every so many time steps, we compute the integrals in those invariants and in (3.1.17) from the numerical results for Ω and monitor the deviations from the invariance conditions and from the above global decay law. These deviations measure the error of the numerical solution and indicate whether the size of the domain is still large enough.

The essence of the numerical scheme formulated above is summarized as follows:

- After choosing a discrete approximation to the differential equations, we select the grid size Δ_s relative to ℓ and the time step Δ_t relative to ℓ/U, by matching the errors due to discretizations, e_s and e_t, with the required degree of accuracy, e_a.
- By matching the error e_A from the boundary data in (3.1.10) with e_a, we determine the size D of the computational domain and then the total number of grid points, N.

- The number of operations for each time increment remains $O(N \ln N)$. This number is dominated by the Poisson solver.
- This scheme is efficient because the errors from different sources are of the same level so that we do not perform excessive computations just to keep the error from one source much smaller than the other errors.

A computer code based on these ideas was developed by Chamberlain and Liu (1985) and employed to study the global merging of vortex filaments by Weston and Chamberlain (1984). Several examples from these references will be described in **Sec. 3.2.1** and **Sec.3.3.1** below.

3.1.3 Approximate Boundary Data for Merging of Type I(b)

An example for this type of merging is the nearly head on collision of two filaments as shown in Fig.3.1(b). In this case, the vortical cores begin to merge along almost their entire length. The initial vorticity distribution, Υ, defined by the core structures of the two filaments, can now be contained in a domain, $\mathcal{G}$, of size L. The domain does not have to be simply-connected. The vorticity decay length, ℓ_d, is much smaller than L. Since ℓ_d characterizes the length scale for order one changes of vorticity, the error e_s in the finite-difference scheme is now measured by Δ_s/ℓ_d instead of the smaller Δ_s/L. Thus, for a given degree of accuracy, e_a, we choose the grid size Δ_s such that

$$\Delta_s/\ell_d \ll 1 \,, \tag{3.1.18}$$

resulting in a much smaller Δ_s/L. It is very inefficient to apply the scheme of the preceding subsection for a computational domain $\mathcal{D}$ with size $D >> L$, since the number of grid points N will be increased by a factor of $(L/\ell_d)^3$ beyond what is really needed. To be more specific, let the vorticity distribution be effectively confined inside a sphere of radius L. Then, in a thin layer outside of the sphere, the vorticity decays exponentially with h^2/ℓ_d^2, where h is the distance to the sphere. Outside a sphere of radius $L+m\ell_d$, the vorticity will be of order $\exp(-m^2)$, which means $O(10^{-4})$ for, say, $m = 3$. Hence, the approximate condition

$$\Omega = 0 \tag{3.1.19}$$

may be imposed on the boundary of a computational domain $\mathcal{D}$ with size $D > L + m\ell_d$. It is sufficient to choose, for example,

$$D = 2L \quad \text{with} \quad \ell_d < L/4 \,. \tag{3.1.20}$$

However, the approximate boundary conditions for the vector velocity potential, using only the first three terms in the power series expansion (3.1.12) in $1/r$, is inaccurate since D is no longer much larger than L. Thus it is necessary to reformulate the numerical boundary conditions for $\mathbf{A}$.

Notice first that the series in (3.1.12) is derived by expanding $|\mathbf{x}' - \mathbf{x}|^{-1}$, in the integrand of the Poisson integral, (3.1.11), in powers of $\mathbf{x}'/r$, where r is the distance from the origin of the vorticity distribution. In the current situation, Fig. 3.1b, the overall extena of the vorticity field is still characterized by L, but the volume of the dominant part of the vorticity field is only a small fraction of L^3. To take advantage of this observation, we modify the power series expansion for the boundary data: The Poisson integral over $\mathcal{D}$ is rewritten as the sum of integrals over M subdomains $\{\mathcal{D}_i\}_{i=1}^{M}$ of $\mathcal{D}$, and then for each i a power series expansion with respect to the center of the $\mathcal{D}_i$ is performed. The detailed steps are as follows:

i) The computational domain is divided into subdomains $\mathcal{D}_i$, $i = 1, \ldots, M$, with

$$\bigcup_{i=1}^{M} \mathcal{D}_i = \mathcal{D} . \tag{3.1.21}$$

The subdomains have roughly the same size, $\ell_i \approx \alpha L$, where $\alpha \ll 1$. For example, when $\alpha = 1/4$, and $D = 2L$, we have $\ell_i \approx D/8$ and $M = 512$.

ii) The vorticity Ω is decomposed as

$$\Omega = \sum_{i=1}^{M} \Omega_i \tag{3.1.22a}$$

with

$$\Omega_i = \begin{cases} \Omega(\mathbf{x}, t) , & \mathbf{x} \in \mathcal{D}_i \\ 0 , & \mathbf{x} \notin \mathcal{D}_i . \end{cases}$$

The vector potential (3.1.11) induced by Ω is then expressed as the sum of the contributions induced by all of the Ω_i,

$$\mathbf{A}(t, \mathbf{x}) = \frac{1}{4\pi} \sum_{i=1}^{M} \int\int\int_{\mathcal{D}_i} \frac{\Omega_i(t, \mathbf{x}')}{|\mathbf{x} - \mathbf{x}'|} d\mathbf{x}' . \tag{3.1.22b}$$

Here $\mathbf{x}$ is on $\partial\mathcal{D}$ and $\mathbf{x}'$ is in $\mathcal{D}_i$ where $\Omega_i \neq 0$ as shown in Fig. 3.3.

iv) For a subdomain $\mathcal{D}_i$ next to the boundary $\partial\mathcal{D}$, we can set $\Omega_i = 0$, i. e., we can omit its contribution to $\mathbf{A}$, because of the exponential decay of Ω. For each subdomain not adjacent to $\partial\mathcal{D}$, we expand the denominator $|\mathbf{x}' - \mathbf{x}|$ in the integrand in a power series with respect to the center $\mathbf{x}_i$ of $\mathcal{D}_i$, i.e., in a power series with respect to

$$\frac{|\mathbf{x}' - \mathbf{x}_i|}{r_i} \leq \frac{\alpha L}{2r_i} << 1 , \tag{3.1.23}$$

where $r_i = |\mathbf{x} - \mathbf{x}_i|$. The Poisson integral for Ω_i can be approximated by the first three terms of the power series in $\alpha L/(2r_i)$ with the coefficients given by the moments of the vorticity in $\mathcal{D}_i$ with respect to its center $\mathbf{x}_i$.

v) The sum of the approximate representations of the Poisson integrals of the Ω_i's in (3.1.22b) then yields the approximate boundary data of $\mathbf{A}$ on $\partial\mathcal{D}$.

It should be noted that the evaluation of an nth moment for each subdomain $\mathcal{D}_i$ and then the sum for $i = 1, \ldots, M$ involve the same number of operations as the computation for the single domain $\mathcal{D}$.

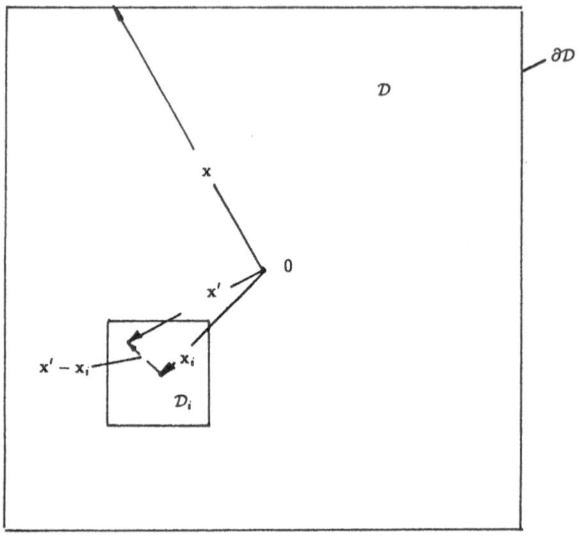

Fig. 3.3. Poisson Integral over a Subdomain $\mathcal{D}_i$

Also note that Ω itself is divergence free but Ω_i in general is not, because it is discontinuous on $\partial\mathcal{D}_i$. Therefore the integral invariants presented in **Sec.1.2** are not applicable to Ω_i. For example, $\langle\Omega_i\rangle$ may not vanish and the matrix of first moments, $\langle\mathbf{x}\Omega^i\rangle$, may not be antisymmetric and time invariant. It is necessary to evaluate all the moments numerically, that is, three components for $\langle\Omega^i\rangle$, nine components for the first moments, and eighteen components for the second moments. The contributions of all these components must be included in the vector potential induced by Ω^i on $\partial\mathcal{D}$.

On the other hand the integral invariants *are* applicable to the moments of vorticity in all of $\mathcal{D}$ with respect to a common center, say the origin, and hence can serve as checks for the accuracy of the scheme. The moments of Ω_i with respect to the origin can be computed readily from the values of the moments and lower moments with respect to the center of the subdomain by translating the origin of the coordinate system to the center of the subdomain. For example, $< \Omega^i >$ is independent of the position of the center and for the first and second moments we have

$$\langle\mathbf{x}\Omega^i\rangle = \mathbf{x}_i\langle\Omega^i\rangle + \langle(\mathbf{x}-\mathbf{x}_i)\Omega^i\rangle \tag{3.1.24a}$$

and

$$\langle\mathbf{x}^2\Omega^i\rangle = \mathbf{x}_i^2\langle\Omega^i\rangle + 2\mathbf{x}_i\cdot\langle(\mathbf{x}-\mathbf{x}_i)\Omega^i\rangle + +\langle(\mathbf{x}-\mathbf{x}_i)^2\Omega^i\rangle\ . \tag{3.1.24b}$$

Using formulae (3.1.24a and b) and similar ones for higher moments, we compute the moments of vorticity in $\mathcal{D}$ with respect to the origin and then compare them with the integral invariants in **Sec. 1.2** to assess the accuracy of the numerical results. In particular, one obtains information whether the size of the computational domain should be increased for the subsequent time steps or not.

This completes the formulation of a numerical scheme tailored for problems of type Ib with a vorticity distribution effectively confined in a domain of size L which is much larger than the effective decay length scale ℓ_d. Applications of this scheme to two and three dimensional problems will be presented in **Sec. 3.2.1** and **3.3.1** respectively.

3.1.4 Numerical Scheme and Approximate Boundary Data for a Local Merging

For the problem of local merging, such as the intersection of two slender filaments, a merging region, $\mathcal{M}_i$, contains small segments of the filament(s) with overlapping cores. For these merging processes of type II, a typical size, ℓ_m, of the merging region, which in many cases is of the order of the core size δ, is assumed to be much smaller than the typical length scale ℓ of a filament, i. e.,

$$\ell_m/\ell \ll 1, \quad \text{say} \quad \ell_m/\ell < 1/10. \tag{3.1.25}$$

We mention an example of 1/10 to emphasize that we are not dealing with a moderately small ratio $\ell_m/\ell = 1/2$ or $1/3$ although this could be accepted as a "small ratio" in many applications of asymptotic analyses. It is necessary to emphasize the smallness of the ratio because for a merely moderately small ratio the merging problem could be solved on a supercomputer, albeit extremely time consuming, by using the numerical scheme for a global merging of type Ib.

When the ratio is as small as 1/10 or smaller, we do need a different numerical scheme built on the observation that away from a local merging region, the characteristics of slender filaments persist and hence the behavior of the filaments should follow closely the asymptotic theory derived in **Sec 2.3**. The asymptotic solution fails only locally in the regions, $\mathcal{M}_i$, where the cores of the filaments are merging.

It is tempting to devise a numerical scheme that resolves the merging process on a fine grid, while it uses the asymptotic theory for the segments of filaments away from the merging region(s). Thus, numerical solutions of the N-S equations are required only for the disjoint domains, $\mathcal{D}_i$, containing $\mathcal{M}_i$. The size of these computational domains should be much larger than the size ℓ_m of $\mathcal{M}_i$, but it may be much smaller than the overall size of a filament on account of (3.1.25). The asymptotic solution is expected to remain valid not only outside of the computational domains but also in the outer portions of $\mathcal{D}_i$ so that there is a matching layer in $\mathcal{D}_i$ next to its boundary $\partial\mathcal{D}_i$ but away from $\mathcal{M}_i$. In the layer, the asymptotic solution has to match the numerical solution in $\mathcal{D}_i$. Such a composite numerical scheme was presented by Krause, Liu and Ting (1986).

In the composite scheme, the code generated by Liu, Tavantzis and Ting (1985) is employed to compute the asymptotic solution of the flow field outside of all $\mathcal{D}_i$'s, while a N-S solver is needed to account for the flow field inside each $\mathcal{D}_i$. For the N-S solver, approximate boundary data for both $\mathbf{A}$ *and* Ω on $\partial\mathcal{D}_i$ are needed since the vorticity does not vanish outside of $\mathcal{D}_i$. Also one needs to find the contribution of the flow field inside of $\mathcal{D}_i$ on the motion of the centerline(s) of the filament(s) outside. Thus for each time step, the asymptotic solution outside of $\mathcal{D}_i$ is coupled with the numerical solution inside. In order to explain clearly how to handle the matching and coupling of the asymptotic and numerical solutions, we describe the procedure in detail for a typical local merging problem as illustrated in Fig. 3.4 in which two filaments are going to intersect in two local regions, $\mathcal{M}_i$, $i = 1,2$.

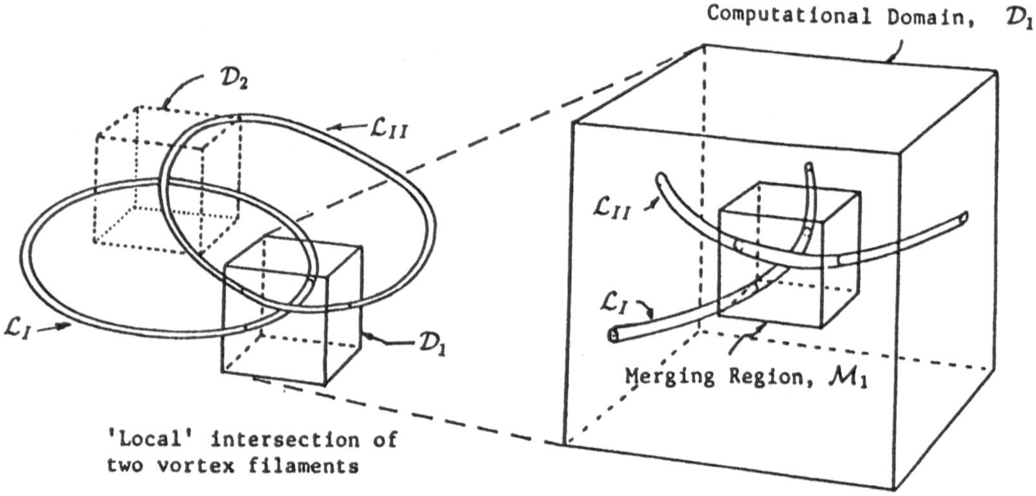

Fig. 3.4. Enlargement of a Local Merging Region $\mathcal{M}_i$

At the instant $t = 0$, prior to merging, the asymptotic solution provides the initial data, the location of the two centerlines $\mathcal{C}_I$, $\mathcal{C}_{II}$ of the two filaments, the associated vorticity, $\Omega_I + \Omega_{II}$, the vector potential $\mathbf{A}_I + \mathbf{A}_{II}$ and the velocity, $\mathbf{v}_I + \mathbf{v}_{II}$. (See **Sec. 2.4** and compare the sketch in Fig. 3.4 with the numerical example shown in Fig. 2.17.) The local merging regions $\mathcal{M}_i$ and the computational domains $\mathcal{D}_i$ are thereby specified. During the merging stage, $t > 0$, the vorticity, velocity and vector potential can be decomposed as:

$$\Omega = \Omega_I + \Omega_{II} + \Omega^* \ , \qquad \mathbf{v} = \mathbf{v}_I + \mathbf{v}_{II} + \mathbf{v}^* \ , \tag{3.1.26}$$

and

$$\mathbf{A} = \mathbf{A}_I + \mathbf{A}_{II} + \mathbf{A}^* \ ,$$

with

$$\Omega^* = \nabla \times \mathbf{v}^* \quad \text{and} \quad \mathbf{v}^* = \nabla \times \mathbf{A}^* \ . \tag{3.1.27}$$

Here Ω^* , $\mathbf{v}^*$ and $\mathbf{A}^*$ represent the corresponding interaction terms, i. e., the deviations from the sum of the solutions of the two filaments disregarding the effects of distortions in the local merging regions. These interaction terms are zero initially,

$$\Omega^* = 0 \;, \quad \mathbf{v}^* = 0 \quad \text{and } \mathbf{A}^* = 0 \quad \text{at } t = 0 \;, \tag{3.1.28}$$

and remain small in the early stage of merging. They appear because of the non-linearity of the vorticity evolution equation (1.1.3) and grow to order one in a local merging region when the sum of the core radii, becomes comparable to the distance of the filament centerlines.

It should be kept in mind that the numerical scheme to be proposed is based on the following two conjectures:

A) The features of two distinct vortex filaments persist outside of the merging regions, $\mathcal{M}_i$.
B) At each time step it is possible to define smooth continuations of the filament centerlines through the merging regions.

In the early stage of merging, a reasonable procedure for these continuations is to trace the local maxima of vorticity in $\mathcal{M}_i$. This may not work for the entire duration of merging. At the end of this section we shall mention examples for which the simple minded approach of tracing the extrema of $|\Omega|$ will fail during the merging period. A procedure that is applicable for all local merging problems remains to be devised at this time. The procedure built on the above two conjectures should therefore be viewed as an exploratory one to study the complexity of the merging process. We now describe the procedure in detail.

Under conjecture A) and from the definition of the interaction term, we expect Ω^* to decay exponentially in h/δ where h is the distance from $\partial\mathcal{M}_i$. Therefore, we may impose

$$\Omega^* = 0 \quad \text{or} \quad \Omega = \Omega_I + \Omega_{II} \quad \text{on} \quad \partial D_i \;. \tag{3.1.29}$$

The formulation of a higher order boundary condition for $\mathbf{A}^*$, i. e., more accurate than $\mathbf{A}^* = 0$, is essential for an efficient N-S solver. The formulation will be described later in step (v) of the proposed procedure.

We assume that at time t_n the centerlines C_I and C_{II} of the two filaments, their core structure Ω_I and Ω_{II}, the induced vector potentials $\mathbf{A}_I$ and $\mathbf{A}_{II}$ and the interaction terms Ω^*, $\mathbf{A}^*$ are known. The procedure for updating the flow field to the next time step t_{n+1} is as follows.

(i) Update the position of the centerline segments C_I' and C_{II}' of the filaments *outside* of the domains $\mathcal{D}_i$ using the velocity defined by the asymptotic theory of the two filament plus the velocity induced by the vorticity interaction term Ω^* in $\mathcal{D}_i$. The latter is defined by the far field expansion of the corresponding velocity potential $\mathbf{A}^*$.
(ii) Update the vorticity distribution Ω in $\mathcal{D}_i$ by solving a discrete version of the vorticity evolution equation (1.1.3).
(iii) Locate the maxima of $|\Omega|$ in a series of grid planes in $\mathcal{M}_i$ and thus locate the two lines $\mathcal{L}_I$ and $\mathcal{L}_{II}$ of maximum $|\Omega|$ in each domain $\mathcal{D}_i$.
(iv) By treating the lines $\mathcal{L}_I$ and $\mathcal{L}_{II}$ as the fictitious extensions of C_I' and C_{II}', two closed centerlines C_I and C_{II} are obtained for t_{n+1}. The core structure

Ω_I associated with $\mathcal{C}_I$ is defined by the asymptotic solution of a filament including the extension $\mathcal{L}_I$ disregarding the presence of the nearby segment $\mathcal{L}_{II}$. Similarly the core structure associated with $\mathcal{C}_{II}$ is defined. This allows us to compute the interaction term $\Omega^* = \Omega - (\Omega_I + \Omega_{II})$ at t_{n+1}.

(v) Compute the vector potential **A** on the boundary of $\mathcal{D}_i$ using the vector potential induced by the filaments $\mathcal{C}_I$ and $\mathcal{C}_{II}$, with core structures Ω_I and Ω_{II}, *and* the contribution of the vorticity interaction term Ω^* in $\mathcal{D}_i$. The former, $\mathbf{A}_I$ and $\mathbf{A}_{II}$, are given by the composite solution given in **Sec. 2.3.4**, uniformly valid inside the vortical cores and away from them. The latter, A^*, is given in (1.1.16) and (1.1.17) in terms of only the first few moments of Ω^*. Thus, the approximate boundary data on each of the computational domains $\mathcal{D}_i$ are specified.

(vi) Solve for the vector velocity potential A^* from (1.1.9) using a fast Poisson solver and hence obtain the velocity field in $\mathcal{D}_i$ for t_{n+1}.

(vii) Compute the minimum distance between the two lines of maximum $|\Omega|$, $\mathcal{L}_I$ and $\mathcal{L}_{II}$. If their minimum distance increases to an amount significantly larger than the core size, the feature of two distinct vortex filaments reappears. Then the effects of local merging disappear and the post intersection period commences.

This completes a cycle of one time step. Step (i) and part of (v) can be accomplished by adopting the code of Liu, Tavantzis and Ting (1985) for the asymptotic description of a filament. Steps (ii), (iii), (vi), and the remaining part of (v) can be accomplished by modifying the code for numerical solutions to the N-S equations of Chamberlain and Weston (1984). By coupling these two codes and adding the identification and connection procedure of steps (iv) and (vii), a numerical code is being developed for the study of the 'local' intersection of two filaments from pre-intersection to the post-intersection period.

It should be noted that during the merging stage, the distance between the two lines of maximum vorticity, $\mathcal{L}_I$ and $\mathcal{L}_{II}$, can diminish to zero so that the curves intersect each other or coincide over a finite length and for a finite duration. If this takes place sometime during the merging process, a computation based upon the above scheme would have to be terminated, because the procedure for continuing $\mathcal{C}_I$ and $\mathcal{C}_{II}$ within a merging region would fail. The problem of reconnection of the centerlines after such a complication remains to be addressed in detail and a rule for the reconnection remains to be devised.

In this respect we mention in particular several numerical examples of *global* merging of vortex filaments, to be presented in **Sec. 3.3.2**. These problems may be considered as local merging problems with moderately small ratios ℓ_m/ℓ, (in contrast to (3.1.25)). In the numerical investigations, which include earlier work by Chamberlain and Weston (1984) and the very recent study by Ishii et al (1989), the initial flow fields feature two or four identical vortex rings oriented symmetrically. All these examples have the following characteristics in common:

a) In the early stage of merging, two adjacent rings begin to touch each other in a single merging region in which the maximum vorticity vector in one ring is equal in magnitude and almost in the opposite direction to that in the other ring.

b) Due to symmetry, the merging process is dominated by the cancellation of vorticity resembling that in the merging of a pair of two dimensional vortices of opposite strength.

c) The topology of the iso-surfaces of $|\Omega|$ changes completely during the merging process. The evolution of the iso-surfaces creates the impression that those initially separated vortex rings link to each other to form a single filament after a finite duration. The same conclusion is drawn from the iso-surfaces of the magnitude of vector potential, $|\mathbf{A}|$.

It is clear that these examples belong to a special class of problems for which the merging process is dominated by the cancellation of vorticity due to anti-symmetry and reconnection of the centerlines. The qualitative features of vorticity cancellation and reconnection of centerlines will appear whenever the initial core sizes are much smaller than the radii of curvature, i. e., $\delta \ll \ell$. They are local mergings but the above numerical scheme will not be applicable unless the rules for the reconnection of the centerlines as implied by conjecture B are formulated.

Since the coding of the above scheme for a local merging problem is still in progress, the next subsections cover only numerical studies of global merging problems of type I(a) and I(b).

3.2 Merging of Two Dimensional Vortices

According to the classifications in **Sec.3.1.1**, the merging of two dimensional vortices is always a global merging problem. The applications of the corresponding numerical schemes formulated in **Sec.3.1.2** and **3.1.3** to two dimensional problems are presented in **Sec.3.2.1**. In particular, we focus on the merging of viscous vortices and the roll-up of thin vortical layers.

In **Sec. 3.2.2** we employ the integral invariants for two dimensional viscous flow to formulate *the rules of merging* for closely interacting vortices and to explain the physical meaning of an optimum Lamb vortex. Then we discuss the superposition of Lamb vortices moving according to the asymptotic theory for nonoverlapping vortices (vortices far apart from each other) as an approximation for a viscous vortical field. When this representation is employed, even if the cores of adjacent vortices overlap considerably, we call it *the extended asymptotic solution.*

In **Sec. 3.2.3** we present a different approach. The sum of Lamb vortices with their velocities treated as unknowns are considered as as an *approximate solution* in the sense that the N-S equations are satisfied approximately under a minimum principle. The minimum principle in turn defines the unknown velocities. This scheme was employed to study the merging of vortices and the results compare favorably with corresponding finite difference solutions reported in **Sec.3.2.1**. This scheme and numerical examples are presented in **Sec. 3.2.3**.

We now restate some relevant results for two-dimensional flows from **Sec. 1.1 - 1.3**, namely, the governing equations, the integral invariants and the far field behaviors. To study the interaction and merging of vortices in two dimensional space, the xy plane, it is natural to use the vorticity ζ and the stream function ψ as dependent variables. For an incompressible fluid, they are governed by the equations

$$\zeta_t + \boldsymbol{V} \cdot \nabla\zeta = \nu\,\Delta\zeta \tag{3.2.1}$$

and

$$\Delta\psi = -\zeta \tag{3.2.2}$$

with the velocity

$$\boldsymbol{V} = \hat{\imath}\psi_y - \hat{\jmath}\psi_x \ . \tag{3.2.3}$$

The prescribed initial vorticity distribution,

$$\zeta(0,x,y) = \zeta_0(x,y) \tag{3.2.4}$$

is assumed to be of bounded support or to decay exponentially in the radial coordinate $\sigma = \sqrt{x^2+y^2}$. This condition is certainly fulfilled when ζ_0 is concentrated around a finite number of points (X_i, Y_i), $i = 1, \ldots, N$ so as to model the vorticity distribution of N vortices. Due to the exponential decay of the vorticity, the induced velocity vanishes at infinity for $t \geq 0$ and thus

$$\zeta(t,x,y) = o(\sigma^{-m}) \tag{3.2.5}$$

for all m and

$$|\boldsymbol{V}| \to 0 \ . \tag{3.2.6}$$

The initial data for $\boldsymbol{V}$ are uniquely defined by (3.2.2), (3.2.3) and (3.2.6).

Equations (3.2.1-6) define an initial value problem in two-dimensional space. It was noted in **Sec.1.2.4** that the solution obeys Poincare's relations (1.2.28). They are the time-invariance constraints on the total strength and first moments of the vorticity,

$$\langle\zeta(t,x,y)\rangle = \langle\zeta_0(x,y)\rangle = \Gamma_0 \ , \tag{3.2.7}$$

$$\langle x\,\zeta(t,x,y)\rangle = \langle x\;\zeta_0(x,y)\rangle = C_1 \ , \tag{3.2.8}$$

$$\langle y\,\zeta(t,x,y)\rangle = \langle y\;\zeta_0(x,y)\rangle = C_2 \tag{3.2.9}$$

and the linear growth equation for the polar moment,

$$\langle\sigma^2\zeta(t,x,y)\rangle = 4\nu\Gamma_0 t + D_3 \ , \tag{3.2.10}$$

where the constant D_3 is defined by the initial data,

$$D_3 = \langle r^2\zeta_0(x,y)\rangle \ . \tag{3.2.11}$$

Here $\langle f\rangle$ denotes the area integral of f over the xy plane.

Note that (3.2.7-10) are also valid for solutions of the linear diffusion equation, which is (3.2.1) without the nonlinear convection term.

Using the above invariants, we arrive at the far field behavior of the stream function in polar coordinates, σ, θ,

$$\psi(t,\sigma,\theta) = -\frac{\Gamma_0}{2\pi}\ln\sigma + \frac{1}{2\pi\sigma}[C_1\cos\theta + C_2\sin\theta] + \frac{1}{4\pi\sigma^2}[F_3(t)\cos 2\theta + H_3(t)\sin 2\theta] + O(\sigma^{-3}) , \tag{3.2.12}$$

where $F_3 = \langle(x^2 - y^2)\zeta\rangle$ and $H_3 = 2\langle xy\zeta\rangle$. This equation provides the boundary data on the computational domain for a total merging of type I(a).

For a total merging of type I(b), we divide the computational domain $\mathcal{D}$ into M subdomains $\mathcal{D}_i$, $i = 1,\ldots,M$ and write

$$\psi = \sum_{i=1}^{M}\psi_i \tag{3.2.13}$$

where ψ_i is the stream function induced by ζ_i. The latter is the vorticity distribution in $\mathcal{D}_i$, which is equal to ζ in $\mathcal{D}_i$ and to zero elsewhere in analogy to (3.1.22). Note that $\psi_i = 0$ when $\mathcal{D}_i$ are adjacent to the boundary $\partial\mathcal{D}$ where $\zeta_i = 0$ is an exponentially accurate approximation.

We can then express ψ_i by its power series expansion

$$\psi_i = -\frac{\langle\zeta_i\rangle}{2\pi}\ln\sigma_i + \frac{1}{2\pi\sigma_i}[\langle x_i\zeta_i\rangle_i\cos\theta_i + \langle y_i\zeta_i\rangle\sin\theta_i] + \frac{1}{4\pi\sigma_i^2}[\langle(x_i^2 - y_i^2)\zeta_i\rangle\cos 2\theta_i + 2\langle x_i y_i\zeta_i\rangle\sin 2\theta_i] + O(\sigma_i^{-3}) . \tag{3.2.14}$$

Here x_i and y_i are Cartesian coordinates and σ_i, θ_i are the corresponding polar coordinates with the origin located at the center of $\mathcal{D}_i$. Because the vorticity distribution ζ_i is discontinuous on $\partial\mathcal{D}_i$ the invariance conditions (3.2.7 - 10) are not applicable to ζ_i. The moments, $\langle\zeta_i\rangle$, $\langle x_i\zeta_i\rangle$, etc., in (3.2.14) are in general time dependent and must be evaluated at each instant.

3.2.1 Numerical Simulation of Vortex Merging and the Roll-up of Thin Shear Layers

In the numerical simulation of merging problems, the computational domain $\mathcal{D}$ is allowed to move so that the dominant part of the vorticity distribution is always located near the center of $\mathcal{D}$. Here we shall explain how to define the velocity of $\mathcal{D}$ and how to simplify the far field condition (3.2.12) depending on whether the total strength of the vorticity distribution in (3.2.7) zero or not.

3.2.1.1 Nonzero Circulation, $< \zeta > \neq 0$

In this case we can use the invariances (3.2.8 and 9) of the first moments to define the center of vorticity $X_{C.G.}$, $Y_{C.G.}$ as

$$X_{C.G.} = \frac{\langle x\zeta\rangle}{\langle\zeta\rangle} = \frac{C_1}{\Gamma_0} , \quad Y_{C.G.} = \frac{\langle y\zeta\rangle}{\langle\zeta\rangle} = \frac{C_2}{\Gamma_0} . \tag{3.2.15}$$

and we conclude that *the center of vorticity is stationary*. By locating the origin at the center of vorticity from the outset, we obtain $C_1 = C_2 = 0$. The computational domain $\mathcal{D}$ will remain stationary and centered at the origin. The boundary values of ψ on $\partial\mathcal{D}$ are then given by its far field behavior (3.2.12) in the simplified form,

$$\psi(t,\sigma,\theta) = -\frac{\Gamma_0}{2\pi}\ln\sigma + \frac{1}{4\pi\sigma^2}[F_3(t)\cos 2\theta + H_3(t)\sin 2\theta] + O(\sigma^{-3})\,. \qquad (3.2.16)$$

Thus in the far field, we see a point vortex of constant strength Γ_0 and two quadrupoles of temporally varying strength but no doublet. The point vortex is located at the origin, i.e., at the center of vorticity.

3.2.1.2 Zero Circulation $<\zeta>=0$ but $<\mathbf{x}\zeta>\neq 0$

With $\Gamma_0 = 0$, we cannot define the center of vorticity by (3.2.15). Instead, we shall define the center of an optimum doublet in terms of the far field behavior (3.2.12), which now becomes

$$\begin{aligned}\psi(t,\sigma,\theta) = {} & \frac{1}{2\pi\sigma}[C_1\cos\theta + C_2\sin\theta] \\ & + \frac{1}{4\pi\sigma^2}[F_3(t)\cos 2\theta + H_3(t)\sin 2\theta] + O(\sigma^{-3})\,,\end{aligned} \qquad (3.2.17)$$

where the first term represents a doublet of constant strength $|\mathbf{C}| = \sqrt{C_1^2 + C_2^2}$ oriented along the direction $C_1\hat{\imath} + C_2\hat{\jmath}$ and again two quadrupoles of temporally varying strength located at the origin. If both C_1 and C_2 are nonzero, we can represent the leading two terms by a doublet of the same strength and orientation but moving with the velocity, $\dot{X}_D(t)$, $\dot{Y}_D(t)$, such that

$$\psi(t,\bar{\sigma},\bar{\theta}) = \frac{1}{2\pi\bar{\sigma}}[C_1\cos\bar{\theta} + C_2\sin\bar{\theta}] + O(\bar{\sigma}^{-3}) \qquad (3.2.18a)$$

with

$$\dot{X}_D = \frac{C_1\dot{F}_3 + C_2\dot{H}_3}{2|\mathbf{C}|^2}\,, \qquad \dot{Y}_D = \frac{-C_2\dot{F}_3 + C_1\dot{H}_3}{2|\mathbf{C}|^2}\,. \qquad (3.2.18b)$$

Here $\bar{\sigma}$ and $\bar{\theta}$ denote the polar coordinates relative to the center $X_D(t)$, $Y_D(t)$. From (3.2.10), we see that the polar moment is conserved since $\Gamma_0 = 0$.

To get additional information for this case, we express the two dimensional space $\mathbb{R}^2$ as the union of $\mathcal{S}^+$, $\mathcal{S}^-$ in which ζ is nonnegative and nonpositive respectively. That is

$$\mathbb{R}^2 = \mathcal{S}^+ \cup \mathcal{S}^- \qquad (3.2.19a)$$

$$\text{with} \quad \zeta \geq 0\,,\ \zeta \leq 0 \text{ and } \zeta = 0 \quad \text{for} \quad \mathbf{x} \in \mathcal{S}^+\,,\ \mathcal{S}^- \text{ and } \mathcal{S}^+ \cap \mathcal{S}^- \qquad (3.2.19b)$$

We denote the nonnegative and nonpositive vorticity distribution by $\zeta^{\pm}$, i.e.,

$$\begin{aligned}\zeta^{\pm} &= \zeta\,, \quad \mathbf{x} \in \mathcal{S}^{\pm}\,, \\ \zeta^{\pm} &= 0\,, \quad \mathbf{x} \in \mathcal{S}^{\mp}\,,\end{aligned} \qquad (3.2.20)$$

and then study the moments of $\zeta^{\pm}$. The condition $\langle\zeta\rangle = 0$ becomes

$$\langle\zeta^+\rangle = -\langle\zeta^-\rangle = \Gamma^+(t) \ , \tag{3.2.21}$$

which is in general not conserved. Its rate of change is given by,

$$\langle\zeta^+\rangle_t = \int_{\partial S^+} \hat{n}\cdot[-\mathbf{V}\zeta + \nu\nabla\zeta]\, ds \ , \tag{3.2.22}$$

where $\hat{n}$ denotes the outward unit normal vector and s the arc length. From the definitions (3.2.19)-(3.2.21), we have $\zeta = 0$ and $\hat{n}\cdot\nabla\zeta \le 0$ along ∂S^+ and hence the total strength of positive vorticity cannot increase

$$\frac{d\Gamma^+}{dt} \le 0 \ . \tag{3.2.23}$$

The equality sign holds only if

$$\hat{n}\cdot\nabla\zeta = 0 \quad \text{along } \partial S^+ \ . \tag{3.2.24}$$

3.2.1.3 Antisymmetric vorticity field

Now we consider the special case that the vorticity field is antisymmetric with respect to an axis, say the y-axis. In addition to $\langle\zeta\rangle = 0$, we have

$$\zeta(t,x,y) = -\zeta(t,-x,y) \ , \quad \zeta(t,0,y) = 0 \ . \tag{3.2.25}$$

Let $\langle\ \rangle^+$ and $\langle\ \rangle^-$ denote the area integrals over the right and left half plane respectively. Then (3.2.7-9) and (3.2.24) yield

$$\langle\zeta\rangle^+ = -\langle\zeta\rangle^- \neq 0 \ , \quad \langle x\zeta\rangle^+ = \langle x\zeta\rangle^- = C_1/2 \ , \quad \langle y\zeta\rangle^+ = -\langle y\zeta\rangle^- \ , \ C_2 = 0 \ . \tag{3.2.26}$$

Note that we have assumed that $\langle\zeta\rangle^+ \neq 0$, but that ζ may change signs in $x > 0$. Because of the antisymmetry, we need to analyze only the flow field on a half plane say the right half, $x \ge 0$. The computational domain $\mathcal{D}$ will be bounded on its left by the y-axis with the boundary condition,

$$\zeta(t,0,y) = 0 \ , \quad \psi(t,0,y) = 0 \ . \tag{3.2.27}$$

In order to contain the vorticity field away from the remaining boundary of $\mathcal{D}$, we move $\mathcal{D}$ with a velocity Y^+ parallel to the y-axis. To be more specific, we choose the domain $\mathcal{D}$ to be, $0 \le x \le D$ and $|y - Y^+| \le D$ with $D >> L$ and identify Y^+ as the y-coordinate of the center of vorticity on the right half plane (also that on the left half plane), i.e.

$$Y^+(t) = Y^-(t) = \langle y\zeta\rangle^+ \ / \ \langle\zeta\rangle^+ \ . \tag{3.2.28}$$

Note that this definition is meaningful only for an antisymmetric vorticity field described by (3.2.25) with $< \zeta >^+ \neq 0$. Equation (3.2.28) can be considered as the equation for the ordinate of the center of the vorticity field which behaves as a doublet in the far field. But this equation differs from (3.2.18) for the center of an optimum doublet. The latter yields $Y_D(t) =< yx\zeta >^+ /C_1$. The difference between these two definitions will be elaborated later in **Sec. 3.2.3.**

For the merging of a vortex pair of opposite strength, we assume in addition to the antisymmetry (3.2.25) that the vorticity on either side of the y-axis does not change sign. We assume that the vorticity on the right side of the y-axis is positive, i.e., $\zeta \geq 0$ for $x \geq 0$. Noting that $\zeta_x(t, 0, y) \geq 0$ we conclude from (3.2.23) that the total strength of positive (negative) vorticity has to decrease (increase)

$$\Gamma_t^+ = \langle\zeta\rangle_t^+ < 0 \,, \tag{3.2.29}$$

It follows that the x-coordinate of the center of positive vorticity has to move away from the y-axis since

$$\dot{X}^+ = [\frac{\langle x\zeta\rangle^+}{\langle\zeta\rangle^+}]_t = -\frac{C_1\Gamma_t^+}{2(\Gamma^+)^2} > 0 \tag{3.2.30}$$

If in addition there is no significant diffusion of vorticity across the y-axis, for example, when the vorticity is negligible near the line $x = 0$, we have

$$\zeta_x(t, 0, y) = 0 \quad \text{in addition to} \quad \zeta(t, 0, y) = 0 \,. \tag{3.2.31}$$

This is true in the model problem for the roll-up of a thin trailing vortical layer behind an airplane wing of high aspect ratio. This model problem will be discussed in detail in **Sec.3.2.1.5**. Under (3.2.31) we obtain from the vorticity evolution equation that $\langle\zeta\rangle_t^+ = 0$ or

$$\langle\zeta(t, x, y)\rangle^+ = \langle\zeta_0(t, x, y)\rangle^+ = \Gamma^+(0) = \text{ constant} \tag{3.2.32}$$

and then conclude that the center of vorticity on the right half plane is stationary with

$$X^+ = \langle x\zeta\rangle^+ / \langle\zeta\rangle^+ = C_1/\Gamma^+(0) = \text{ constant.} \tag{3.2.33}$$

We now apply the far field behavior and the equation of the center of a vortical field to define the approximate boundary conditions and the motion of the center of the computational domain in the numerical schemes formulated in **Sec.3.1** for global mergings of type I(a) and I(b). The corresponding computational codes are then employed to study the merging of several viscous vortices in **Sec.3.2.1.4** and the roll up of a thin trailing vortical layer in **Sec.3.2.1.5**.

3.2.1.4 Merging of Several Vortices

Numerical studies of the merging of several viscous vortices, which were initially far apart from each other were carried out by Lo and Ting (1975, 1976). The numerical results demonstrate the merging of the vortices to a single one with total strength Γ_0, which is the sum of the strengths of the initial vortices, provided that $\Gamma_0 \neq 0$. They merge to a limiting doublet configuration if $\Gamma_0 = 0$. The results also show that the asymptotic solutions for a collection of Lamb vortices remain good approximations even when their vortical cores overlap each other. Furthermore, in the long time limit these *extended asymptotic solutions* yield the correct solutions for the final stages of merging. Lo and Ting employed the numerical scheme of Wu and Thompson (1972). At each time step the Poisson equation (3.2.2) for the stream function was solved by numerical evaluation of the Poisson integral. Thus a special construction of boundary data for ψ was not needed, but the number of computational steps per time step was order N^2 instead of $N \ln N$. This scheme is very inefficient. Nevertheless, the numerical results of Lo and Ting (1975) are useful in verifying the accuracy of the more efficient schemes for global merging of type I(a) and I(b), which employ approximate boundary conditions for ψ in conjunction with a fast Poisson solver, (see **Sec. 3.1.2** and **3.1.3**).

When the vortices are not too far apart from each other, we say that the merging of vortices commences. To be more precise, we apply the criterion (2.4.11) for the practical region of valid of the asymptotic theory and set the lower bound for the ratio of the distance d_{ij} between two vortices to the sum of their core radii $\delta_i + \delta_j$ equal to 2. Let $t = t_c$ be the instant when $d_{ij}/(\delta_i + \delta_j) = 2$. Using the asymptotic solution at $t = t_c$ as the initial datum, a numerical solution of the N-S equations is constructed to simulate the flow field for $t > t_c$. To illustrate the process of merging we quote several numerical examples dealing with the merging of vortices of equal strength and core size. Additional examples can be found in Lo and Ting (1975) and Liu and Hsu (1984).

The merging of n identical vortices was studied for $n = 2, 3$ and 4. The vortex centers are distributed symmetrically on a circle of radius ρ around the origin. The initial vorticity distribution for each vortex is that of a Lamb vortex with core size δ_0. If we choose δ_0 as the unit length scale and $\delta_0^2/4\nu$ as the time scale then the numerical solution of the Navier-Stokes equations begins at $t = t_c = 1$ with $\rho = 2$ and Re $= \Gamma/\nu = 100$.

Due to the diffusion of vorticity, the vortical cores grow and overlap more and more as time increases. The n points of locally maximal vorticity move gradually away from the vortex centers and spiral inwards. Eventually, say at $t = t_m$, the n points of local maximum coincide at the origin. We say that the n vortices merge to a single one and lose their individual identity for $t \geq t_m$ and note that the origin is the location of the center of vorticity for all $t \geq 0$ according to (3.2.15).

Shown in Fig. 3.5 are the trajectories of the points of maximum vorticity given by the numerical solutions for $n = 2, 3, 4$. The dotted lines are the corresponding trajectories defined by *the extended asymptotic solution* which will be explained in **Sec. 3.2.2**. Shown in Fig. 3.6 is the decay of the maximum vorticity given by these two solutions. We see from Fig. 3.5 and Fig. 3.6 that the asymptotic solution is in good agreement with the numerical solution even when the core size $\delta(t) = [4\nu(t+1)]^{1/2}$ is of the order of the distance between the adjacent vortex centers. Thus the practical region of validity of the asymptotic solution can be extended from $\delta/\ell \sim 1/4$ to $\delta/\ell \sim 1$, where ℓ denotes the distance between two adjacent vortex centers. In contrast, under the inviscid theory for n point vortices, the vortex locations will be rotating along the circle of radius ρ forever with zero core size.

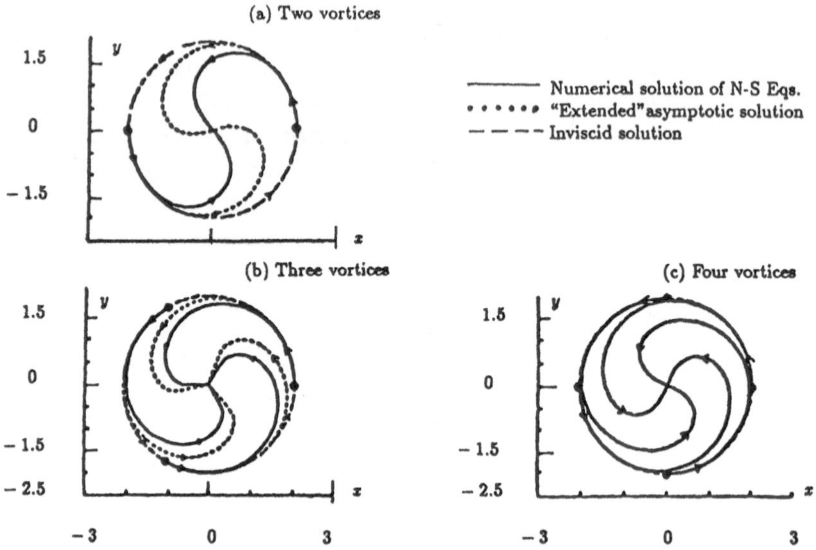

Fig. 3.5. Trajectories of the location of maximum vorticity for the merging of n vortices (Ting and Liu 1986, Ting 1986)

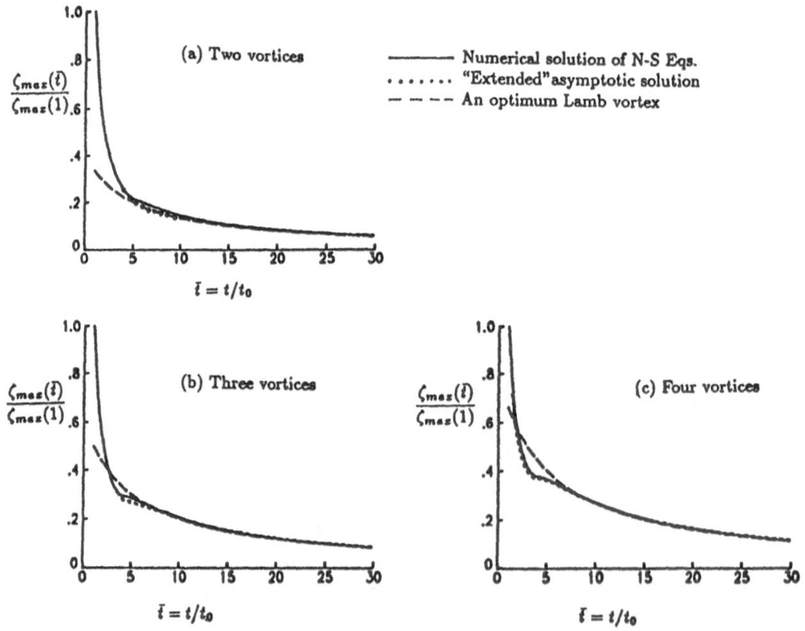

Fig. 3.6. The decay of maximum vorticity, for the merging of n vortices (Ting and Liu 1986, Ting 1986)

The contour lines of constant vorticity for two and three vortices, $n = 2, 3$ are shown in Fig. 3.7 and Fig. 3.8 respectively. At each instant, four contour lines at

0.995, $e^{-1/2}$, e^{-1}, and $e^{-3/2}$ times the maximum vorticity are shown. The initial contour lines of the n nonoverlapping Lamb vortices are displayed in Figs. 3.7a and 3.8a. The successive merging of the contour lines of n vortices are shown in Fig. 3.7 b-c and Fig. 3.8 b-c. For $t \geq t_m$ with t_m equal to 5.5 and 4.6 for $n = 2, 3$ respectively, the n local maximum vorticity points converge to the origin and the character of three distinct vortices disappears. Thus ends the first stage of merging. Figures 3.7 f-h and 3.8 f-h then show the gradual circularization of the contour lines beginning from the innermost and continuing towards the outermost ones as t increases. This is the second stage of merging. In Fig. 3.7h and 3.8h for $n = 2, 3$, all four contour lines are nearly circular and coincide with those of an optimum Lamb vortex with strength. Thus we come to the final stage of merging of n vortex to a single one. Also note that in Fig. 3.6, the decay of maximum vorticity given by the optimum Lamb vortex and that by the extended asymptotic solution are in good agreement with the numerical solution for $t > t_m$. The meaning of the optimum Lamb vortex and the rule of merging will be explained in **Sec.3.2.2**. Also we shall explain why the *extended asymptotic solution* gives good description of not only the early stage of merging but also the final stage of the recircularization of the constant vorticity lines.

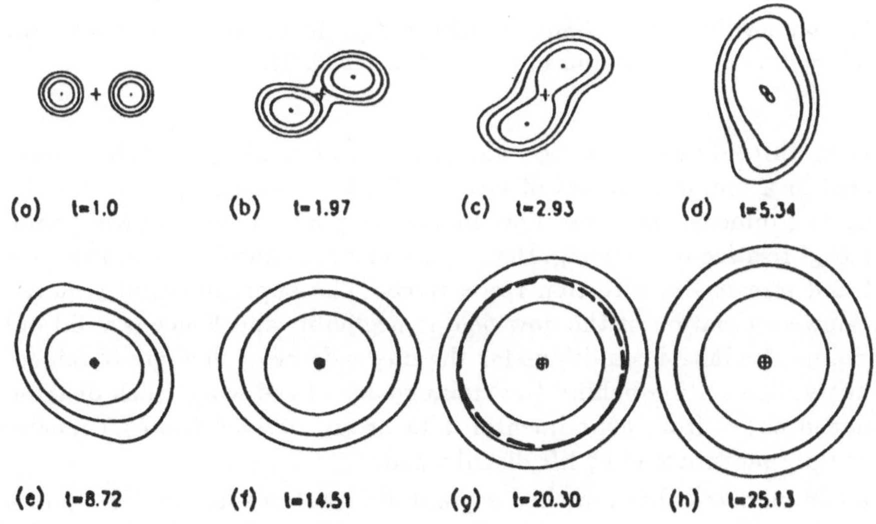

Fig. 3.7. Contour lines of constant vorticity ζ/ζ_{max} in the merging of 2 vortices, — numerical solution of N-S equations (Ting and Liu 1986, Ting 1986)

3.2.1.5 Roll up of a thin trailing vortical layer

The evolution and control of thin vortex wakes trailing airplane wings are of great practical interest for reasons of flight safety, especially during landing and takeoff operations (Olson, Goldberg and Rogers (1971), Donaldson (1971)).

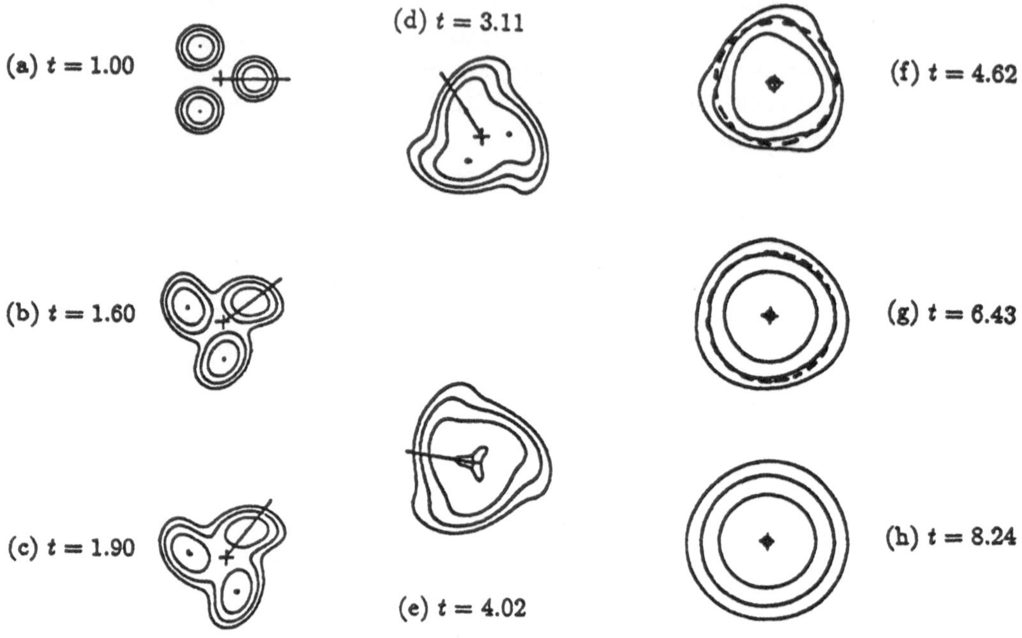

Fig. 3.8. Contour lines of constant vorticity ζ/ζ_{max} in the merging of 3 vortices, — numerical solution of N-S equations (Ting and Liu 1986, Ting 1986)

The steady three-dimensional flow downstream of a wing of high aspect ratio submerged in a uniform stream of velocity $U_\infty\hat{k}$ is usually approximated by an unsteady two-dimensional cross flow in the yz plane moving downstream with velocity $U_\infty\hat{\imath}$ relative to the wing. Here x, y and z are coordinates in the spanwise, vertical and streamwise direction respectively. The approximation requires that the streamwise variation of the flow field is negligible and hence is valid only far downstream. The initial conditions for the unsteady cross flow are obtained from the vorticity distribution behind the trailing edge of the wing. Such distributions are obtained either from experimental data or are derived from an idealization based on the spanwise load or lift distribution.

It should be noted that matched asymptotic solutions for the roll up of an inviscid vortex sheet were presented by Guiraud and Zeytounian (1977, 1979, 1982). Also, a viscous core structure for a tightly rolled up vortex sheet was presented by Hall (1961). Later, the structure of decaying thin vortical layers was matched to the outer solution of an inviscid vortex sheet by Guiraud and Zeytounian (1982) and by Moore (1973).

Here, we will analyze trailing wakes of finite thickness sufficiently far behind the wing so that the shear layer structure can be resolved with sufficient accuracy in the computations. Before describing the numerical solutions, we point out some salient feature of this problem. Since the flow field is symmetric with respect to the vertical plane passing through the center line of the airplane, the yz plane, the vortical field is an odd function of x. In the corresponding unsteady cross flow, the vorticity field $\zeta(t,x,y)$ is antisymmetric with respect to the y-axis. Near the y-axis, the vorticity is weak initially and becomes weaker as t increases (or in going downstream) because the vortical layer is rolling away from the y-axis. The vorticity field fulfills (3.2.31), i. e., $\zeta_x = 0$ and $\zeta = 0$ on $x = 0$. We obtain from (3.2.32 - 33) that the total strength and the center of vorticity on the right half of the xy plane is stationary, i. e.,

$$\Gamma^+(t) = <\zeta>^+ = <\zeta_0>^+ = \Gamma_0 \tag{3.2.34a}$$

and

$$X^+ = \frac{<x\zeta>^+}{<\zeta>^+} = \frac{<x\zeta_0>^+}{\Gamma_0} \tag{3.2.34b}$$

where $< >^+$ denotes the area integral over the half plane $x \geq 0$.

From the classical theory of wings of finite span (see for example Kármán and Burger 1963), the linear strength $\gamma(x)$ of the vortical layer at the initial station $z = z_0$ can be related to the load or lift per unit span, $\Lambda(x)$ as follows,

$$\gamma(x) = \int_{-\infty}^{\infty} \zeta_0 dy\ , \quad \text{with} \quad \gamma(x) = 0 \quad \text{for } x > S\ ,$$

$$\Lambda(x) = \rho U_\infty \int_x^S \gamma(x')dx'\ , \quad \text{for } 0 \leq x \leq S$$

and

$$\Lambda(0) = \rho U_\infty \Gamma_0\ ,$$

where S denotes the half span. Note that Γ_0 is equal to the circulation around the root section of the wing ($x = 0$). The special case of an elliptic load distribution is of great importance because it yields the minimum induced drag for a given span and total lift. In this case, Λ and γ are given by

$$\Lambda(x) = \Lambda(0)\sqrt{1 - x^2/S^2}\ , \tag{3.2.35a}$$

$$\gamma(x) = \Gamma_0 x H(S^2 - x^2)/\sqrt{1 - x^2/S^2}\ , \tag{3.2.35b}$$

where H denotes the Heaviside function. From (3.2.34b) and (3.2.35 a, b), we obtain the stationary spanwise coordinate of the center of vorticity with $x \geq 0$,

$$X^+ = \frac{\pi}{4}S\ , \tag{3.2.35c}$$

for $z \geq z_0$. This is a well known result for an inviscid vortex sheet induced by an elliptic loading. Here we show that it is also true for a viscous thin vortical layer.

We now apply the numerical schemes presented in the preceding sections for global mergings to simulate the vortex wakes evolving from specified spanwise load distributions. The initial vorticity distribution is generated by placing a series of Lamb vortices along the computational grid line corresponding to the wing trailing edge. The strengths of the Lamb vortices are determined from the change in the level of the load distribution between grid points. The core radius of each lamb vortex is equal to one-half the thickness of the wake. The initial thin wake is then replaced by the sum of the Lamb vortices.

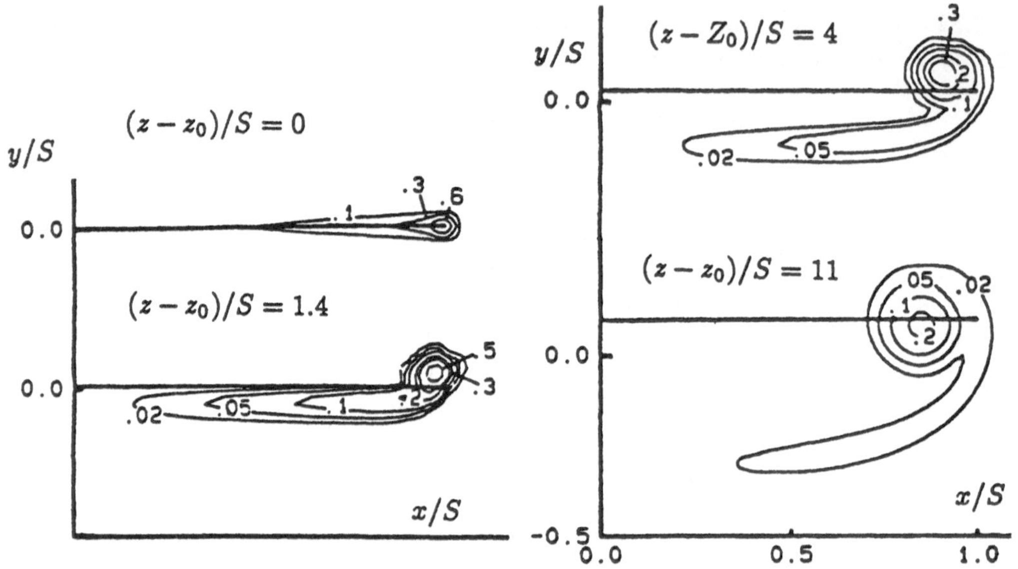

Fig. 3.9. Vortex roll up for elliptic span load at $Re = 20,000$ (Weston and Liu 1982)

The first case considered is the above elliptic load distribution. Calculations for this load distribution with a wake thickness of 5% of the semispan were made by Weston and Liu (1980) using the numerical scheme given in **Sec. 3.1.2** for a global merging of type I(a). The initial vorticity field is illustrated in Fig. 3.9 using contours of constant vorticity. These calculations were performed at a Reynolds number (Γ_0/ν) of 20,000. Due to the symmetry of the flow field with respect to the line $x = 0$, it was sufficient to consider only the right half-plane, $x \geq 0$. Since the linear extension of the vorticity distribution is about the half span S, the size D of the computational domain, $\mathcal{D}$, for global merging of type I(a) should be much larger than S. In Weston and Liu (1982), the size of $\mathcal{D}$ is chosen to be $4S$ and the center of the computational domain moves with the center of vorticity with respect to the half space $x \geq 0$. Figures 3.9a-d show contour lines of constant vorticity in several cross-sectional planes downstream from the wing. The vortex sheet rapidly rolls up into a single, nearly circular vortex with an eventual lateral position near $x/S = \pi/4$ as predicted by the analytical result (3.2.35c). The lack of smoothness in the contour lines for the tip region is observable. This problem is a result of a computational grid that is too coarse to resolve the flow, but could not be finer in this computation, due to the restricted computer memory available at the time the calculation was performed. Approximately 300 grid points across the semispan are required in order to properly resolve the flow, while only 50 grid points across the semispan were used in order to fit the entire computational domain into the

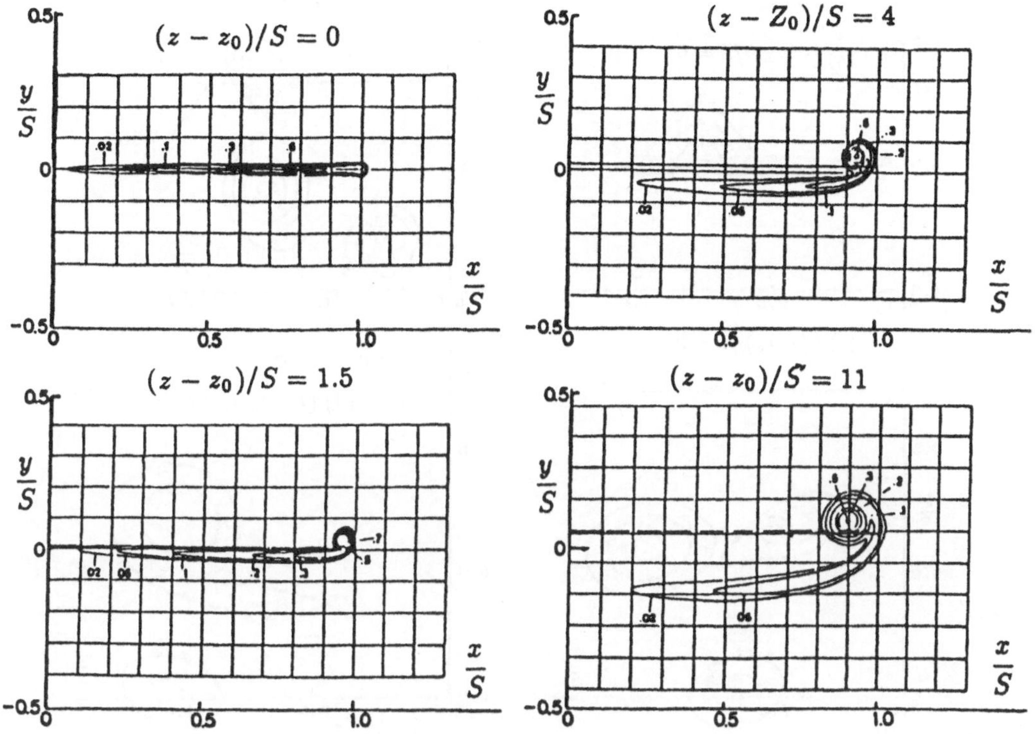

Fig. 3.10. Vortex roll up for elliptic span load at $Re = 40,000$ (Weston, Ting and Liu 1986)

computer memory (nonvirtual machine). This difficulty was subsequently eliminated by in calculation using the scheme given in **Sec. 3.1.3** for global merging of type I(b). Here the computational domain can be reduced to a thin rectangle containing the vorticity distribution in a region $\mathcal{G}$ of width S and thickness $S/20$. Consequently 200 grid points could be placed on the semispan in that calculation.

A more severe initial vorticity distribution is obtained with an initial wake thickness of 1.0 percent of S and a Reynolds number of 40,000 requiring initially as much as 500 grid points across the semispan. The resulting initial vorticity distribution is shown in Fig. 3.10(a). Parts (b) to (d) of Fig. 3.10 illustrate the subsequent roll-up of the vorticity. The division of the computational domain into subdomains is indicated in each figure, with the size of each subdomain chosen to be 10 percent of the semispan. The extent of the domain boundary is maintained at a size that keeps at least two subdomains between any region of significant vorticity and the closest boundary except near the y-axis. The vortex development at the tip is more rapid than in the previous case, and results, as expected, in a smaller vortex with a higher level of vorticity. The lateral position of the main vortex again approaches $x = \pi/4$.

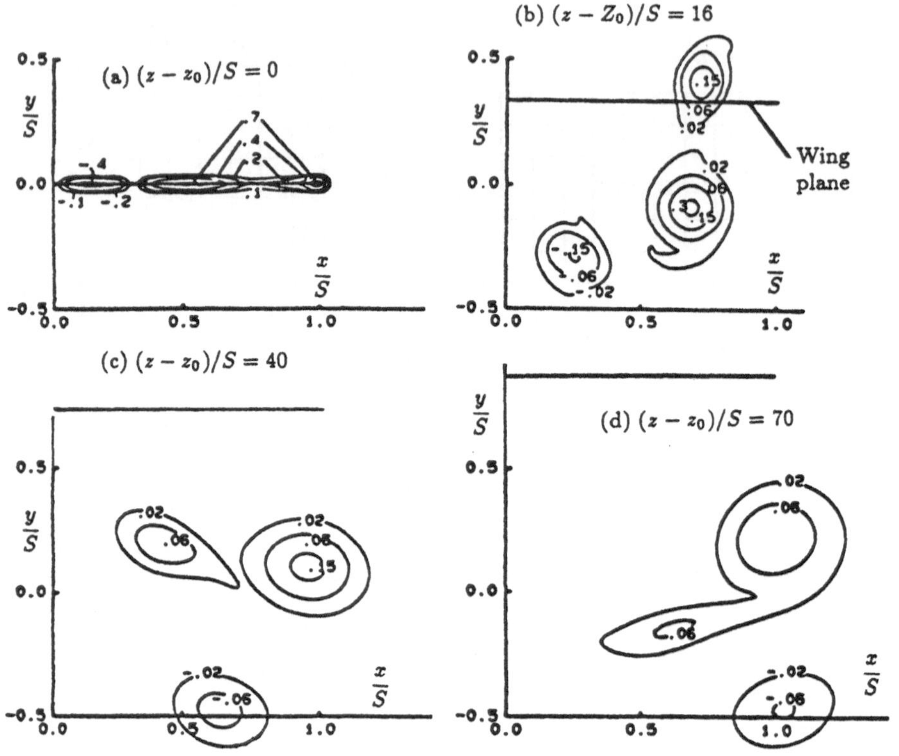

Fig. 3.11. Numerically simulated vorticity distribution in the wake of a transport aircraft (Weston and Liu 1982)

Another example presented by Weston and Liu (1982) is for a load distribution similar to that of a transport aircraft using flaps on landing. Note that the appearance of negative values of vorticity is associated with the reduction in lift around the fuselage-wing juncture. Part (a) of Fig. 3.11 displays the initial vorticity distribution for a wake thickness of 1.5 percent at a Reynolds number of 20,000. This example requires 500 grid points across the semispan in order to properly resolve the flow. Parts (b) through (d) of Fig. 3.11 show the subsequent evolution of this vortex system in time or downstream distance and demonstrate the ultimate merging of the vortices from the wing tip and flap tip. Experimental investigations were carried out using a 3% scale model of a B-747 aircraft with the inboard flaps deflected to give a spanwise load distribution similar to that used in the numerical model. The visualized vortex positions are qualitatively similar to the calculated ones shown in Fig.3.11(b) and (c) at $16S$ and $40S$ downstream of the initial station. The comparison of the numerical simulation with the experimental vortex wake can be found in Weston and Liu (1982) and Krause, Liu and Ting (1985).

3.2.2 Rules of Merging of Vortices

When we apply the Poincare relations (3.2.7-10) to two solutions of the initial value problem (3.2.1-6) that have different initial data we arrive at the following three statements:

"*If there are two initial data having the same total strength and first moments, the corresponding solutions retain the same strength and first moments for* $t \geq 0$."

(3.2.36a)

"*If in the above case the total strength is nonzero, then the solutions also have the same center of vorticity for* $t \geq 0$."

(3.2.36b)

"*If two initial data have the same total strength and polar moment, then the two solutions retain the same total strength and match in their polar moment for* $t \geq 0$".

(3.2.36c)

It was explained in **Sec. 2.2** that, on a long time scale, one can approximate the solution of our initial value problem, (3.2.1-6), by a similarity solution, $\Gamma\zeta^*$, i. e.,

$$\zeta(\bar{r},t) = \Gamma\zeta^*(\bar{r},t+t_0)\,\{1+O[(t+t_0)^{-1}]\}\ . \tag{3.2.37}$$

for any $t_0 > 0$, where

$$\zeta^*(\bar{r},t) = \frac{1}{4\pi\nu t}\,e^{-\bar{r}^2/(4\bar{\nu}t)}\ , \tag{3.2.38}$$

represents a Lamb vortex of unit strength created at the instant $t = 0$ with zero initial core radius. The similarity solution $\Gamma\zeta^*(\bar{r},t+t_0)$ represents a Lamb vortex with the same total strength Γ as that of the actual flow field. Its core size is

$$\delta_S(t) = [4\nu(t+t_0)]^{1/2}\ . \tag{3.2.39}$$

Since $\delta_S = 0$ at $t = -t_0$, we say that the Lamb vortex in (3.2.37) was created at $t = -t_0$ and that its initial "age" (at the instant $t = 0$) is t_0. This initial age is a free parameter of in the similarity solution. We define an optimum time shift $t_0 = t^*$ by the condition that ζ approaches the corresponding similarity solution in the fastest possible manner, i. e.,

$$\zeta(\bar{r},t) = \Gamma\,\zeta^*(\bar{r},t+t^*)\,\{1+O[(t+t_0)^{-2}]\}\ . \tag{3.2.40}$$

This is achieved when the coefficient C_1 of the second term in the series representation (2.2.42) of ζ vanishes (see also (2.2.45)). The condition $C_1 = 0$ yields,

$$(4\nu t^*)\Gamma = 2\pi\int_0^\infty r^2\zeta_0\,r\,dr\ , \tag{3.2.41}$$

where $\Gamma = \langle\zeta_0\rangle$. Recall that t^* should be positive. This is certainly true when ζ_0 does not change sign.

The right side of (3.2.41) represents the polar moment of the initial vorticity distribution. The left side is equal to $\Gamma\delta_0^2$ which is the polar moment of $\Gamma\zeta^*$ at $t = 0$. Hence, (3.2.41) is equivalent to the condition of matching the initial polar moment of the actual distribution with that of the leading order approximation,

$$\langle r^2 \Gamma\zeta^*\rangle|_{t=0} = \Gamma\delta_S^2(0) = \langle r^2\zeta_0\rangle \ . \tag{3.2.42}$$

It follows from statement (3.2.36c) that under (3.2.42) the polar moment of the optimum similarity solution matches that of the exact solution (2.2.34) for all $t \geq 0$.

We should mention a relevant result obtained by Kleinstein and Ting (1971) regarding the solution of a linear diffusion equation, $\zeta_t = \nu\Delta\zeta$ with nonaxisymmetric initial data $\zeta_0(x, y)$. The solution soon loses its asymmetry and approaches the optimum axisymmetric similarity solution $\Gamma\zeta^*$ in the sense of (3.2.40). Thus, $\Gamma\zeta^*$ matches the total strength and polar moment of $\zeta_0(x, y)$ and the center of $\Gamma\zeta^*$ is located at the center of gravity of $\zeta_0(x, y)$. In the next section we shall observe the same behavior for the nonlinear initial value problem of the merging of several vortices to a single one.

We now summarize the properties of an optimum similarity solution (or a Lamb vortex with the optimum initial core size):

"*A similarity solution, $\Gamma\zeta^*(\bar{r}, t+t_0)$, is specified by two parameters, its strength Γ and its initial age t_0 (or its initial core size $\sqrt{4\nu t_0}$) with $t_0 > 0$*". (3.2.43a)

"*All the similarity solutions preserve the total strength and the first moments (the location of the center of vorticity) of the exact solution.*" (3.2.43b)

"*The optimum similarity solution, with $t_0 = t^*$, in the sense of (3.2.40), has the same polar moment as the initial data at time $t = 0$ and hence has the same the polar moment as the exact solution for all times.*" (3.2.43c)

By combining statements (3.2.36a, b, c) on two vorticity fields with different initial data and (3.2.43a,b,c) on an optimum single Lamb vortex, we come up with the rule of merging of n vortices:

"*In the long time limit, n initially nonoverlapping vortices merge to an optimum Lamb vortex. Its total strength is given by the initial value of the circulation around the n vortices, its center remains at the initial center of vorticity and its optimum age, t^*, is defined by matching of its polar moment with that of the n vortices.*" (3.2.44)

Tests of the rule of merging can be found in Figs. 3.6 and 3.7. They show that the numerical solutions of N-S equations are in good agreement with the solutions predicted by the corresponding optimum Lamb vortex for $t > t_m$ and the agreement improves as t increases.

In the following subsection, we will express an *extended asymptotic solution*, which is an asymptotic solution used beyond its region of validity, as sum of Lamb vortices and employ the rule of merging to explain why the solution is a good approximation not only to the early stage but also to the final stage of merging.

3.2.2.1 Extended asymptotic solutions

Let us consider an initial vorticity field which can be represented approximately by a sum of n Lamb vortices. Let their strengths, initial ages and the coordinates of their centers be denoted by Γ_k, t_k^* and $X_k(0), Y_k(0)$ for $k = 1, \ldots, n$.

If these n vortices are far apart from each other, the asymptotic solution of the flow field is given by the sum of the Lamb vortices for $t \geq 0$. The vorticity field is

$$\zeta(x,y,t) = \sum_{k=1}^{n} \Gamma_k \zeta^*(x - X_k, y - Y_k, t + t_k^*) = \sum_{i=1}^{n} \frac{\Gamma_k}{\pi \delta_k^2(t)} e^{-r_k^2/\delta_k^2(t)} \tag{3.2.45}$$

where

$$r_k^2 = (x - X_k)^2 + (y - Y_k)^2$$

and

$$\delta_k^2(t) = 4\nu(t + t_k^*) \ .$$

The corresponding velocity field is

$$\begin{aligned} \boldsymbol{V}(x,y,t) &= \sum_{k=1}^{n} \boldsymbol{V}_S(x - X_k, y - Y_k, t + t_k^*) \\ &= \sum_{k=1}^{n} \frac{\Gamma_k}{2\pi r_k^2} \left[-(y - Y_k)\hat{\imath} + (x - X_k)\hat{\jmath}\,\right] \left[1 - e^{-r_k^2/\delta_k^2(t)}\right] \end{aligned} \tag{3.2.46}$$

The velocity of the k-th vortex center is

$$\dot{X}_k(t)\hat{\imath} + \dot{Y}_k(t)\hat{\jmath} = \boldsymbol{V}(X_k, Y_k, t) \tag{3.2.47}$$

for $k = 1, \ldots, n$. We note that the vorticity and velocity defined by (3.2.45 - 46) are uniformly valid in the xy plane and agree with the classical inviscid solution for n point vortices in the region away from the vortex centers.

The asymptotic solution (3.2.45-46) remains valid so long as each core size δ_i is much smaller than the distance, d_{ij}, to an adjacent vortex center, i.e.,

$$\delta_i / d_{ij} \leq \alpha << 1 \qquad \text{for } i,j = 1 \ldots n \text{ and } j \neq i. \tag{3.2.48}$$

In practice we choose $\alpha = 1/4$ as noted in **Sec. 2.4**. When condition (3.2.48) is no longer satisfied, the ith and jth vortical cores are overlapping each other and the merging of vortices commences. When we continue to use the asymptotic solution in this case, we call it the *extended asymptotic solution.* We then study its accuracy by comparison with the numerical (finite difference) solution of the merging process described in **Sec.3.2.1**.

Numerical results for the merging of n identical vortices symmetrically located on a circle centered at the origin were described in **Sec. 3.2.1.4** for $n = 2, 3, 4$.

The corresponding results given by the *extended asymptotic solutions* were also included in Figs. 3.5 and 3.6.

As shown in Fig. 3.5, the trajectories of n points of maximum vorticity, given by the extended asymptotic solution, spiral inward and end at the origin at an instant t_m which depends on n. To illustrate this process, we consider $n = 2$. The points of maximum vorticity are located along the radial line joining the two vortex centers and hence may be found by computing the roots of

$$\zeta_r(t,r,0) = \frac{2\Gamma}{\pi\delta^4}\left[(a-r)\,e^{-(r-a)^2/\delta^2} - (r+a)\,e^{-(r+a)^2/\delta^2}\right] = 0\ .$$

For $t < t_m$ there are two roots, $r = 0$ for the local minimum and $r^*(t) < a$ for the maximum with $r^*(t_m) = 0$. For $t > t_m$, there is only one root, $r = 0$, where ζ is the maximum. The instant t_m is defined by the condition $r^*(t_m) = 0$ or $\zeta_{rr} = 0$ at $t = t_m$ and $r = 0$. The result for the case in Fig. 3.5a is $t_m = a^2/(2\nu) = 8$. This value is greater than the value 5.5 predicted by the finite difference solution but the difference gets smaller for larger n.

The decay of the maximum vorticity given by the extended asymptotic solution is also shown in Fig. 3.6. The difference between the extended asymptotic solution and the finite difference solution is noticeable only during the second stage of merging, that is, for the period from the disappearance of n local maxima $t \sim t_m$ to the recircularization of contour lines around the origin ($t \sim t_f$). The difference during this period is smaller for larger values of n. For $t > t_f$, the extended asymptotic solution is again in good agreement with the numerical solution.

In Figs. 3.7 and 3.8 we see the merging of $n = 2, 3$ initially distinct vortices to a single one with nearly circular contours of constant vorticity. The $|\Omega|$-contours gradually become circularized beginning with the innermost line and progressing to the outer ones as time goes on. The dotted lines are the corresponding contour lines predicted by an optimum Lamb vortex defined by the rule of merging (3.2.44). The contour lines of the optimum Lamb vortex and those given by the numerical solution become indistinguishable in the last time frame in Figs. 3.7 and in 3.8. This agreement was pointed out right after (3.2.44). We note here that the contour lines given by the extended asymptotic solutions also become indistinguishable with those by the numerical solutions. The fact that the extended asymptotic solution is valid in the early stage of merging is to be expected because it is within the practical region of validity of the asymptotic solution. The extended solution becomes accurate again for large $t > t_m$ can be explained by the fact that the solution obeys the linear diffusion equation and hence conserves the total strength and first moments and matches the polar moment of the numerical solution. When the latter approaches that of an optimum single Lamb vortex for $t > t_f$, the extended asymptotic solution converges to the same limit.

The fact that the *extended asymptotic solution* remains quite accurate for all t except for the finite period of the second stage of merging, $t_c < t < t_f$, prompts us to recast the sum of Lamb vortices as an *approximate solution* to the N-S equations. In this process, we identify the source of errors and derive an improvement. This

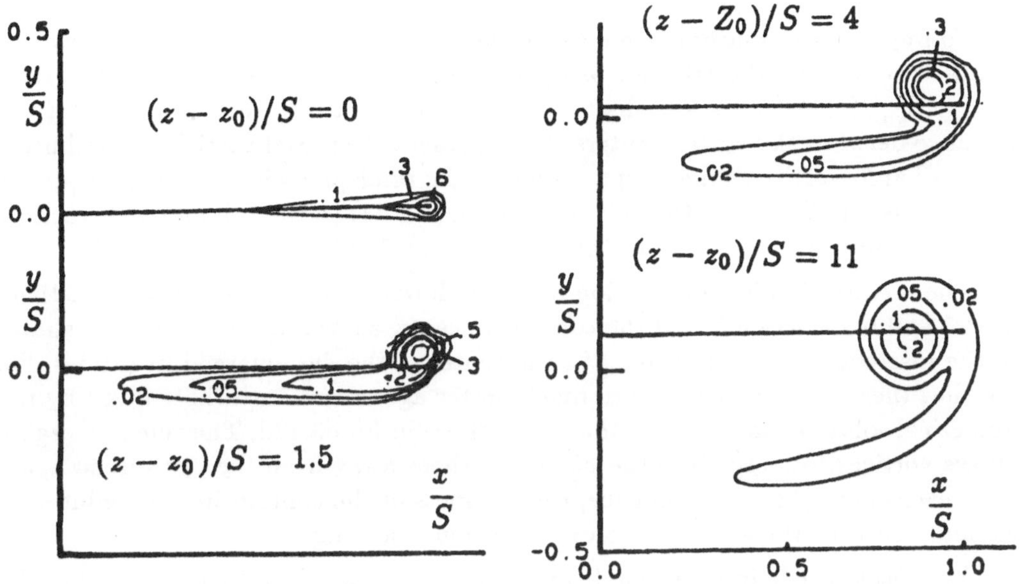

Fig. 3.12. Contour lines of constant vorticity ζ/ζ_{max} in the roll up and decay of a trailing vortex sheet — "extended" asymptotic solution. (Ting 1986)

is the topic of **Sec. 3.2.3**, in which we point out to what extent the *approximate* solution compensates for the deficiencies of the *extended asymptotic* solution.

To end this subsection, we mention another application of the extended asymptotic solution and the rule of merging. They were employed by Ting (1986) and Ting and Liu (1986) to simulate the roll-up and decay of a thin trailing vortical layer behind an aircraft wing. An estimate of the accuracy of the simulation was obtained by comparing the results with those of the numerical solutions presented in **Sec.3.2.1.5**. Recall that in that subsection the initial data for the numerical simulation of a trailing vortical layer were a collection of Lamb vortices which might overlap the adjacent ones. The same collection is now used to initialize the *extended asymptotic solution* (3.2.45 - 47). The initial data used by Weston et al. are a collection of 100 vortices equally spaced along the semispan S with initial core sizes equal to 2.5% of the semispan. The Reynolds number is 20,000 based on the circulation Γ_0 around the root section, $x = 0$. Figure 3.12 shows the contour lines at distances of $1.4S$, $4S$ and $11S$ downstream from the wing. They are similar to those of numerical solutions to the N-S equations shown in Fig. 3.9. The contour lines of high vorticity show the core structure of an *eye* known as the tip vortex. At the station $z = 11S$, more than half of the initial vortices, accounting for 86%

of Γ_0 are packed in and around the *eye*. The contour lines of constant vorticity near the *eye* become more wavy for larger z/S. This is due to the fact that near the *eye* many vortices have been closely packed together and subjected to intense merging.

To improve the solution, we employ the rule of merging (3.2.44) to allow two Lamb vortices, say the ith and mth, to merge into a single one when the ratio $\Lambda_{km} = d_{km}^2/(\delta_k^2 + \delta_m^2)$ is less than a given critical value. Here d_{km} stands for the distance between the vortex centers. We represent the initial vorticity distribution by 160 Lamb vortices equally spaced along the semispan with core radii equal to 2.5% semispan. We apply the rule of merging (3.2.44) to the kth and mth vortices when the ratio Λ_{km} is less than the critical value 1/32.

The results of this modification are as follows: At the station $z = 11S$, there are 94 Lamb vortices left and the contour lines are shown in Fig. 3.13. Also shown in the insert are the contour lines in the center of the *eye* enlarged five times. We see that these contour lines are in much better agreement with those given by the numerical solutions in Fig. 3.9d than were those in Fig. 3.12d. The rule of merging allows vortices to merge into the *eye* while those away from the *eye* spread apart from each other. As a consequence, the waviness of the contour lines is reduced in comparison with the extended solution without merging.

As mentioned before, we shall improve the *extended asymptotic solution* by introducing a new criterion for the velocities of the vortex centers. The velocity of the vortex centers are adjusted so that the solution fulfills the N-S equations approximately under a minimum principle. The new solution to be described in the following subsection will be called the *approximate solution*.

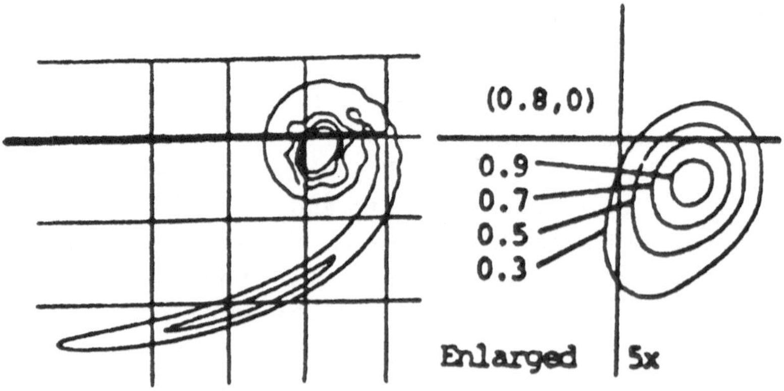

Fig. 3.13. Contour lines of constant vorticity at station $z = 11S$ — initial 160 Lamb vortices merged to 96 vortices (Ting and Liu 1986)

3.2.3 Approximate solution of N-S Equations using superposition of Lamb vortices

To prepare for the construction of our *approximate solution*, we use $\zeta^*(x,y,\delta(t))$ to denote the vorticity distribution of a Lamb vortex of unit strength and centered at the origin, i.e.

$$\zeta^*(x,y,\delta(t)) = \frac{1}{\pi\delta^2(t)}\, e^{-r^2/\delta^2(t)} \tag{3.2.49}$$

with

$$\delta^2(t) = 4\nu t \quad \text{and} \quad r^2 = x^2 + y^2 \ .$$

The Lamb vortex is created at $t = 0$ since the core size δ vanishes at that time. We denote the corresponding stream function by $\psi^*(x,y,\delta)$ and obtain the induced velocity

$$\boldsymbol{V}^*(x,y,\delta) = \psi_y^*\hat{\imath} - \psi_x^*\hat{\jmath} = \frac{1}{2\pi r^2}[-y\hat{\imath} + x\hat{\jmath}]\,[1 - e^{-r^2/\delta^2(t)}] \ . \tag{3.2.50}$$

We then construct an *approximate solution* to the initial value problem, (3.2.1)–(3.2.6), by linear superposition of n Lamb vortices as follows:

$$\zeta(t,x,y) = \sum_{k=1}^{n} \Gamma_k\, \zeta_k^* \tag{3.2.51a}$$

and

$$\psi(t,x,y) = \sum_{k=1}^{n} \Gamma_k\, \psi_k^* \ , \tag{3.2.51b}$$

where

$$\zeta_k^* = \zeta^*(x - X_k(\tau)\ ,\ y - Y_k(\tau)\ ,\ \delta_k(\bar{t} + t_k^*)) \tag{3.2.52a}$$

and

$$\psi_k^* = \psi^*(x - X_k(\tau)\ ,\ y - Y_k(\tau)\ ,\ \delta_k(\bar{t} + t_k^*)\ ) \ . \tag{3.2.52b}$$

The formal introduction of two time variables t, τ will be explained shortly. The strength Γ_k, initial vortex center $X_k(0)$, $Y_k(0)$ and age t_k^* for $k = 1\ldots n$ are chosen to fit the initial data $\zeta_0(x,y)$. The representation of the *approximate solution*, (3.2.52a, b), by a superposition of n Lamb vortices is identical to that for the *extended asymptotic solution*, (3.2.45 - 46). They differ only in the velocities of the vortex centers. For the latter, the velocities of the centers are prescribed by the asymptotic theory for nonoverlapping vortices, (3.2.47) and are in error for overlapping vortices. Now the velocities, $\dot{X}_k(\tau)$ and $\dot{Y}_k(\tau)$ are treated as $2n$ unknowns to be defined such that (3.2.51a,b) and (3.2.52a,b) give the "best" approximate solution to the N-S equations, i. e., they are derived from a minimum principle. This is described in the following.

Note that the dependence of the solution (3.2.51a -52b) on t appears indirectly through the changing core sizes and the moving vortex centers. To show this difference explicitly, we have denoted the time variable t in the core size δ_k and the position of the center (X_k, Y_k) by $\bar{t}$ and τ respectively. Thus the time derivative of ζ_k^* is written as,

$$\partial_t \zeta_k^* = (\partial_{\bar{t}} + \partial_\tau)\zeta_k^* = \left[\partial_{\bar{t}} - (\dot{X}_k(\tau)\partial_x + \dot{Y}_k(\tau)\partial_y)\right] \zeta_k^* \tag{3.2.53}$$

We note that the vorticity ζ_k^*, stream function ψ_k^* and the velocity $\boldsymbol{V}_k^*$ fulfill the following equations,

$$\partial_{\bar{t}}\zeta_k^* = \nu\Delta\zeta_k^* \,, \tag{3.2.54a}$$

$$(\boldsymbol{V}_k^* \cdot \nabla)\zeta_k^* = 0 \,, \tag{3.2.54b}$$

$$\Delta\psi_k^* = -\zeta_k^* \tag{3.2.55a}$$

and

$$\boldsymbol{V}_k^* = (\psi_k^*)_y\,\hat{\imath} - (\psi_k^*)_x\,\hat{\jmath} \,. \tag{3.2.55b}$$

We observe that the stream function ψ in (3.2.52b) fulfills the Poisson equation (3.2.2) with an inhomogeneous term given by ζ in (3.2.51a) and that the velocity **V** defined by (3.2.3) yields

$$\boldsymbol{V}(t,x,y) = \sum_{k=1}^{n} \Gamma_k\, V_k^* \,. \tag{3.2.56}$$

The far field conditions (3.2.5) and (3.2.6) are also fulfilled by ζ and $\boldsymbol{V}$ of (3.2.51) and (3.2.56). Only the nonlinear vorticity diffusion equation, (3.2.1), remains to be checked. By using (3.2.53), (3.2.1) becomes

$$\sum_k [\boldsymbol{V}(t,x,y) - \dot{X}_k(\tau)\hat{\imath} - \dot{Y}_k(\tau)\hat{\jmath}] \cdot \nabla(\Gamma_k\zeta_k^*) = \sum_k \Gamma_k[\nu\Delta\zeta_k^* - \partial_{\bar{t}}\zeta_k^*] \,. \tag{3.2.57}$$

Due to (3.2.54a, b), the right side of (3.2.57) vanishes, while the left side reduces to

$$F(t,x,y) = \sum_k \Gamma_k [(\sum_{\ell \neq k} \Gamma_\ell \boldsymbol{V}_\ell^*) - \dot{X}_k(t)\hat{\imath} - \dot{Y}_k(t)\hat{\jmath}] \cdot \nabla\zeta_k^* \tag{3.2.58a}$$

and (3.2.57) becomes

$$F(t,x,y) = 0 \,. \tag{3.2.58b}$$

The function F vanishes in the domain of zero vorticity but it cannot vanish for all x, y in the domain where $\zeta \neq 0$. To fulfill (3.2.58b) approximately, we seek $\dot{X}_k(t)$, $\dot{Y}_k(t)$, $k = 1 \ldots n$ such that the function

$$H(t, \dot{X}_1 \ldots \dot{X}_n \,,\, \dot{Y}_1 \ldots \dot{Y}_n) = \langle F^2(t,x,y) \rangle = \min. \tag{3.2.59}$$

for $t \geq 0$, where $\langle \quad \rangle$ again denotes the area integral over the xy plane. In this sense, we say that ζ and ψ in (3.2.51 and 52) are the "best" approximate solutions of the N-S equations. At each instant t, $H(t, \dot{x}_1, \ldots \dot{y}_n)$ is minimized by $\dot{X}_1 \ldots \dot{Y}_n$, if and only if the following $2n$ linear equations for $\dot{X}_m$ and $\dot{Y}_m$ are fulfilled:

$$\sum_m (a_{km}\dot{X}_m + e_{km}\dot{Y}_m) = C_k \quad \text{and} \quad \sum_m (b_{km}\dot{Y}_m + e_{mk}\dot{X}_m) = -D_k\ , \tag{3.2.60}$$

for $k = 1 \ldots n$ with,

$$a_{km} = \Gamma_k\Gamma_m\langle\partial_x\zeta_k^*\ \partial_x\zeta_m^*\rangle\ , \quad e_{km} = \Gamma_k\Gamma_m\langle\partial_x\zeta_k^*\ \partial_y\zeta_m^*\rangle \quad ,$$

$$C_k = \sum_m c_{km}\ , \quad D_k = \sum_m d_{km}\ , \tag{3.2.61}$$

$$c_{km} = \Gamma_k\Gamma_m\langle\partial_x\zeta_k^*\ (\partial_x\zeta_m^* \sum_{\ell\neq m}\Gamma_\ell\partial_y\psi_\ell^* - \partial_y\zeta_m^* \sum_{\ell\neq m}\Gamma_\ell\partial_x\psi_\ell^*)\rangle$$

and b_{km} and d_{km} are the same as a_{km} and c_{km} respectively with ∂_x interchanged with ∂_y. These coefficients are elementary functions of X_k, X_m, Y_k, Y_m, δ_m and δ_k, listed in Appendix A.4. In particular, we have $a_{kk} = b_{kk} = 2\Gamma_k^2/(\pi\delta_k^4)$ and $e_{kk} = 0$. Since ζ^* decays exponentially on the length scale δ, all the nondiagonal elements, $a_{km} \cdots e_{km}$, contain a factor $\exp(-\Lambda_{km}^2)$ where $\Lambda_{km}^2 = |\mathbf{X}_k - \mathbf{X}_m|^2/(\delta_k^2 + \delta_m^2)$. The n pairs of equations can be rearranged as

$$GZ = H \tag{3.2.62}$$

Here Z and H denote respectively the column matrices of $\{\dot{X}_1, \dot{Y}_1, \cdots, \dot{X}_n, \dot{Y}_n\}$ and $\{C_1, D_1, \cdots, C_n, D_n\}$ and G denotes the corresponding $2n \times 2n$ matrix. Due to the fact that the minimum principle is applied to the square of a linear function in the $\{\dot{\mathbf{X}}_1, ..., \dot{\mathbf{Y}}_1, ...\}$, G is symmetric and positive definite. Since the nondiagonal elements vanish as $\exp(-\Lambda_{km}^2)$, G is dominated by its diagonal elements. Equation (3.2.62) can be solved readily for $\dot{X}_k$ and $\dot{Y}_k$, $k = 1 \cdots n$, the velocities of the n Lamb vortices in the representation of the *approximate solution*, (3.2.51a - 52b).

The ratio Λ_{km}^2, in the exponential factor, has been used to characterize the interaction between the k-th and m-th vortices. When $\Lambda_{km}^2 >> 1$, these two vortices are isolated from each other. When $\Lambda_{km}^2 << 1$, they have merged to a single one. For a k-th vortex, that is far apart from the others such that $\Lambda_{km}^2 >> 1$, for $m \neq k$, both the $(2k-1)$-st and $2k$-th equation of (3.2.62) are decoupled from the remaining $2n - 2$ equations and yield the asymptotic classical result for the velocity of the k-th vortex center given in (3.2.47). The nondiagonal elements of G and the c_{km}, d_{km} for $k \neq m$ in H account for the effects of interactions between overlapping adjacent vortical cores. Thus we observe the following :

- For a vortex that is far apart from the others, the velocity of its center given by the *approximate solution* reduces to that by the asymptotic solution.
- For overlapping vortices, the velocities of the vortex centers are determined by minimizing the effect of the nonlinear interaction terms in (3.2.58a).

The above observations also explain why the simulation of a thin trailing vortical layer in **Sec. 3.2.2** by the extended asymptotic solution together with the rule of merging agrees quite well with the finite difference solution even in the later stage of the roll up process. The reason is that the core of a vortex will overlap at most with one or two adjacent vortices at a time while the velocity of a vortex center is the resultant of the velocities induced by all the vortices, over 100 of them used in the modeling of the trailing vortex layer.

Since the probability of a simultaneous merging of more than two vortices is much less than that of only two, the merging of two vortices is clearly the simplest situation, yet the fundamental building block in the evolution of a vortical field simulated by many vortices. We shall test the accuracy of our *approximate solution* in the following studies of the merging of two vortices of the same strength and that of opposite strength.

Merging of two identical vortices

In Fig. 3.5a we compare the trajectories of the locations of maximum vorticity for two identical vortices with strength $\Gamma = 25$ and Re $= 100$ which were initially ($t = 1$) centered at (2,0) and $(-2, 0)$ with core size $\delta_0 = 1$. With the initial value of the parameter $\Lambda^2 = 8$, the two vortices are considered to be far apart from each other and each one is represented by a Lamb vortex. Under the inviscid theory, the two point vortices will spin around the origin at constant angular velocity forever. The trajectories of the two points of maximum vorticity given by the extended asymptotic solution spiral inward and combine to a single one at the origin to end the first stage of merging. These trajectories agree only qualitatively with those given by the finite difference solutions described in **Sec. 3.2.1**.

Figure 3.5a is reproduced in Fig. 3.14 with the addition of the trajectories given by the *approximate solution*. These additional curves are much closer to those of the finite difference solution than those of the extended asymptotic solution. Figure 3.7 showed the contour lines of $\zeta/\zeta_{\max} = 1/\sqrt{e}$, $1/e$ and $1/\sqrt{e^3}$ of the finite difference solution. The heavy dotted lines show the lines based on the optimum single vortex. Hence, the merging to a single vortex is nearly completed by $t = 25$.

Shown in Fig. 3.15 is the rotation of the major axis of vorticity during the merging process. As expected, the inviscid solution yields a straight line while the approximate solution is in considerably better agreement with the finite difference solution than the extended asymptotic solution. It should be noted that in the final phase of merging, $t > 20$, the contour lines are approaching circles and then it becomes rather difficult (or less accurate) to determine the orientations of the major and minor axes. However, the contour lines are no longer sensitive to the orientations of the axes which only account for the small deviations of the contour lines from circles. At that stage, the mean radii of the contour lines given by all three solutions, the numerical, *extended asymptotic* and *approximate* solutions, are in good agreement with those predicted by the optimum Lamb vortex.

Figure 3.16 shows the contour lines of the *approximate solution* based upon the minimum principle. They agree with those in Fig.3.7 in size and in the orientation of the principal axes, but they differ in details. In particular, the numerical results are not symmetric with respect to the principal axes while the *approximate* results are. Also shown in dotted lines are the corresponding axes based on the extended asymptotic solution. We see that the contour lines of the latter will be completely out of phase with the finite difference solution.

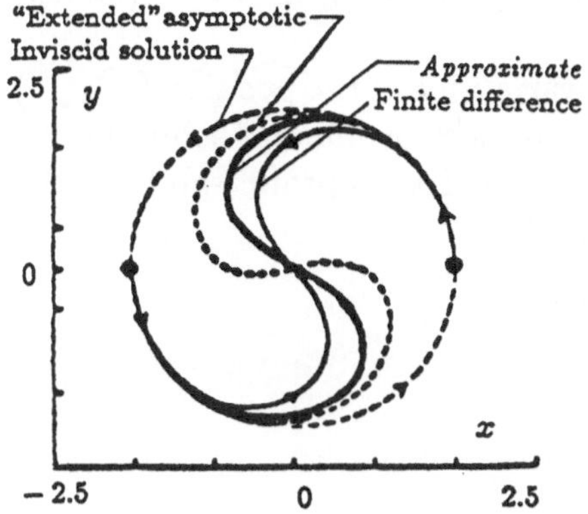

Fig. 3.14. Trajectories of the locations of the maximum vorticity in the merging of two identical vortices (Ting and Liu 1986, Ting 1986)

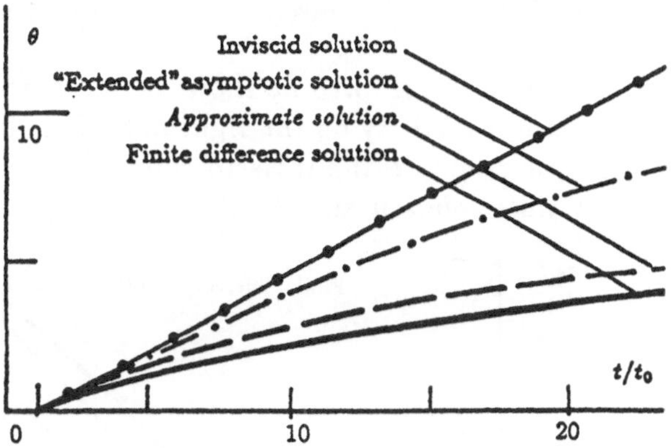

Fig. 3.15. Rotation of the major axis of vorticity distribution during the merging of two identical vortices (Ting and Liu 1986)

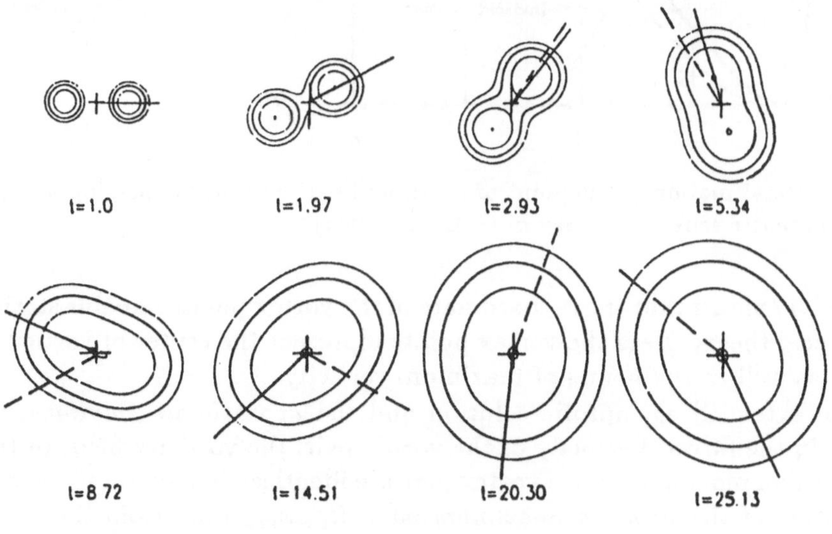

Fig. 3.16. Contour lines of constant vorticity in merging of two vortices — approximate solution based on a minimum principle (Ting and Liu 1986)

Cancellation of two vortices of opposite strength

Under the inviscid theory, a pair of point vortices of opposite strength, $\pm\Gamma$, located at $(\pm a, 0)$ moves with constant velocity $\Gamma/(4\pi a)$ in the direction of the y-axis. The general behavior of the cancellation of a counter-rotating vortex pair was described at the beginning of **Sec. 3.2** for the antisymmetric vorticity distribution $\zeta(t,x,y) = -\zeta(t,-x,y)$ and $\zeta > 0$ for $x > 0$. As time proceeds, the strength of the vorticity on the right half plane $\Gamma^+ = \langle\zeta\rangle^+$ decreases, while the total strength of the first x-moment remains constant. As a consequence, the x-coordinate of the center of vorticity increases:

$$\Gamma_t^+ < 0\,, \quad \langle x\zeta\rangle = 2\langle x\zeta\rangle^+ = C_1 \quad \text{and} \quad (x_c)_t = [C_1/(2\Gamma^+)]_t > 0\,. \qquad (3.2.63)$$

In Fig. 3.17 and Fig. 3.18 we show the temporal variations of the position (x^+, y^+) of maximum vorticity on the right half plane based on four different solutions. As before, we compare the inviscid solution, the extended asymptotic solution, the *approximate solution* and the finite difference solution.

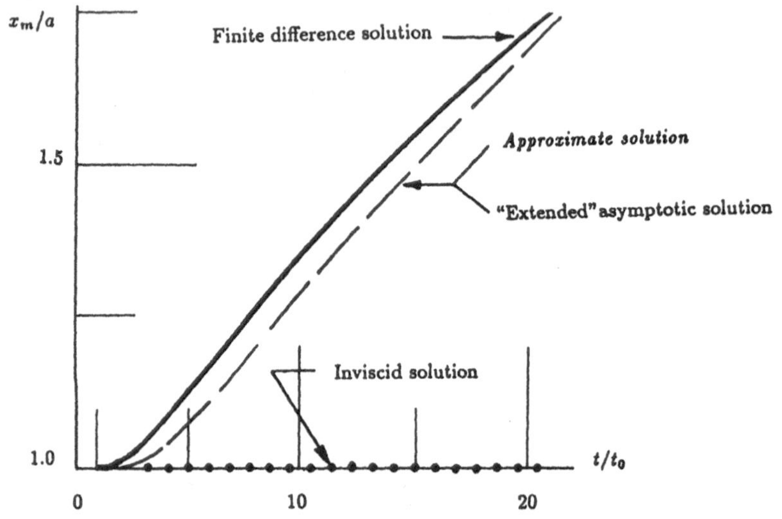

Fig. 3.17. Lateral motion of the point of maximum vorticity in the merging of a pair of vortices of opposite sense (Courtesy of G. C. Liu 1986)

Figure 3.17 shows that the x-coordinate of the vortex points remains stationary in the inviscid theory. Here the vortex points represent the center of vorticity in a half plane as well as the points of maximum vorticity.

For the extended asymptotic solution and the *approximate solution* , which differ only in the forward velocity of the vortex pair, the vorticity fields in the coordinate system moving with the vortex pair are identical. There is a nonvanishing lateral motion of the *points of maximum vorticity*, x_m, away from the center of the Lamb vortex. The motion is induced by the cancellation of vorticity in the half plane $x > 0$ by its negative image in $x < 0$, which in turn is due to diffusion of vorticity across the y-axis. This lateral motion of x_m is in good agreement with that given by the finite difference solution. The distance between the point of maximum vorticity and its negative image increases by almost 75% as the time, t, increases from 1 to 20. During this period, the vorticity fields of the two Lamb vortices have cancelled each other substantially since the ratio of the distance between the vortex centers to the sum of their radii decreases from the initial value of 2 to $1/\sqrt{5}$.

Figure 3.18 shows the forward motion of the vortex pair. Note that the scale of the ordinate for y_m is tenfold that for x_m in Fig. 3.17. The forward velocity remains constant only for the inviscid solution. The velocity decreases in the other three solutions. The approximate solution is in good agreement with the finite

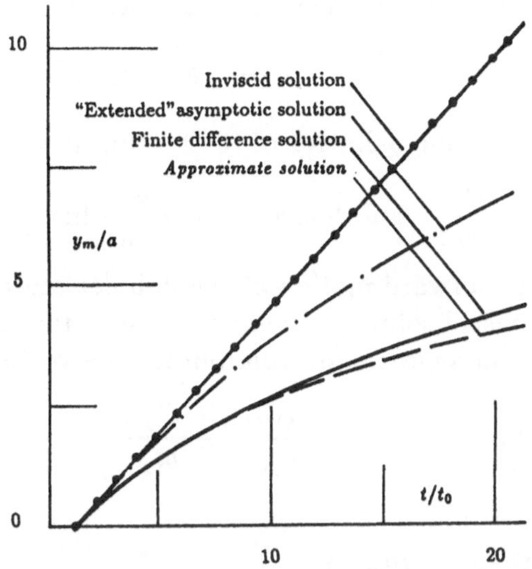

Fig. 3.18. Forward motion of the point of maximum vorticity in the merging of a pair of vortices of opposite strength (Courtesy of G. C. Liu 1986)

difference solution while the extended asymptotic solution becomes less and less accurate as t increases.

We hope that the above examples will encourage further applications and/or improvement of the method of *approximate solutions* (3.2.51a - 52b). We hope that they will stimulate further studies on procedures for optimum representation of an initial vorticity distributions by a finite number of Lamb vortices and on efficient implementation of the rule of merging of vortices.

A basic question arises from the above studies: We note that the solution of the initial value problem (3.2.1 - 4) with zero total strength, $<\zeta_0>=0$, should behave in the far field like a doublet of constant strength and orientation. Here we have defined the orientation by the direction of the first moment vector, $<\mathbf{x}\zeta>\times\hat{k}$, see (1.2.12). The velocity of the vorticity field, or that of the doublet, $(\dot{X}_D, \dot{Y}_D)$, can be defined by (3.2.18). We can always choose the coordinate axes such that the doublet is oriented in the direction of the x-axis, for example, by locating the centers of the initial vortex pair of strengths, $\Gamma_1 > 0$ and $\Gamma_2 = -\Gamma_1$, on the x-axis at $(\pm a, 0)$. Due to antisymmetry, the center of the doublet will remain on the $y-$axis, i. e., $X_D = 0$, and the forward motion of the vortex pair is in the downward direction. There is an intriguing question concerning the long time behavior of the motion of the doublet. Will the doublet center $(0, Y_D)$ approach a stationary point, $(0, Y_\infty)$? That is,

$$\text{Does} \quad \lim_{t\to\infty} Y_D(t) \quad \text{exist ?} \tag{3.2.64}$$

In the inviscid theory, the answer is *no*. The vortex pair will move at constant speed and Y_D will increase linearly in t,

$$\dot{Y}_D = \Gamma_2/(4\pi a) \qquad \text{and} \qquad Y_D = t\Gamma_2/(4\pi a), \tag{3.2.65}$$

where $\Gamma_2 = -\Gamma_1$ and $2a$ denotes the distance between the vortex centers.

The velocity given by the extended asymptotic solution is

$$\dot{Y}_D = \frac{\Gamma_2}{4\pi a}[1 - e^{-4a^2/\delta^2}] \quad \text{with} \quad \delta^2 = 4\nu(t + t_0), \tag{3.2.66}$$

where $\delta = \delta_1 = \delta_2$. The long time behavior of the motion of the doublet center is

$$\dot{Y}_D \sim \frac{a\Gamma_2}{4\pi\nu t} \quad \text{and hence} \quad Y_D \sim \frac{a\Gamma_2}{4\pi\nu}\ln t \tag{3.2.67}$$

as $t \to \infty$ or $\delta^2 \gg 4a^2$. The forward motion of the doublet center, although much slower than that in the inviscid solution, does not have a stationary limit point.

The velocity given by the *approximate solution* for the vortex pair is

$$\dot{Y}_D = \dot{Y}_1 = \frac{\Gamma_1\Gamma_2\,(d^*_{11,2} + d^*_{12,1})}{\Gamma_1\, b^*_{11} + \Gamma_2\, b^*_{12}} \tag{3.2.68}$$

where

$$b^*_{11} =< ((\zeta^*_1)^2_y >= 1/(2\pi\delta^4), \tag{3.2.69a}$$

$$b^*_{12} =< (\zeta^*_1)_y(\zeta^*_2)_y >= b^*_{11}e^{-2a^2/\delta^2} \tag{3.2.69b}$$

$$d^*_{11;2} =< (\zeta^*_1)_y \mathbf{v}^*_2 \cdot \nabla\zeta^*_1 >= \frac{b^*_{11}}{4a\pi}[1 - e^{-8a^2/(3\delta^2)}] \tag{3.2.69c}$$

$$d^*_{12,1} =< (\zeta^*_1)_y \mathbf{v}^*_1 \cdot \nabla\zeta^*_2 >= \frac{b^*_{11}}{a\pi}[e^{-2a^2/\delta^2} - e^{-8a^2/(3\delta^2)}] \tag{3.2.69d}$$

They come from (3.2.60) for two vortices, $n = 2$, of opposite strength,† $\Gamma_2 = -\Gamma_1$ and the same core size δ. The long time behavior of $\dot{Y}_D$ is

$$\dot{Y}_D \sim \frac{a\Gamma_2}{12\pi\nu t} \quad \text{and hence} \quad Y_D \sim \frac{a\Gamma_2}{12\pi\nu}\ln t \tag{3.2.70}$$

as $t \to \infty$ or $\delta^2 \gg 2a^2$. The leading term is one third of that given by the extended asymptotic theory. Again the forward motion of the doublet center does not have a stationary limit point.

We note that in the long time, the size of the vortical field, L, increases as $\sqrt{t}$. If the forward displacement of the doublet center, $|Y_D|$, increases as $\ln t$, it stays well inside the vortical field, since $L \gg |Y_D|$. Furthermore, when we say "in the far field", it implies "at a large distance of length scale $R \gg L$". In the length scale R, the vortical flow field reduces to a doublet centered at $(0, 0)$. This statement is not true if Y_D increases linearly in t for all t according to the inviscid theory.

Intuitively, we could define the forward velocity of the vortex pair by the average velocity on the right half plane weighed by the vorticity distribution,

$$\dot{Y} = \frac{< v\zeta >^+}{< \zeta >^+} = \frac{\Gamma_2}{\pi\sqrt{\pi}\delta\Phi(\bar{a})}\Big[-e^{-\bar{a}^2} + \frac{1}{\sqrt{2}}e^{-2\bar{a}^2} + \frac{(1-2\bar{a}^2)\sqrt{\pi}}{2\bar{a}}\Phi(\bar{a}) - \frac{(1-4\bar{a}^2)\sqrt{\pi}}{4\bar{a}}\Phi(\sqrt{2}a)\Big], \tag{3.2.71a}$$

† If two vortices are of the same strength, then (3.2.68) defines the circumferential velocity of a vortex center.

where $\bar{a} = a/\delta$ and Φ denotes the probability function. The long time behavior of the doublet center is

$$\dot{Y}_D \sim \frac{2(\sqrt{2}-1)}{3}\frac{a\Gamma_2}{4\pi\nu t} \quad \text{and hence} \quad Y_D \sim \frac{0.828 a\Gamma_2}{12\pi\nu}\ln t. \tag{3.2.71b}$$

Once more the doublet moves forward as $\ln t$ in the same manner as that predicted by the former two solutions, (3.2.67), and (3.2.70), while the coefficient in (3.2.71b) differs from that of the *approximate solution (3.2.70)* by a factor of 0.828.

Due to the symmetry of The vorticity distributions in both the extended asymptotic and the *approximate solution* are symmetric with respect to the horizontal axis joining the two Lamb vortex centers. Therefore, the vertical coordinates, Y_m, Y^+, Y_D, of the point of maximum vorticity, the center of positive vorticity and of the equivalent doublet in the far field, all coincide, i.e., $Y_m = Y^+ = Y_D$. This may not be true for the finite difference solution, i. e., $Y_m \neq Y_D$, because the solution in general does not have a horizontal line of symmetry for $t > 0$. This also explains why we say the definition of the doublet velocity by (3.2.71a) is intuitive. If we define the center of the doublet by the condition, $< (y-Y)\zeta >^* = 0$, the velocity $\dot{Y}$ is given by the equation,

$$\dot{Y} < \zeta >^* = < (y-Y)\zeta_t >^* = < v\zeta >^* - \bar{\nu} < (y-Y)\zeta_x >^* \ . \tag{3.2.72}$$

The second term on the right-hand side of the equation does vanish when ζ is an even function of $y - Y$, as is the case for a superposition of two Lamb vortices. However, in general, the second term does not vanish for $t > 0$, and (3.2.72) differs from (3.2.71a).

Using dimensional analysis, we can show that the forward velocity of the doublet given by the numerical solution also behaves as t^{-1} for large t. Therefore, we expect $Y_D \sim c_1 \ln t + c_2$ where the constants c_1 and c_2 can be determined by a careful numerical solution of the N-S equation. Using the same argument, we can show that the axial velocity of a circular vortex ring, which appears in the far field as a doublet oriented in the axial direction, will behave as $t^{-3/2}$ and hence the axial position of the doublet, $Z(t)$ *will* approach a stationary position Z_∞ as $t \to \infty$. Again, the value Z_∞ has to be determined numerically. In the next subsection we present several numerical simulations of three dimensional merging problems.

3.3 Merging or Intersection of Filaments

In order to illustrate how to implement the numerical schemes formulated in **Sec.3.1**, that is, how to choose the appropriate computational domain and how to generate the boundary data for different types of initial vorticity distributions, we consider first the axisymmetric problems in **Sec. 3.3.1** and then the fully three-dimensional problems in **Sec. 3.3.2**

3.3.1 Merging of Coaxial Vortex Rings

In an axisymmetric flow all the vortex lines are coaxial circles. The vorticity vector can then be expressed in terms of its circumferential component, ϖ, by

$$\Omega = \hat{\theta}\varpi(t,\sigma,z) = [-\sin\theta\,\hat{\imath} + \cos\theta\,\hat{\jmath}]\varpi(t,\sigma,z)\ , \tag{3.3.1}$$

where σ, θ, z denote the radial, circumferential and axial coordinates respectively. The vector potential $\mathbf{A}$ is also a circumferential vector,

$$\mathbf{A} = \hat{\theta}\,\psi(t,\sigma,z)/\sigma\ . \tag{3.3.2}$$

Here ψ represents the stream function in the meridian plane, the σz plane with $\sigma \geq 0$. The stream function is related to the vorticity ϖ by the axisymmetric Poisson equation,

$$\psi_{\sigma\sigma} + \psi_{zz} - \psi_\sigma/\sigma = -\sigma\varpi\ . \tag{3.3.3}$$

The vorticity, ϖ, is governed by the evolution equation,

$$\varpi_t - \psi_z\left(\frac{\varpi}{\sigma}\right)_\sigma + \frac{\psi_\sigma}{\sigma}\varpi_z = \nu\left[\varpi_{\sigma\sigma} + \varpi_{zz} + \left(\frac{\varpi}{\sigma}\right)_\sigma\right]\ . \tag{3.3.4}$$

For Ω to decay rapidly, we require ϖ to decay rapidly in $r = \sqrt{\sigma^2 + z^2}$. It was pointed out in **1.2.3** that Ω is divergence-free so that all the consistency conditions of are fulfilled automatically. The three time invariants of first moments (1.2.12) reduce to one nontrivial invariant (1.2.23) of the axial component of $< \mathbf{x} \times \Omega >$. It is

$$2\pi \int_{\infty}^{\infty}\int_0^{\infty} \sigma^2\varpi(t,\sigma,z)\,d\sigma dz = E_3 = constant, \tag{3.3.5a}$$

because $\sigma\varpi = x\omega_2 - y\omega_3$. In the far field, the flow behaves as a doublet with constant strength E_3 oriented along the z-axis. For later reference, we define here the circulation around an isolated vortex ring by the area integral of vorticity in a meridian plane,

$$\Gamma(t) = \int_{-\infty}^{\infty} v_3(t,0,z)\,dz = \int_{-\infty}^{\infty}\int_0^{\infty} \varpi(t,\sigma,z)\,d\sigma\,dz\ . \tag{3.3.5b}$$

Given these preliminaries, we are ready to study the merging of vortex ring(s). Because of axisymmetry, we will always have a **global** merging problem. In **Sec.3.3.1.1** we consider the self-merging of a vortex ring, which occurs when its core size becomes comparable to the ring radius. This is as an example for a global merging of type I(a). In **Sec.3.3.1.2**, we consider the merging of two rings. Their distance becomes comparable to their core radii, while both radii are still much smaller than the radii of the rings. This serves as an example for a global merging of type I(b).

3.3.1.1 Type I(a), Vorticity concentrated near the origin

The self-merging of a vortex ring is a global merging problem of type I(a). The vorticity distribution is concentrated within a sphere of radius $L = O(R)$ centered at the origin moving in the axial direction with the velocity of the center of vorticity in the meridian plane. The far field is defined by the condition $r >> R$. The vorticity decays exponentially with r and hence with σ and z. The computational domain, $\mathcal{D}$, in the σz plane can be a square of side $2mR$ with $0 \leq \sigma \leq 2mR$ and $|z| \leq mR$. The number m must be chosen so that the error in the boundary data generated through a far field expansion is within the required accuracy of the numerical solution. Due to symmetry and the regularity condition along the z-axis we have $\varpi = 0$ and $\psi = \psi_\sigma = 0$ on $\sigma = 0$. By following the scheme formulated in **Sec.3.1.2**, the boundary condition for the vorticity is $\varpi = 0$ on $\partial\mathcal{D}$. The boundary data for ψ but not on the z-axis can be generated from (1.2.25) for the vector potential $\mathbf{A}$ and then from (3.3.2) for ψ, but it is easier to derive the far field behavior of ψ directly from the Poisson integral solution of (3.3.3),

$$\psi = \frac{\sigma}{4\pi} \int_{-\infty}^{\infty} \int_{0}^{\infty} \int_{0}^{2\pi} \frac{\varpi(t, \sigma', z')}{\rho} \sigma' \cos\theta \, d\theta \, d\sigma' \, dz' \tag{3.3.6}$$

where $\rho^2 = (z - z')^2 + \sigma^2 + \sigma'^2 - 2\sigma\sigma' \cos\theta$. To obtain the behavior of ψ at large $r = (z^2 + \sigma^2)^{1/2}$, Liu and Ting (1982) expanded $1/\rho$ in the integrand in powers of $1/r$ and obtained the following expansion,

$$\begin{aligned}\psi(\sigma, z, t) = \frac{1}{4r}\left(\frac{\sigma}{r}\right)^2 \Big\{ \frac{E_3}{2\pi} + \frac{3}{r}\left(\frac{z}{r}\right)\langle\varpi\sigma'^2 z'\rangle + \frac{3}{2r^2}\Big[\big(5\left(\frac{z}{r}\right)^2 - 1\big)\langle\varpi\sigma'^2 z'^2\rangle \\ + \left[\frac{5}{4}\left(\frac{\sigma}{r}\right)^2 - 1\right]\langle\varpi\sigma'^4\rangle\Big]\Big\} + O(r^{-4}) ,\end{aligned} \tag{3.3.7}$$

where

$$\langle\varpi\sigma^m z^n\rangle = \int_{-\infty}^{\infty} \int_{0}^{\infty} [\, \varpi(t, \sigma, z)\, \sigma^{m-1} z^n \,]\, \sigma \, d\sigma \, dz \, . \tag{3.3.8}$$

Here we use $< >$ to denote the area integral over the half σz plane. Because of the axisymmetry, m will always be even and hence the term inside the square brackets in (3.3.8) is equal to $z^n(x^2+y^2)^{(m-2)/2}[x\omega_2 - y\omega_1]$ and the area integral, in (3.3.8), represents a linear combination of $(m + n - 1)$st moments of vorticity in $\mathbb{R}^3$.

It was noted in **Sec.3.1** that the deviation of an integral invariant evaluated numerically at an instant t from its initial value is smaller for a larger computational domain. Therefore, we use the integral invariant (3.3.5a) to measure the error due to the finite size of the domain, or to decide when we should increase the size of the computational domain.

Since the numerical integration of the vorticity evolution equation (3.3.4) is much faster than that of the Poisson equation (3.3.3), the efficiency of the scheme can be further enhanced by using the *two-domain method*. One solves the finite difference vorticity evolution equation on a large domain $\mathcal{D}_2$ but the Poisson equation only on a smaller subdomain $\mathcal{D}_1 \subset \mathcal{D}_2$. The four moments of the vorticity needed in (3.3.7) are evaluated over $\mathcal{D}_2$. The one which should be time invariant is compared with the initial value to measure the error due to the finite size of $\mathcal{D}_2$. The stream function in the complementary domain $(\mathcal{D}_2 - \mathcal{D}_1)$ and the boundary data on $\partial\mathcal{D}_1$ are defined by the far-field representation (3.3.7).

Numerical results presented by Liu and Ting (1982) demonstrate that the *two-domain method* provides an accuracy comparable to that of the single domain code operating on the larger domain $\mathcal{D}_2$. On the other hand, it consumes only little more computational time than the single domain code operating on the smaller domain, $\mathcal{D}_1$. We conclude that the *two-domain method*, coupled with the development of appropriate far-field representations for the flow field, is an efficient numerical scheme for the analysis of unsteady unbounded viscous flow.

To end this subsection, we note that the merging of two moderately slender rings to a single one was treated as a global merging problem of type I(a) by Ting (1981). Recently, the self-merging of a vortex ring in long time to a similarity solution in the low Reynolds number limit was reported by Berezovskii and Kaplanskii (1988).

3.3.1.2 Type I(b), Vorticity concentrated near a circular center line

The merging or collision of two coaxial vortex rings with their initial core radii, $\delta_i(0)$, $i = 1, 2$, much smaller than their ring radii, $R_i(0)$, are examples of global merging problems of type I(b). The vorticity field associated with both rings can be effectively contained inside a torus whose cross-sectional radius $a = O(\delta_i)$ is much smaller than the radius $R = O(\ell)$ of its center line. In general, the torus containing its center line moves with the axial velocity of the center of the vorticity field in the σz plane. The plane, $z = Z(t)$ containing the centerline of the torus remains stationary if the two vortex rings are of the same size but of opposite circulation. Equations defining the center of the vorticity in the meridian plane will be provided later. The assumption of $a \ll R$ implies that the vorticity decays exponentially in ρ_+, where ρ_+ is the minimum distance from a reference point $Q(\sigma, z)$ in the flow field to the center line of the torus. The maximum distance is denoted by ρ_- as shown in Fig. 3.19. Also shown are the minimum and maximum distances, τ_+ and τ_-, from the point $Q(\sigma, z)$ to a circular vortex line of radius σ' in the plane $z = z'$.

The computational domain $\mathcal{D}$ in the meridian plane can be a square with side $2ma$ centered at $(R(t), Z(t))$, provided that $R > ma$. Otherwise, the lower boundary , $\sigma = R - ma$, of $\mathcal{D}$ shall be replaced by $\sigma = 0$. In order to demonstrate the choice of the computational domain and the generation of boundary data suitable for different types of initial vorticity distributions, we discuss a numerical example from Liu and Ting (1982) for the head on collision of two vortex rings of identical shape but with opposite circulations. The vorticity field is an odd function of z. We set $Z(t) = 0$ and identify $R(t)$ as the radial coordinate of the point of maximum $|\varpi|$. At $t = 0$, we have $R(0)$ equal to the initial radius R_0 of the vortex rings.

Again, we adjust the number m, which characterizes the size of $\mathcal{D}$, so as to achieve a required accuracy. Depending on the order of magnitude of R/a relative to m, i.e., $R/a = O(m)$ or $R/a >> m$, two different expansions for the Poisson integral must be developed to obtain an accurate far field description. We will explain these expansions explicitly for the one-domain method. The analogous procedures for the *two-domain method* will then become quite obvious.

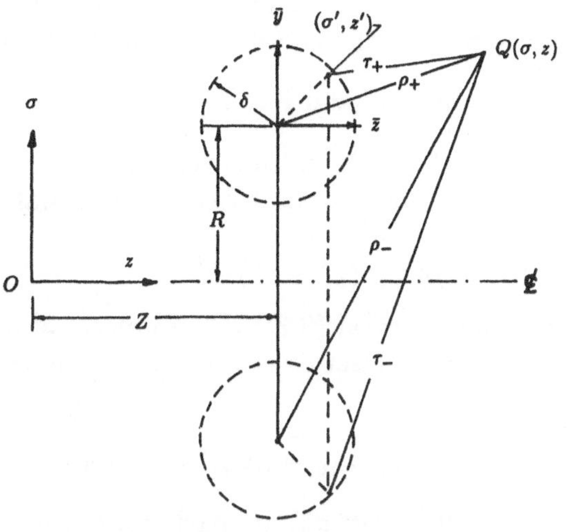

Fig. 3.19. Coordinate system for an axisymmetric vorticity distribution

Size of Computational Domain $ma = O(R)$

In this case we can choose the computational domain $\mathcal{D}$ for the finite difference solution of the Poisson equations (3.3.4) to be a rectangle,

$$|z - Z(t)| \leq ma \quad \text{and} \quad R(t) + ma \geq \sigma \geq R(t) - ma \geq 0 \tag{3.3.9}$$

with a sufficiently large m, say $m = 4$, and impose $\varpi = 0$ on $\partial\mathcal{D}$ since the decay length is $\ell_d = O(a)$. In case $R < ma$, we replace the lower boundary by the axis of symmetry, $\sigma = 0$, on which we have $\varpi = 0$ and $\psi = \psi_\sigma = 0$. The area of $\mathcal{D}$ is of the order $4R^2$ while the area of a computational domain for the preceding case of type I(a) was $O(2m^2R^2)$. Thus the number of grid points in $\mathcal{D}$ is reduced from that for type I(a) by a factor of $m^2/2$. To obtain far field boundary data for ψ as $\rho_+ >> a$, we cannot use the former expansion (3.3.8) for the Poisson integral in (3.3.6), because r/ℓ or r/R is now order one instead of being much greater than one.

Taking advantage of the axisymmetry, we first carry out the integration in θ, and obtain the stream function for a vortex ring of cross sectional area $d\sigma'\, dz'$, (see Lamb (1932) p.237). The Poisson integral (3.3.6) then becomes

$$2\pi\psi(t, \sigma, z) = \int_{-\infty}^{\infty}\int_{0}^{\infty} d\sigma'\, dz'\, (\tau_1 + \tau_2)\, G(\lambda)\, \varpi(t, \sigma', z') \tag{3.3.10a}$$

where

$$\tau_\pm = \sqrt{(z - z')^2 + (\sigma \mp \sigma')^2}\ , \quad \lambda = \frac{\tau_- - \tau_+}{\tau_- + \tau_+} = \frac{4(R + \bar{y}')(R + \bar{y})}{(\tau_- + \tau_+)^2} \tag{3.3.10b}$$

and

$$G(\lambda) = F(\lambda) - E(\lambda)\ . \tag{3.3.10c}$$

Here F and E are the complete elliptic integrals of the first and second kind respectively and we have introduced the coordinates $(\bar{y}, \bar{z})$ relative to (R, Z) by

$$\sigma = \bar{y} + R \quad \text{and} \quad z = \bar{z} + Z \ .$$

Next we extend the integrals in (3.3.10), which cover the half σz plane with $\sigma \geq 0$, to the entire $\bar{y}\bar{z}$–plane using the fact that ϖ decays exponentially in terms of the scaled radius

$$\rho_+^2/a^2 = (\bar{y}^2 + \bar{z}^2)/a^2 \ . \qquad (3.3.11)$$

Then we expand $\tau_\pm$, λ and the integral (3.3.10a) in powers of $a/\rho_\pm = O(\delta/\ell) = O(1/m)$. The result is again a power series expansion for the stream function of the form

$$2\pi\,\psi(t,\sigma,z) = \langle\varpi\rangle^* a_{00} + \langle\varpi\bar{z}\rangle^*\, a_{10} + \langle\varpi\bar{y}\rangle^* a_{01} + O(\delta^2/\ell^2)\ , \qquad (3.3.12)$$

where $\langle f\rangle^*$ stands for $\int\int_{-\infty}^{\infty} f(\bar{y},\bar{z})\,d\bar{y}\,d\bar{z}$. In particular, $\langle\varpi\rangle^*$ is equal to the strength of the axisymmetric vortical field defined by (3.3.5) and in case of the merging of two vortex rings of opposite circulations, we have $\langle\varpi\rangle^* = 0$.

The coefficients in (3.3.12) are

$$a_{00} = (\rho_+ + \rho_-)\,G(\lambda_0) \qquad (3.3.13a)$$

$$a_{10} = (\frac{\bar{z}}{\rho_+} + \frac{\bar{z}}{\rho_-})\,[-G(\lambda_0) + 2\lambda_0 G'(\lambda_0)] \qquad (3.3.13b)$$

$$a_{01} = -(\frac{\bar{y}}{\rho_+} - \frac{\bar{y}+2R}{\rho_-})\,G(\lambda_0)$$
$$+ [\frac{\rho_+ + \rho_-}{R} + 2(\frac{\bar{y}}{\rho_+} - \frac{\bar{y}+2R}{\rho_-})]\,\lambda_0\, G'(\lambda_0)\ , \qquad (3.3.13c)$$

where $\lambda_0 = 4R\sigma/(\rho_+ + \rho_-)^2$. Although only the first two terms in ψ are given in (3.3.12), the original numerical computations of Liu and Ting (1982) included two more terms, so that the error of the boundary data was $O(\delta^4/\ell^4)$.

Size of Computational Domain $ma << R$

In this case the computational domain $\mathcal{D}$ can again be a square of size $2ma$ centered at (R, Z). With the size of $\mathcal{D}$ being now much smaller than R, the leading order solution in $\mathcal{D}$ should be a two-dimensional solution since the far-field expansion parameter m^{-1} is independent of δ/R. The far-field behavior of ψ on $\partial\mathcal{D}$ is obtained from (3.3.10) with $\delta/\rho_1 = O(1/m)$ and $\Lambda = \rho_1/\rho_2 = O(ma/R)$ as small independent parameters. The first three terms, also given by Liu and Ting (1982), are

$$\frac{2\pi}{\sqrt{\sigma R}}\,\psi(t,\sigma,z) = \langle\varpi\rangle^* b_{00} + \langle\varpi\bar{z}\rangle^* b_{10}$$
$$+ \langle\varpi\bar{y}\rangle^* b_{01} + o(\frac{\delta^j}{\ell^j}\Lambda^n \log\Lambda)\ ,\ j+n=2\ , \qquad (3.3.14)$$

where

$$b_{00} = \ln \frac{4}{\Lambda} - 2 \,,$$

$$b_{10} = \frac{\bar{z}}{\rho_+^2}(1 - \frac{1}{2R}) - \frac{\bar{z}}{\rho_-^2} \,,$$

$$b_{01} = \frac{\bar{y}}{\rho_+^2}(1 - \frac{1}{2R}) - \frac{\bar{y} + 2R}{\rho_-^2} + \frac{b_{00}}{2R} \,.$$

Note that the leading term containing $\ln(4/\Lambda)$ represents the local two-dimensional result. Again, in the numerical solutions of Liu and Ting (1982), the expansion (3.3.14) was carried out up to terms of $O(\Lambda^n \log \Lambda\, \delta^j/\ell^j)$ with $j + n = 4$. It should be mentioned that the integral invariant corresponding to (1.2.23) becomes

$$\langle (R + \bar{y})^2 \varpi \rangle = R^2 \langle \varpi \rangle^* + 2R \langle \varpi \bar{y} \rangle^* + \langle \varpi \bar{y}^2 \rangle^* = \text{constant}. \tag{3.3.15}$$

Interaction of coaxial vortex rings

The present scheme was applied to study the interaction of a pair of coaxial vortex rings. A particular example is the interaction of identical vortex rings of opposite strength in a "head-on collision" as sketched in Fig. 3.20. The initial vorticity distribution at $t_0 = 1$ is given by a superposition of two nonoverlapping vortex rings with the core structure of a Lamb vortex,

$$\begin{aligned} \varpi(r, z) =& \frac{\Gamma_1}{\pi} \exp\{ -[(z + Z_0)^2 + (r - R_0)^2]\} \\ &+ \frac{\Gamma_2}{\pi} \exp\{ -[(z - Z_0)^2 + (r - R_0)^2]\} \end{aligned} \tag{3.3.16}$$

The length and the time scales are chosen so that the initial core size $\delta_0 = \sqrt{4\nu} = 1$. We set $\Gamma_1 = -\Gamma_2 = 16\pi$, $R_0 = 20$ and $Z_0 = 2$. The two vortex rings are of opposite strength and their centers are located in the σz plane at $(20, \pm 2)$. The initial distance between the center lines of the rings is equal to 4, i. e., $4\delta_0$, so that they are considered to be nonoverlapping. As t increases, $R(t)$ increases. Consequently the condition of $ma = O(m\delta) << R$ is realized for $t \geq 1$. Recall that $(R, Z_\pm)$ denote the locations of the maximum $|\varpi|$ associated with the two rings.

The computed trajectory of the center of the vortex ring in the left half plane $z < 0$, i. e., $R(t)$ vs $-Z(t)$, is shown in Fig. 3.21 along with the predictions of inviscid theory and extended asymptotic analysis. The variations of the maximum vorticity, $|\varpi|_{\max}$, predicted by these three solutions are shown in Fig. 3.22 while the corresponding circulations or integrals of vorticity over the left half plane, $\Gamma^-(t)$, are plotted versus the axial displacement, $-Z(t)$, in Fig. 3.23. The ratio of the circulation to its initial value , $\Gamma^-(t)/\Gamma_2$, serves as a measure of the extent of vorticity cancellation as the two vortex rings overlap each other.

For an inviscid slender vortex ring, we represent the core structure by a Lamb vortex with the core radius defined by the formula $\delta_{\text{inv}}(t) = \delta_0\sqrt{R_0/R(t)}$ to account for the stretching of the ring. The inviscid theory for the head on collision

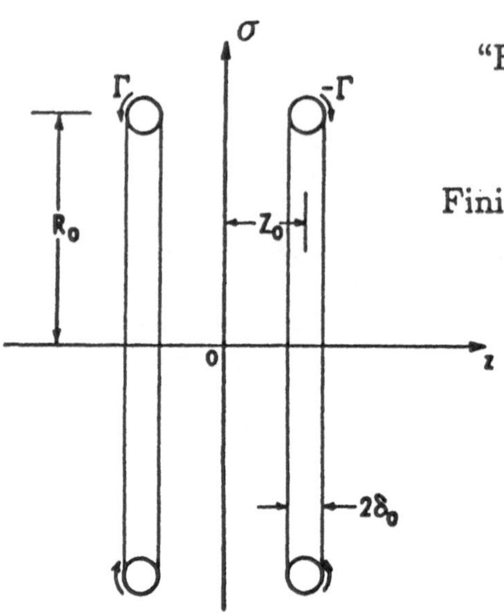

Fig. 3.20. Initial geometry for two coaxial vortex rings

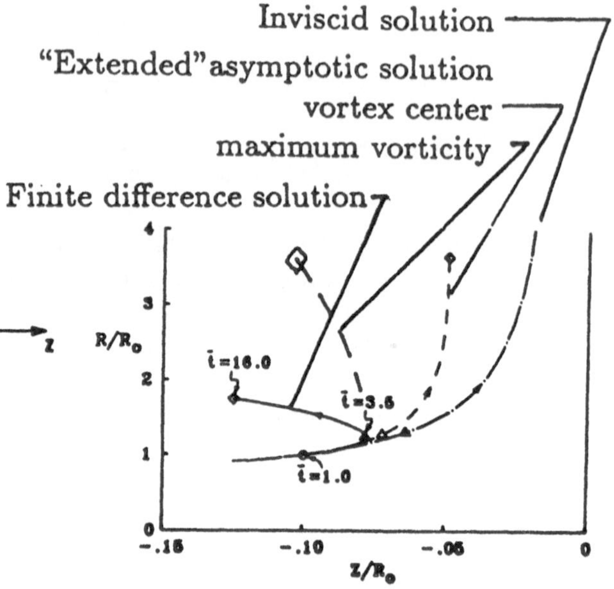

Fig. 3.21. Path of the Colliding vortex ring (Liu and Ting 1982)

of two vortex rings with small vortical cores predicts that : the circulation is conserved , $\Gamma^-(t)/\Gamma_2 = 1$, the core radius $R(t)$ increases with t, the distance between the rings, $2Z(t)$, decreases to zero and the maximum $|\varpi(t)|$ increases as the core size decreases due to stretching. The prediction on $|\varpi(t)|_{\max}$ is incorrect for $t > 1$. The first three predictions become erroneous for a larger t, certainly for $t > 3.5$, when the cancellation of vorticity in the overlapping region becomes substantial. This we see from Fig. 3.23. At $t = 3.5$, the circulation given by the numerical solution is already reduced to 75% of its initial value.

For a slender vortex ring given by the extended asymptotic solution, the core structure is that of a Lamb vortex, with its core radius $\delta_{\text{asy}}(t) = \delta_0\sqrt{tR_0/[t_0R(t)]}$ to account for diffusion and stretching. The effect of diffusion dominates that of stretching so long as $R(t)/t$ decreases as t increases. This is the case for our example. As the core size decreases the maximum vorticity, $|\varpi|_{\max}$ increases as $1/\delta_{asy}$. As shown in Fig. 3.22, the maximum begins to deviate from the numerical solution when $t/t_0 = 2$. The relative deviations are 12.6, 74, 168 and 233 percent at $\bar{t} = t/t_0 = 4, 8, 12$, and 16, respectively. The deviations represent the cumulative effects of the nonlinear convection terms which are not accounted for in the extended asymptotic solution.

In the early stage, $t \sim 1$, the rings are sufficiently far apart relative to their core size, hence the trajectories given by the three solutions in Fig. 3.21 are in good agreement. As t increases, the rings are approaching each other while the ring radius is increasing , i. e., $\dot{Z}(t) < 0$ and $\dot{R}(t) > 0$. The trajectory given by the finite difference solution deviates from the other two at about $t = 3.5$, when the

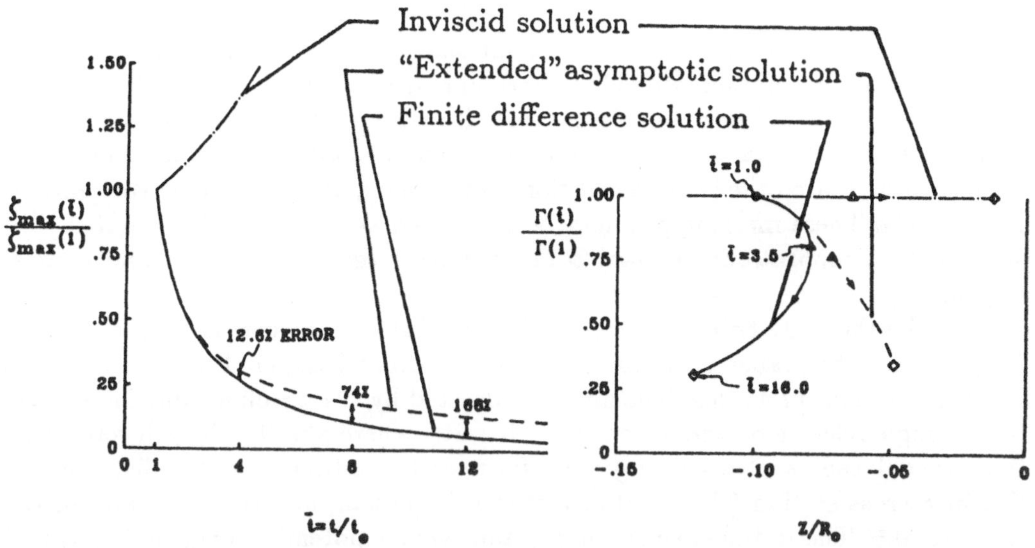

Fig. 3.22. Decay of the Maximum Vorticity (Liu and Ting 1982)

Fig. 3.23. Decay of the circulation and the half σz plane (Liu and Ting 1982)

distance between the two rings $2Z(t)$ reaches its minimum and begins to increase. Recall that $\pm Z(t)$ denote the axial axial coordinates of the two maxima of $|\varpi|$. The reversal of the axial velocity $\dot{Z}(t)$ is caused by the cancellation of vorticity in the overlapping vortical cores and by the interaction of nonlinear convection terms. At $t = 3.5$, the core radius predicted by the extended asymptotic solution is already about 88% of the semi-distance between the core centers. Because of the overlap of the vortical cores, the trajectory of the point of maximum vorticity, $(R(t), -Z_{asy}(t))$, differs from that of the center of the ring, $(R(t), Z_-(t))$, in their z-coordinates and their difference increases with t. These two trajectories are shown in Fig.3.21. Note that the reversal of the trajectory of the point of maximum vorticity brings the trajectory to better agreement with that of the numerical solution at least qualitatively for $t > 3.5$. From Fig. 3.23, we see that the deviation of the circulation given by the extended asymptotic solution from that of the numerical solution is within 10% even when $t = 16$. This is not surprising because the rate of the decrease of circulation is measured by the line integral of vorticity gradient along the positive σ-axis, $(z = 0, \sigma \geq 0)$. There the effect of the nonlinear convection terms on the vorticity gradient is weak initially $(t \sim 1)$, reaches a maximum and then decreases as the points of maximum vorticity begin to move away from each other while the vorticity in the overlapping region vanishes due to cancellation. The deficiency of the extended asymptotic solution for large t lies in the inaccurate prediction of the radial velocity, $\dot{R}(t)$. An improvement for large t that accounts for the nonlinear interaction terms and corrects the radial velocity is needed. We next discuss problems without axisymmetry.

3.3.2 Numerical Modeling of Merging of Vortex Filaments

The numerical solutions of the N-S equations modeling the merging of vortex filaments were first obtained by Chamberlain and Liu (1985) and by Chamberlain and Weston (1984). They considered the self-merging of an elliptical vortex ring with a core size comparable to the minimum radius of curvature of its center line. They also simulated the oblique collision and merging of two and four vortex rings respectively. These merging problems were treated as problems of Type I(a) with sizes of the computational domain $\mathcal{D}$ being much larger than those of the vortical regions.

Initially there are two vortex rings with equal strengths, core radii and toroidal (ring) radii. The planes containing their center lines (ring circles) intersect each other at an angle of 45 deg. The angle is bisected by the zx plane and the centers of the ring circles lie on the y axis at ± 1.5 units, where a unit of length represents the toroidal radius of each ring. The form of the initial vorticity distribution, $|\Upsilon|$, in a cross section (the core) of a ring is Gaussian, and the core radii of the rings are 0.5. The initial surfaces of constant vector potential magnitude $|\mathbf{A}|$ and constant vorticity magnitude, $|\Omega|$, are shown in Fig. 3.24 (a) and (b) respectively. The constant $|\Omega|$ surfaces show the distinct features of two vortex rings with little interaction, while the constant $|\mathbf{A}|$ surfaces clearly indicate the effect of interaction. This demonstrates the fact that the vorticity decays exponentially while its induced velocity and vector potential fields decay much slower in inverse powers of the distance.

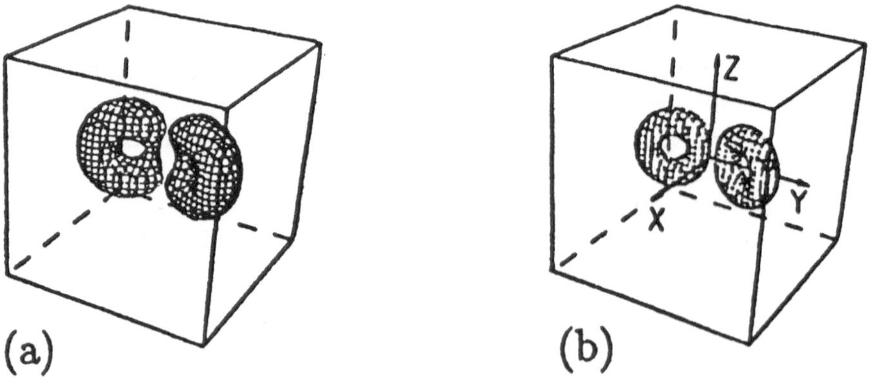

Fig. 3.24. Initial isosurfaces of vector potential $|\mathbf{A}|$ in (a) and vorticity $|\Omega|$ in (b) (Chamberlain and Liu 1985)

Perspective views of surfaces of constant magnitude of vector potential, $|\mathbf{A}|$, at several stages are shown in Fig. 3.25 (a) - (d). (The corresponding top, side and front views are presented in Chamberlain and Liu (1985).) The computational domain is a cube centered about the origin with sides 8 by 8 by 8, while the initial vorticity distribution is contained effectively in a box of sides $4 \times 3 \times 2$. Figure 3.25

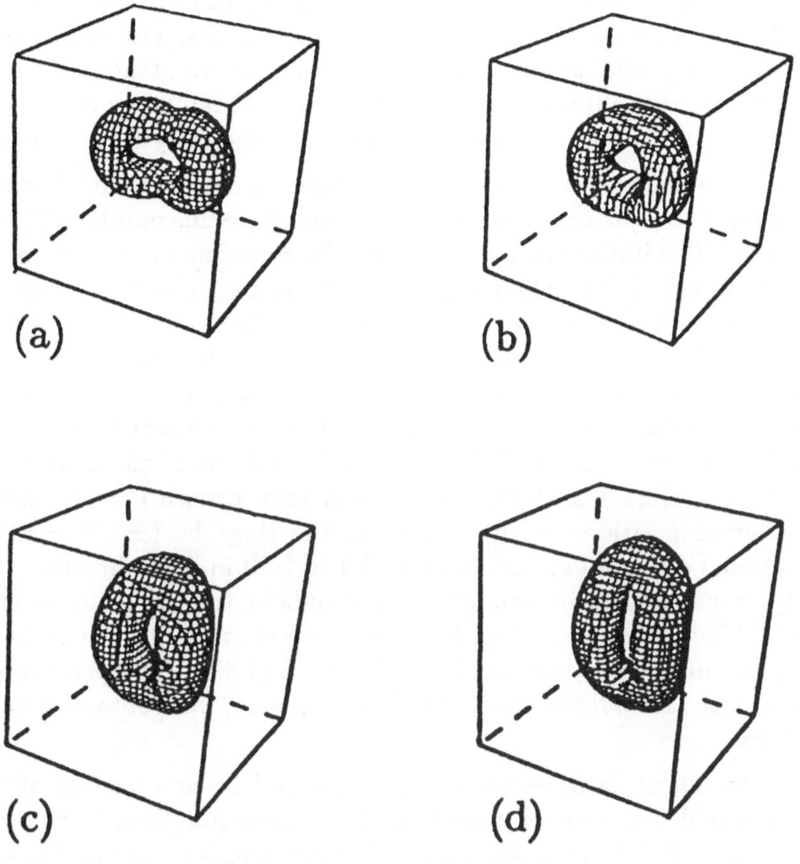

Fig. 3.25. Isosurface of vector potential magnitude at $t|\Omega_0|_{max} = 10.4$, (a); 20.2, (b); 44.9 (c); and 60.1 (d). (Chamberlain and Liu 1985)

shows the subsequent merging of the two into a single distorted oblong ring, and the exchange of the roles of the major and minor axes of the ring. This behavior is in good qualitative agreement with experimental visualizations by Fohl and Turner (1975) and Oshima and Asaka (1975, 1977).

To achieve an accurate resolution of the merging of constant $|\Omega|$ surfaces, the above global merging problems were reconsidered by Ishii et al (1989) as problems of type I(b) (instead of as type I(a)) because the decay length δ is much smaller than a typical size of the vorticity distribution, ℓ, by a factor of at least 4. The authors obtained the evolution of the isosurfaces of constant vorticity magnitude for the self merging of a 4:1 elliptical vortex ring and for the global merging of two vortex rings. The calculation used a cubic grid of 131^3 points, a second order central difference scheme in space and the explicit Euler method for the time evolution with $\Delta t = 0.0125$.

A total of 26^3 subdomains, with 5^3 grid points per subdomain, was employed to evaluate the values of **A** at the boundaries using a series representation of the Poisson integral. However, by taking advantage of the fact that the length and time scales of **A** on the boundary are much larger than those within the vortical region these computations were simplified considerably. The vector potential was computed at only 6700 points on the boundary (instead of all 67604 points), while the values of **A** at the intermediate positions were obtained by cubic interpolation. Furthermore, the boundary data were updated only every 5 time steps.

Here we quote the descriptions of the numerical results by Ishii et al. (1989) for the merging of two vortex rings for two different Reynolds numbers $\Gamma_0/\nu = 628$ and $\Gamma_0/\nu = 62.8$. The initial vorticity distribution represents a pair of circular vortex rings of identical shape. Their center lines (the ring circles) are coplanar, lying in the xy plane. The rings have the same center line radius $R_0 = 4.0$, circulation $\Gamma_0 = 20\pi$ and a Gaussian vorticity distribution with core radius $\delta_0 = 1.0$. The direction of the circulation is chosen so that the velocities of the ring circles are in the direction of the z-axis. In case 1, the kinematic viscosity is $\nu = 0.1$, leading to a Reynolds number of $Re = 628$. The initial positions of the centers of the rings are $(X, Y, Z) = (\pm 4.0, \pm 4.0, 0.0)$. In case 2, they choose $\nu = 1$, and $Re = 62.8$ and the initial positions of the centers of the rings be $(\pm 4.25, \pm 4.25, 0.0)$. The computational domain extends from -20.0 to 20.0 in each direction. Case 2 is of interest because the initial configuration simulates the experiments of Oshima et al. (1988). The three views, top, front and side views, of the surfaces of constant vorticity magnitude are presented by Ishii et al (1989). The perspective views of surfaces of constant $|\Omega|$ for case 2 shown in Fig. 3.26, were obtained from Dr. Ishii, ICFD Tokyo.

From the initial data, we see that the $y-$ and $z-$components of vorticity, ω_2 and ω_3, are odd functions of x while ω_1 is an even function. In the yz plane, we have $\omega_2 = \omega_3 = 0$, i. e., $\Omega = \omega_1 \hat{\imath}$ on $x = 0$. Therefore, a vortex line or an isosurface crossing over the yz plane has to be orthogonal to the plane. The vorticity vector in the segments of these two rings closest to each other are dominated by the components in the yz plane and are of opposite sense. The decrease of vorticity magnitude due to the mutual vorticity cancellation of the vorticity fields of these two rings are most enhanced in the region close or common to these two segments. As time increases, the rings turn towards and approach each other while their core size increases. The mean forward motion of the rings is accounted for by the translation of the computational domain. The two initially disjoint isosurfaces of vorticity magnitude (at 30% of the maximum $|\Omega|$) begin to join across the yz plane where $|\Omega| = |\omega_1|$. We see the formation of a third hole by the isosurface in addition to the original two. Part of the vortex lines in the isosurface coming from the far side of one ring, say from $x > X$ crosses over the yz plane at right angle and is continued or connected to the corresponding part on the $x < 0$ side. The remaining part of vortex lines in the isosurface remain on the same side $x > 0$, is continued to the segment separating the third hole and the hole on the $x > 0$ side, and sustains the shape of a ring on the $x > 0$ side. The vorticity magnitude in the third hole is lower than the value of the isosurface because of the cancellation of the vorticity belonging to the overlapping segments of the two rings. As time increases the percentage of the vortex lines in the isosurface crossing over the yz plane increases. Eventually all the vortex lines in the isosurface cross over the plane

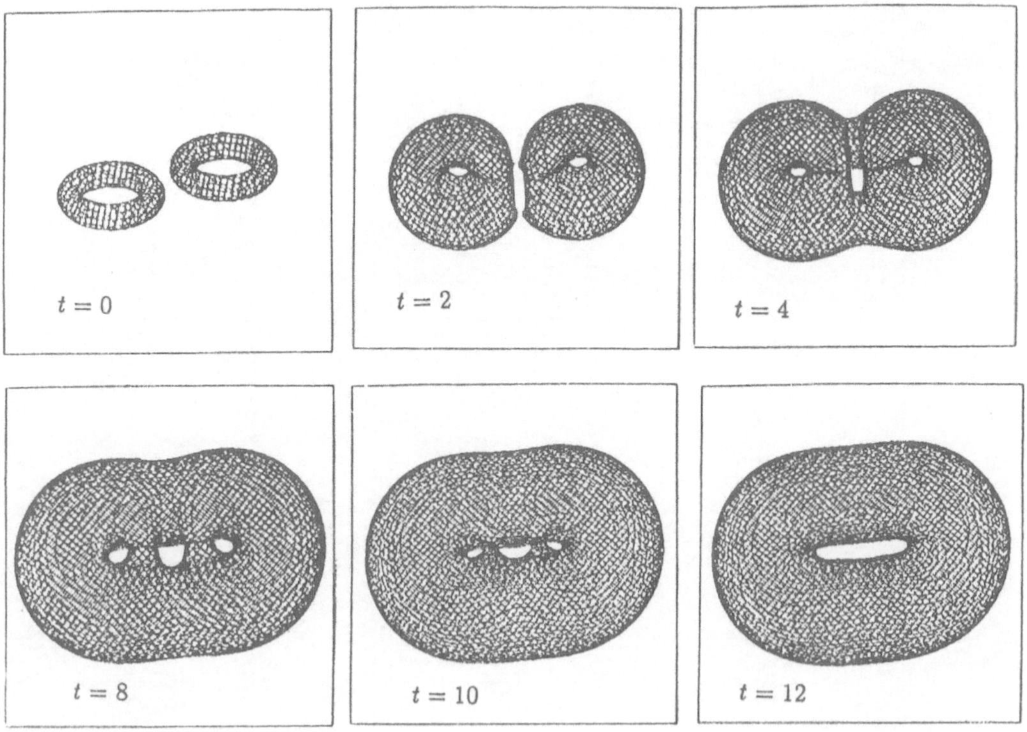

Fig. 3.26. Merging of two vortex rings, perspective views of surfaces of constant $|\Omega|$. (Courtesy of Dr. Ishii, ICFD Tokyo, 1989)

from one ring to the other and the three holes formed by the isosurface merge to a single one. Thus, the isosurface with $|\Omega| = 30\% \ |\Omega|_{\max}$ gives the impression of the merging of two rings to a single ring or filament. This is the same type of behavior as those seen in the experimental vortex ring pairs by Oshima et al. (1988). The corresponding experimental results are reproduced in Fig. 3.27 using the originals kindly provided to us by Dr. Oshima.

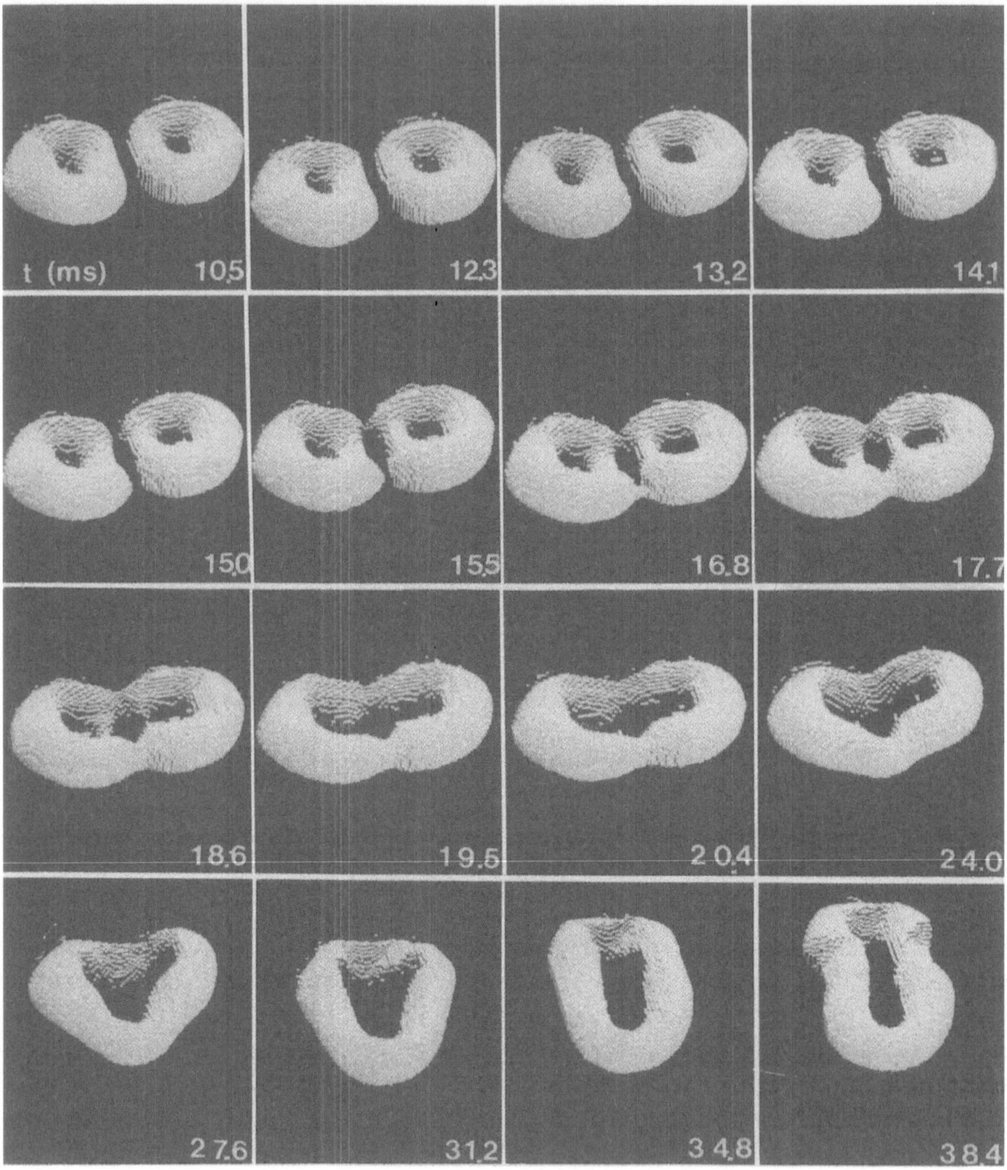

Fig. 3.27. Merging of two vortex rings observed by Oshima et al (1988)

4. Closing Remarks

In the preceding chapters we studied viscous vortical flows in free space under the following assumptions:

(1) The vorticity field is far away from boundaries or body surfaces, i. e., for any region with vorticity of the order U/ℓ or larger, its distance to a boundary is much larger than its typical length scale, ℓ.
(2) The flow field is incompressible without body force.

In chapter 2 we presented matched asymptotic solutions of slender vortex filaments under the additional assumptions:

(3) There is a small reference length scale δ for the vortical core size with $\delta/\ell = \epsilon \ll 1$.
(4) The strength of a filament is $O(1)$, i. e., the vorticity in its core is $O(\epsilon^{-2})$.
(5) The slender filaments are submerged in a background potential flow.
(6) All other length scales of the filaments are $O(\ell)$, i. e., the minimum radius of curvature of the centerline of a filament, its length L and the distance between two adjacent filaments are $O(\ell)$.

In this chapter we shall mention several problem areas which violate at least one of the assumptions above. For each area, we point out the appropriate modifications of the asymptotic and/or numerical methods, whenever possible.

Assumption (3) defines the slenderness of the filaments, identifies the small expansion parameter ϵ and hence is essential for the matched asymptotic analysis. Assumptions (4) to (6) can be relaxed or modified to describe flows with different physical characteristics and as a result we get different asymptotic solutions. When assumption (6) is modified, for example, by allowing the length of the filament(s) to be much larger than the typical radius of curvature which remains $O(\ell)$, we arrive at the asymptotic solutions for long filaments in which the axial dynamics in the core becomes important. Another interesting problem area emerges when a filament is subjected to small amplitude short wavelength perturbations. The resulting asymptotic solution accounts for the creation of small axial scales by vortex self-interaction. These two types of problems will be studied in **Sec. 4.1.** When assumption (5) is relaxed we have the asymptotic solution for a filament in a background rotational flow of order one vorticity. In this case the leading order solution of the background flow is coupled with the motion of the filament. If assumption (2) is replaced by the tangent plane approximation in dynamical meteorology, we arrive at the theory for the motion of a geostrophic vortex for which the core structure is a decaying Bessel vortex instead of a Lamb vortex. These types of problems are described in **Sec. 4.2.**

When we deal with the interaction of vortex filament(s) with a rigid surface, assumption (1) is dropped. Besides the geometry of the surface we have two additional length scales, the boundary layer thickness and the distance d_S between a filament and the surface. If the surface is moving, then the velocity of the surface scaled by the velocity U of the flow field becomes another parameter. Depending on the configurations of the filament relative to the surface we can have a wide class of problems ranging from weak to strong interactions. For a weak interaction problem, the distance d_S is much larger than the core size and the boundary layer thickness. The strong interaction problems include the merging of segment(s) of a filament with the boundary layer along the surface and the cutting of a filament by the moving surface. For strong interaction problems, the numerical schemes formulated in **Chapter 3** under assumption (1) are not applicable. New numerical schemes suitable for specific types of strong interactions are needed. The complexity of these types of problems will be described in **Sec.4.3.**

4.1 Dynamics of Vortex Filaments with Multiple Axial Length Scales

In this section we report on recent research efforts that use and generalize some of the methods and results described in Chapters 1 – 3 above. **Section 4.1.1** resumes a study by Klein and Majda (1990) that describes the dynamics of small amplitude–short wavelength distortions of a vortex filament. In a particular regime for the perturbation scalings, their theory reveals a direct competition between nonlocal, but linear, effects due to self-induction and nonlinear effects induced by the logarithmically strong binormal propagation. The axial characteristic length of the perturbations is small compared to the radii of curvature of the center line but it is still large compared to the core size. **Section 4.1.2** suggests a theory for long filaments that may provide new insight in problems of vortex breakdown. The approach combines the filament theory of **Sec.2.3** with an asymptotic description of slowly varying axial waves on the vortex core.

4.1.1 Effects with Self-Stretching for Perturbed Vortex Filaments

In numerical simulations, e.g., by Chorin (1982) or Knio and Ghoniem (1990), but also in experimental studies by Maxworthy (1977), one observes the formation and (rapid) growth of hairpin-like distortions on slender vortex rings. The typical length scale of these structures in the axial direction is much shorter than the ring radius but it is large compared to the vortex core size. Once a hairpin starts to grow, the elongation of the filament induces strong vortex stretching. Thus, the evolution of short wavelength perturbations on a broadly curved background filament and the phenomenon of filament self-stretching seem to be closely associated in this process. A similar connection may exist in turbulent flows where vortex stretching and folding are believed to dominate the generation of small flow scales (Chorin (1988),(1989)).

To obtain an understanding of the underlying mechanisms, Klein and Majda analyze a single vortex filament embedded in an outer potential flow that is subject to small amplitude short wavelength perturbations. The work by Callegari and Ting (1978) and an earlier study by Hasimoto (1972) provide the basic tools for the analysis. However, several generalizations and new concepts had to be developed to unravel a direct competition between weakly nonlinear and linear nonlocal effects on the evolution of the disturbances. The nonlocal effects are, in particular, responsible for the self-stretching.

Hasimoto (1972) assumes the "local induction approximation" for vortex filament dynamics, which says that the filament center line, $\mathcal{C}$, moves according to the binormal law

$$\frac{\partial \mathbf{X}}{\partial \bar{t}}(\bar{t}, \tilde{s}) = \kappa\, \hat{b}(\bar{t}, \tilde{s}) \,, \tag{4.1.1}$$

where $\tilde{s}$ is an arclength coordinate. This equation is a consistent leading order approximation for the results in **Sec.2.3**, (2.3.73), if one considers $\ln^{-1}(1/\delta) \ll 1$ as an expansion parameter and introduces the scaled time variable $\bar{t} = \frac{4\pi}{\Gamma \ln(1/\delta)}\, t$. Hasimoto invents a complex "Filament Function", defined by

$$\psi = \kappa \exp(i\Phi) \tag{4.1.2a}$$

where $\kappa(\bar{t}, \tilde{s})$ is the local center line curvature and

$$\Phi(\bar{t}, \tilde{s}) = \int_{\tilde{s}_0}^{\tilde{s}} T(\bar{t}, \tilde{s}')\, d\tilde{s}' + A(\bar{t}) \,, \tag{4.1.2b}$$

is the integral of the torsion along the filament. The function $A(\bar{t})$ is arbitrary and may be chosen for convenience. Its choice does not affect the result for the filament geometry.

By a very elegant transformation, Hasimoto shows that the binormal law (4.1.1) is equivalent to the cubic nonlinear Schrödinger equation

$$\frac{1}{i}\,\psi_{\bar{t}} = \psi_{\tilde{s}\tilde{s}} + \frac{1}{2}|\psi|^2\psi \tag{4.1.3}$$

for the Filament Function. To know about this equivalence is a considerable achievement, since extensive theoretical understanding has been developed in recent years of nonlinear wave equations of the type of (4.1.3) (see e.g., Whitham (1974), Flaschka and Newell (1975), Newell (1975)).

Importantly, however, the effect of vortex stretching is completely absent from (4.1.1), (4.1.2), because only velocity components, $\mathbf{X}_{\bar{t}} \cdot \hat{n}$, in the principal normal direction of the curve lead to changes of the arclength of $\mathcal{C}$. On the other hand, the finite part of the Biot-Savart integral, $\mathbf{Q}^f$, in (2.3.73) can contribute a component $\mathbf{X}_{\bar{t}} \cdot \hat{n} \neq 0$, of the self induced velocity. This yields a potential for vortex stretching even if the outer flow, $\mathbf{Q}_2$, is at rest. On the "Hasimoto-time scale" represented by $\bar{t}$, this contribution of $\mathbf{Q}^f$ appears as an order $O(\ln^{-1}(1/\delta))$ correction to the binormal law (4.1.1). Letting

$$\tilde{\delta} \sim \frac{\Gamma}{4\pi\, \ln(1/\delta)}, \tag{4.1.4}$$

one may write Callegari and Ting's result as

$$\frac{\partial \mathbf{X}}{\partial \bar{t}} = \kappa \hat{b} + \tilde{\delta} \mathbf{v}^*, \tag{4.1.5}$$

where

$$\mathbf{v}^* = \kappa C^* \hat{b} + \mathcal{P}(\mathbf{Q}^f + \mathbf{Q}_2). \tag{4.1.6}$$

The operator

$$\mathcal{P}(\hat{t}) = (I - \hat{t} \circ \hat{t}) \tag{4.1.7}$$

represents the projection into the cross-sectional plane at $\mathbf{X}(\bar{t}, \bar{s})$.

Klein and Majda reconsider the Hasimoto transform and derive from (4.1.5) the *exact* generalized equation

$$\frac{1}{i}\psi_{\bar{t}} = \psi_{\bar{s}\bar{s}} + \frac{1}{2}|\psi|^2\psi - \tilde{\delta}\left\{ i\,\left[(\mathbf{N} \cdot \mathbf{v}^*_{\bar{s}})_{\bar{s}} - \psi \mathbf{v}^*_{\bar{s}} \cdot \hat{t} \right] + \psi \int_{\bar{s}_0}^{\bar{s}} \mathrm{Im}(\psi \bar{\mathbf{N}}) \cdot \mathbf{v}^*_{s'}\, ds' \right\}. \tag{4.1.8}$$

for the Filament Function. Here $\mathbf{N}$ is a complex vector defined by

$$\mathbf{N} = (\hat{n} + i\,\hat{b})\,\exp(i\,\Phi)\,, \tag{4.1.9}$$

with Φ from (4.1.2b) and $\hat{n}, \hat{b}$ being the principal normal and binormal unit vectors. This result generalizes Hasimoto's equation to include the effect leading to vortex self-stretching, but it is not yet satisfactory for the following reasons: First, the equation contains the vector $\mathbf{N}$, which could, in principle, be computed from ψ, but only through an inversion of the Serret-Frenet formulae (2.1.45). Secondly, even if the outer flow is assumed to be at rest, $\mathbf{v}^*$ still contains $\mathbf{Q}^f$, which is given by a complicated nonlinear and nonlocal functional acting on the curve $\mathcal{C}$.

Considerable simplifications arise for small amplitude, short wavelength distortions of a straight background filament. Klein and Majda introduce a center line description of the form

$$\mathbf{X}(\bar{t}, s) = s\,\hat{t}_0 + \tilde{\varepsilon}^2 \mathbf{X}^{(2)}(s/\tilde{\varepsilon}, \bar{t}/\tilde{\varepsilon}^2) + o(\tilde{\varepsilon}^2) \qquad \text{with} \qquad (\tilde{\varepsilon} \ll 1)\,. \tag{4.1.10a}$$

The small parameter, $\tilde{\varepsilon}$, is yet to be defined and $\hat{t}_0$ is a constant unit vector. The perturbation vector

$$\mathbf{X}^{(2)}(\sigma, \tau) = x^{(2)}(\sigma, \tau)\,\hat{n}_0 + y^{(2)}(\sigma, \tau)\,\hat{b}_0 \tag{4.1.10b}$$

is chosen to be normal to $\hat{t}_0$ with $\hat{t}_0, \hat{n}_0, \hat{b}_0$ spanning an orthogonal basis. This ansatz ensures that the curvature due to the perturbation is of order $O(1)$, since

$$\kappa \hat{n} = \mathbf{X}^{(2)}_{\sigma\sigma} + O(\tilde{\varepsilon}) \tag{4.1.11}$$

with

$$\sigma = \frac{s}{\tilde{\varepsilon}} \tag{4.1.12}$$

as the reader may confirm. Thus, the filament moves at an order $O(1)$ velocity on the Hasimoto time scale according to (4.1.5). This, in turn, implies that the relevant evolution time scale for $\mathbf{X}^{(2)}$ must be small of order $\tilde{\varepsilon}^2$, justifying the introduction of the scaled time

$$\tau = \frac{\bar{t}}{\tilde{\varepsilon}^2} \tag{4.1.13}$$

as an independent variable in (4.1.10a). Under (4.2.10a) for the centerline, the arclength coordinate $\tilde{\sigma} = \tilde{s}/\tilde{\varepsilon}$ obeys $\tilde{\sigma} = \sigma(1 + O(\tilde{\varepsilon}^2))$. Using this fact and introducing $\tilde{\sigma}, \tau$ in (4.1.8) while keeping only first order terms in $\tilde{\varepsilon}$ and $\tilde{\delta}$ reduces the equation to

$$\frac{1}{i}\psi_\tau = \psi_{\tilde{\sigma}\tilde{\sigma}} + \tilde{\varepsilon}^2 \left[\frac{1}{2}|\psi|^2\psi - \alpha\, i\,(\mathbf{N} \cdot \mathbf{v}^*_{\tilde{\sigma}})_{\tilde{\sigma}} + O(\tilde{\varepsilon}\alpha)\right] \tag{4.1.14}$$

where $\alpha = \tilde{\delta}/\tilde{\varepsilon}^2$. The distinguished limit

$$\alpha = \frac{\tilde{\delta}}{\tilde{\varepsilon}^2} = O(1) \qquad \text{as} \qquad (\tilde{\varepsilon} \to 0) \tag{4.1.15}$$

yields a regime where the nonlinearity of Hasimoto's filament equation, can directly compete with the nonlocal effects associated with $\mathbf{v}^*$ from (4.1.6). Through (4.1.4), this limit essentially couples $\tilde{\varepsilon}$ and the core size parameter δ except for logarithmic corrections. Observing that the core structure coefficient C^* in (4.1.6) will be constant on the τ-time scale one replaces $\ln(1/\delta)$ in (4.1.4) by $\ln(1/\delta)+C^*$, thereby removing the first term in the equation (4.1.6) for $\mathbf{v}^*$. Furthermore, the detailed analysis of the Biot-Savart integral yields an additional logarithmic correction to the binormal velocity of order $O(\ln(1/\tilde{\varepsilon}))$. The final result for $\tilde{\varepsilon}$ is

$$\alpha\tilde{\varepsilon}^2 = \frac{4\pi}{\Gamma(\ln(2\tilde{\varepsilon}/\delta) + C^*).} \tag{4.1.16}$$

What is left to discuss is the particular form of the self-induced velocity $\mathbf{Q}^f$ in (4.2.6) under the special perturbation scalings in (4.1.10). Klein and Majda show that the finite part of the integral linearizes to

$$\mathbf{Q}^f_{\tilde{\varepsilon}} = \mathcal{I}[\mathbf{X}^{(2)}] \times \hat{t}_0 + \cdots, \tag{4.1.17}$$

where the linear integral operator, $\mathcal{I}[\cdot]$, is defined by

$$\mathcal{I}[w](\tilde{\sigma}) = \int_{-\infty}^{\infty} \frac{1}{|h^3|}\left[w(\tilde{\sigma}+h) - w(\tilde{\sigma}) - hw_{\tilde{\sigma}}(\tilde{\sigma}+h) + \frac{h^2}{2}\mathcal{H}(1-|h|)w_{\tilde{\sigma}\tilde{\sigma}}(\tilde{\sigma})\right] dh. \tag{4.1.18}$$

To obtain the corresponding expression that will appear in the filament equation one has to compute $-i(\mathbf{N} \cdot \mathbf{Q}^f_{\tilde{\varepsilon},\tilde{\sigma}})_{\tilde{\sigma}}$. For a curve with scalings in (4.1.10) it turns out that

$$\mathbf{N} = \mathbf{N}_0 + O(\tilde{\varepsilon}) \qquad \text{with} \qquad \mathbf{N}_0 = (\hat{n}_0 + i\hat{b}_0) = \text{const}, \tag{4.1.19}$$

such that the desired expression becomes $-i\,\mathbf{N}_0 \cdot \mathbf{Q}^f_{\tilde{\varepsilon},\tilde{\sigma}\tilde{\sigma}} + \cdots$. Furthermore, it is shown that

$$\mathbf{N} \cdot \frac{\partial^j \mathbf{X}^{(2)}}{\partial \sigma^j} = \frac{\partial^{j-2} \psi}{\partial \bar{\sigma}^{j-2}} + O(\bar{\varepsilon}) \qquad \text{for} \qquad (j = 2, 3, 4). \tag{4.1.20}$$

Using (1.4.17)–(1.4.20), one finds the surprisingly simple result that for $\mathbf{Q}_2 \equiv 0$,

$$-i \left(\mathbf{N} \cdot \mathbf{v}^*_{\bar{\sigma}}\right)_{\bar{\sigma}} = -\mathcal{I}[\psi]. \tag{4.1.21}$$

The consequence for the Filament equation is

$$\frac{1}{i} \psi_\tau = \psi_{\bar{\sigma}\bar{\sigma}} + \bar{\varepsilon}^2 \left(\frac{1}{2} \psi^2 \psi - \alpha \mathcal{I}[\psi] \right), \tag{4.1.22}$$

which one might call an "*Asymptotic Filament Equation with Self Stretch*".

The local rate of self-stretch of the curve is given by its motion in principal normal direction as

$$\frac{1}{\ell} \dot{\ell} = -\kappa \hat{n} \cdot \mathbf{X}_{\bar{t}} \,, \tag{4.1.23}$$

where $\ell = |\mathbf{X}_{s_0}|$ is the local accumulated stretch of the center line with respect to the arclength s_0 on the initial curve. In the asymptotic regime considered, the stretching rate can be computed as a quadratic functional of the filament function. It is given by

$$\dot{\ell} = \bar{\varepsilon}^2 \dot{\ell}^{(2)} = \bar{\varepsilon}^2 \frac{i}{4} \int_{-\infty}^{\infty} \frac{1}{|h|} \left[\bar{\psi}(\sigma + h) \psi(\sigma) - \psi(\sigma + h) \bar{\psi}(\sigma) \right] dh \,. \tag{4.1.24}$$

The explicit calculation shows that it is only the $\mathcal{I}$-operator term in (4.1.22) that can produce a nonzero value for $\dot{\ell}^{(2)}$.

The result (4.1.24) resolves an apparent contradiction regarding the stretching effect. From the vorticity transport equation in three dimensions it is known that vortex stretching is a quadratically nonlinear effect. However, in the filament equation the stretching term appears as a linear operator, which seems to be paradoxical. Equation (4.1.24) provides the answer by showing that a *quadratic* functional in ψ describes the curve elongation.

The filament equation is interesting from both a physical and a mathematical point of view.

Regarding the physical implications we notice that, formally, the binormal term dominates in (4.1.5) and that it is this term that leads to the cubic nonlinearity. However, its nonlinear influence becomes weak for a curve (4.1.10) to a level where it has to compete with the nonlocal self-induction represented by the $\mathcal{I}$-operator.

As regards the mathematical aspects, the cubic nonlinear Schrödinger equation ($\alpha = 0$) is an integrable equation in the sense of inverse scattering theory (Flaschka and Newell (1975)). The question arises as to whether, or to what extend, the inverse scattering approach remains applicable to the filament equation. Notice that a straight-forward perturbation theory, based on known solutions to the cubic Schrödinger equation is not appropriate, since the cubic and the nonlocal terms appear at the same perturbation order.

Numerical solutions to (4.1.24) as well as several extensions of the theory involving perturbed filaments in straining flows, pairs of co- and counter rotating Filaments and non-straight background curves are currently being developed.

4.1.2 Axial Core Dynamics

The previous section explained a mechanism that can lead to the creation of small scale structures in the geometry of a vortex filament. The driving forces in this mechanism were the local and nonlocal contributions of the Biot-Savart integral to the filament motion. The flow structure in the vortex core was assumed constant on the relevant space and time scales. This assumption rules out a description of a "vortex breakdown" associated with strong changes of the core structure. Two examples of this phenomenon are given in Figs. 4.1 and 4.2 (Krause and Liu (1989), Krause (1990)).

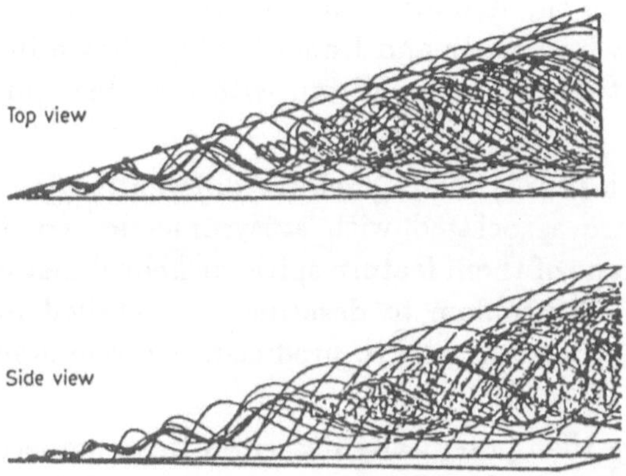

Fig. 4.1. Streamline patterns of stationary vortex breakdown on the upper surface of a delta-wing obtained by direct numerical simulation (Krause and Liu 1989)

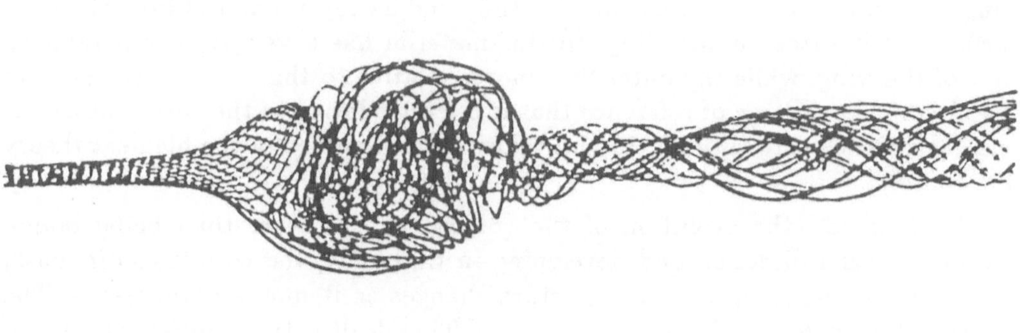

Fig. 4.2. Streamline patterns of a *bubble* type vortex breakdown in an isolated vortex, obtained by direct numerical simulation of the full Navier Stokes equations (Krause 1990)

Figure 4.1 shows the vortex system on the upper side of a delta-wing at a high angle of attack. The pattern of streamlines, which was obtained by means of a direct numerical simulation for steady flow, shows a sudden disruption of the regular curling motion of the particles that emerge from the tip region and are entrained into the leading edge vortex. Figure 4.2 shows streamline patterns from a simulation of a single vortex undergoing an "axisymmetric" or "bubble type" breakdown. Both results are in good agreement with experiments by Payne et al. (1986) and Wentz and Kohlmann (1971). Clearly, in situations as displayed in the figures, an appropriate description of the flow must include axial variations of the vortex core structure.

On the other hand, we know from our discussions in **Sec.2.3** and **4.1** that a slender vortex has its own dynamic behavior involving deformations of its centerline. One might expect that hydrodynamical instabilities of the filament geometry can compete with vortex core instabilities in disrupting a slender vortex. This is true, in particular, when the filament is long enough to allow extensive centerline motion. Computations by Breuer and Hänel (1989), shown in Fig. 4.3 (Courtesy Krause (1990)) exhibit a vortex breakdown with a strong spiral excursion of the vortex center line.

Faler and Leibovich (1977) distinguish six different regimes for vortex breakdown. Two regimes are associated with axisymmetric structures, as shown in Figs.4.1 and 4.2, and four of them feature spiral or helical distortions as in Fig.4.3. It is an open theoretical problem to describe the detailed mechanisms governing the two breakdown patterns and to predict which one dominates under what conditions.

Aiming at a theory that potentially describes this competition, we generalize the filament theory of **Sec2.3** by allowing weak axial variations along the vortical core. Thus, we supplement Callegari and Ting's theory for the evolution of the filament geometry with new dynamical equations for the core structure. This builds in both potential mechanisms for vortex breakdown.

To motivate the scalings involved, consider, e.g., the tip vortex of an aircraft wing. There, the vortical core is filled with boundary layer material from the wing surface. As it sheds off the wing tip this material has a velocity comparable to that of the wing, while the outer flow moves relative to this core at the speed of the aircraft. In a frame of reference that is at rest relative to the surrounding air, one has a vortex core with large axial velocities as assumed in the filament theory in **Sec.2.3**.

Consider now the evolution of the vortical core. Rather than being homogeneous in axial direction and developing in time only, the core is continuously created at the wing tip and its structure changes as it moves downstream. The mathematical difference is that Callegari and Ting dealt with an initial value problem for the filament, whereas for the tip vortex one has to formulate an (initial-) boundary value problem that describes the anchoring of the vortex at the wing tip.

In the scalings of **Sec.2.3** the aircraft velocity gives the order of magnitude of the difference in the axial velocity between the core and the outer flow. It is of order $O(1/\epsilon)$. In times of order $O(1)$, which is the typical evolution time scale of the vortical core, the core material moves a distance of order $O(1/\epsilon)$ relative to

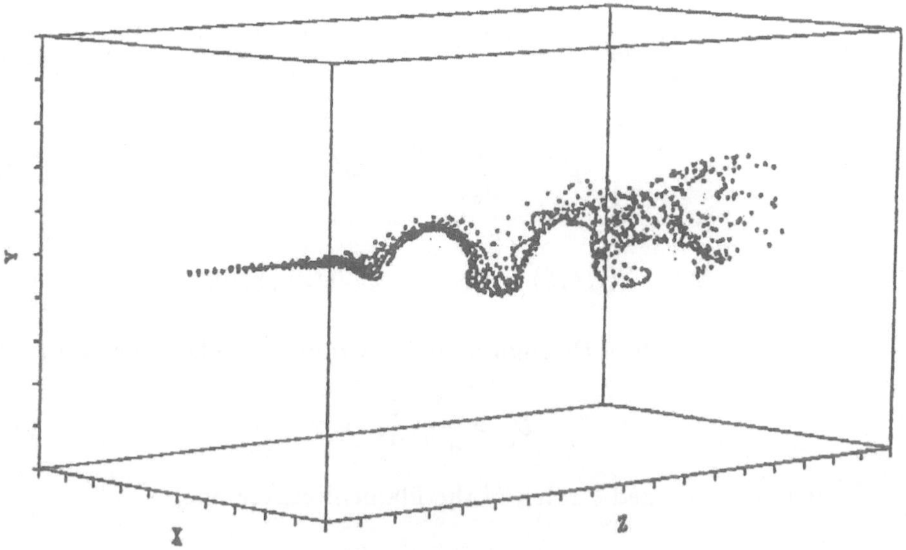

Fig. 4.3. Streamline patterns of a *spiral* type vortex breakdown in an isolated vortex, obtained by direct numerical simulation of the full Navier Stokes equations (Krause 1990)

the surroundings. This introduces a new outer length scale $L = O(\ell/\epsilon)$, where ℓ characterizes the typical radii of curvature of the center line.

From these considerations it seems natural to introduce a new axial variable, $\xi = \epsilon s$, in order to describe the slow variation of the vortical core in downstream direction. The core quantities then have expansions

$$\begin{aligned} v &= \frac{1}{\epsilon} v^{(0)} + v^{(1)} + \epsilon v^{(2)} + \cdots, \\ w &= \frac{1}{\epsilon} w^{(0)} + w^{(1)} + \epsilon w^{(2)} + \cdots \end{aligned} \tag{4.1.25}$$

where

$$\begin{aligned} v^{(0)} &= v^{(0)}(t, \bar{r}, \xi), \\ v^{(1)} &= v^{(1)}(t, s, \bar{r}, \xi) \qquad \text{etc.} \end{aligned} \tag{4.1.26}$$

While the leading order structure has only a slow axial variation, multiple axial variables must be introduced in the higher order functions.

The analysis for the leading order and first order continuity and momentum equations is almost unchanged as compared to that described in **Sec. 2.3**, including the matching conditions and the derivation of the filament velocity. Considerable modifications appear at the second order. We skip the details of the derivation but present next the core evolution equations and discuss the physical meaning of the new terms that appear.

The leading order axial and circumferential velocities obey

$$w_t^{(0)} + w^{(0)} w_{\hat{\xi}}^{(0)} - \frac{1}{\bar{r}} m_{\hat{\xi}}^{(0)} w_{\bar{r}}^{(0)} = -p_{\hat{\xi}}^{(0)} + \frac{\bar{\nu}}{\bar{r}} (\bar{r} w_{\bar{r}}^{(0)})_{\bar{r}} + \frac{\bar{r}^3}{2} \left(\frac{w^{(0)}}{\bar{r}^2} \right)_{\bar{r}} \dot{\Xi}^{(0)} \tag{4.1.27a}$$

$$v_t^{(0)} + w^{(0)} v_{\hat{\xi}}^{(0)} - \frac{1}{\bar{r}^2} m_{\hat{\xi}}^{(0)} (\bar{r} v^{(0)})_{\bar{r}} = -\frac{\bar{\nu}}{\bar{r}} \left[(\bar{r} v_{\bar{r}}^{(0)})_{\bar{r}} - \frac{v^{(0)}}{\bar{r}} \right] + \frac{(\bar{r} v^{(0)})_{\bar{r}}}{2} \dot{\Xi}^{(0)} \tag{4.1.27b}$$

and

$$p^{(0)} = \int_{\infty}^{\bar{r}} \frac{[v^{(0)}]^2}{r'} \, dr' . \tag{4.1.27c}$$

Here

$$2\pi m^{(0)}(t, \bar{r}, \xi) = 2\pi \int_0^{\bar{r}} r' w^{(0)}(t, r', \xi) \, dr' \tag{4.1.28}$$

is the overall axial mass flux through a circle of radius $\bar{r}$. The subscript $\hat{\xi}$ abbreviates

$$\partial_{\hat{\xi}} = \frac{1}{\bar{\sigma}^{(0)}} \frac{\partial}{\partial \xi} \tag{4.1.29}$$

where $\bar{\sigma}^{(0)}$ is the s-averaged stretch of the filament center line,

$$\bar{\sigma}^{(0)}(t, \xi) = \lim_{\epsilon \to 0} \frac{\epsilon}{\Delta \xi} \int_{\frac{1}{\epsilon}(\xi - \Delta\xi)}^{\frac{1}{\epsilon}(\xi + \Delta\xi)} \left| \frac{\partial \mathbf{X}^{(0)}}{\partial s} \right| (t, s, \xi) \, ds . \tag{4.1.30}$$

This limit must be *assumed* to exist and to be independent of $\Delta\xi$ for the multiple scales procedure to be applicable (Keller 1980).

The quantity $\dot{\Xi}^{(0)}$ denotes the s-averaged rate of stretching of the filament

$$\dot{\Xi}^{(0)}(t, \xi) = \frac{1}{\bar{\sigma}^{(0)}} \lim_{\epsilon \to 0} \frac{\epsilon}{\Delta \xi} \int_{\frac{1}{\epsilon}(\xi - \Delta\xi)}^{\frac{1}{\epsilon}(\xi + \Delta\xi)} \frac{\partial}{\partial t} \left| \frac{\partial \mathbf{X}^{(0)}}{\partial s} \right| (t, s, \xi) \, ds . \tag{4.1.31}$$

Going back to (2.3.63) we observe that the above system (4.1.27) is equivalent to Callegari and Ting's equations if $\partial_{\hat{\xi}} \equiv 0$. If there is a large scale axial dependence, the ξ-derivatives introduce nonlinear convection terms and a pressure gradient force, thereby allowing for core waves along the filament. The meaning of the terms with factor $m_{\hat{\xi}}^{(0)}$ becomes clear when we introduce, formally, an averaged radial velocity $\bar{u}^{(2)}$ by

$$(\bar{r} \bar{u}^{(2)})_{\bar{r}} + (\bar{r} w^{(0)})_{\hat{\xi}} = 0 , \qquad \bar{u}^{(2)} \Big|_{\bar{r}=0} = 0 . \tag{4.1.32}$$

Then we find

$$m_{\hat{\xi}}^{(0)} = -\bar{r} \bar{u}^{(2)} \tag{4.1.33}$$

and as a consequence one has $-\frac{1}{\bar{r}} m_{\hat{\xi}}^{(0)} w_{\bar{r}}^{(0)} = \bar{u}^{(2)} w_{\bar{r}}^{(0)}$ showing that these terms essentially account for radial convection.

We emphasize that this extended theory is far from being examined in all detail. Several aspects remain to be addressed, such as the appropriate multiple scales description of the filament centerline and the prescription of boundary data at the anchoring point of the vortex. However, the approach is promising, since it unifies two formerly independent theories that deal separately with axial flow dynamics on a straight vortex and with the motion of a filament with axially homogeneous core structure in three dimensions: From our description it should be clear that the above system is, technically, an extension of Callegari and Ting's analysis. On the other hand, there is a direct connection with the so called "slender vortex approximation" (see Krause (1985), Reyna and Menne (1988) and the references therein).

By introducing $\bar{u}^{(2)}$, one obtains from (4.1.27a,b) and (4.1.32) the system

$$(\bar{r}\bar{u}^{(2)})_{\bar{r}} + (\bar{r}w^{(0)})_{\hat{\xi}} = 0 \,, \tag{4.1.34a}$$

$$w_t^{(0)} + w^{(0)}w_{\hat{\xi}}^{(0)} + \bar{u}^{(2)}w_{\bar{r}}^{(0)} = -p_{\hat{\xi}}^{(0)} + \frac{\bar{\nu}}{\bar{r}}(\bar{r}w_{\bar{r}}^{(0)})_{\bar{r}} + \frac{\bar{r}^3}{2}\left(\frac{w^{(0)}}{\bar{r}^2}\right)_{\bar{r}} \dot{\Xi}^{(0)} \tag{4.1.34b}$$

$$v_t^{(0)} + w^{(0)}v_{\hat{\xi}}^{(0)} + \bar{u}^{(2)}\frac{1}{\bar{r}}(\bar{r}v^{(0)})_{\bar{r}} = \frac{\bar{\nu}}{\bar{r}}\left[(\bar{r}v_{\bar{r}}^{(0)})_{\bar{r}} - \frac{v^{(0)}}{\bar{r}}\right] + \frac{(\bar{r}v^{(0)})_{\bar{r}}}{2}\dot{\Xi}^{(0)}, \tag{4.1.34c}$$

which governs the behavior of $\left[\bar{u}^{(2)}, v^{(0)}, w^{(0)}\right](t, \bar{r}, \xi)$.

For a straight filament with $\dot{\Xi} \equiv 0$ and under the assumption of exactly axisymmetric flow these equations reduce to the unsteady slender vortex equations and $\bar{u}^{(2)}$ attains the physical meaning of the leading order radial velocity. (Notice that this interpretation is not valid in the general case, where the definition (4.1.34) is just a convenient abbreviation and the description of the radial velocities becomes more involved.)

A suitably adapted combination of the small amplitude-short wavelength theory of Klein and Majda, **Sec.4.1.1**, and the axial core dynamics description in this section seems to be a promising approach to the vortex breakdown problem.

4.2 Vortices in Complex Background Flows - Rotational or Geostrophic

In this section, we shall extend the matched asymptotic analyses presented in **Chapter 2** for filaments in a background potential flow to that in a more complex flow. In **Sec. 4.2.1** we consider vortices in a background rotational flow of order one vorticity. In **Sec. 4.2.2** we study geostrophic vortices.

4.2.1 Background rotational flow of order one vorticity

The asymptotic analysis for small vortices or slender vortex filaments were presented in **Chapter 2** under the assumption that background flow was irrotational. Under this assumption, the motion and decay of the filaments induce only higher order terms in the background flow. Therefore, we were able to concentrate our effort on explaining in detail the evolution of the core structure and its coupling with the velocity of the centerline of a filament. However, this assumption is not essential for the success of a matched asymptotic analysis.

We now consider slender filament(s) in a background rotational flow. Again we use U and ℓ to denote the reference speed and length of the background velocity field $\mathbf{Q}$ and δ^* to denote the typical core size. The ratio of the two length scales $\epsilon = \delta^*/\ell$ serves as the small parameter for the matched asymptotic analysis.

The scaled solution for the flow field away from a filament is order one. in particular, the scaled vorticity is order one, i.e.,

$$\Omega_b(t, \mathbf{x}) = \nabla \times \mathbf{Q} = O(U/\ell) , \tag{4.2.1}$$

instead of zero. The Reynolds number has been assumed to be order ϵ^{-2}, therefore, the outer solution obeys the Euler equations up to order ϵ^2. The vorticity evolution equation is

$$\Omega_t + \mathbf{u} \cdot \nabla \Omega - \Omega \cdot \nabla \mathbf{u} = \epsilon^2 \bar{\nu} \Delta \Omega . \tag{4.2.2}$$

This equation reduces to $\Omega \equiv 0$ for all orders of ϵ when the outer solution is initially irrotational. Since the velocity induced by a filament contributes to the leading order velocity in the outer region, (4.2.2) shows that the leading order vorticity field in the outer region will be unsteady due to the motion of the filament, even if the initial background velocity field $\mathbf{Q}(0, \mathbf{x})$, without the filament(s) represents a steady rotational flow. Thus the leading outer solution is coupled with the inner solution(s) of its filament(s). Due to the difference in the length scales, we shall use an Euler solver for the leading (and first) order outer solution in conjunction with the method of matched asymptotics to define the core structure and the velocity of the filament(s).

With the strength of a filament being order one, i.e., $\Gamma/(U\ell) = O(1)$, the velocity and vorticity in the core (the inner region) are order ϵ^{-1} and ϵ^{-2} respectively. The moving curvilinear coordinate system and the expansion scheme for the inner solution introduced in **Chapter 2** remain applicable. In particular, we mention the expansion (2.4.14e) for the vorticity,

$$\Omega(t, \bar{r}, \theta, s) = \epsilon^{-2} \Omega^{(0)}(t, \bar{r}, \theta) + \epsilon^{-1} \Omega^{(1)} + \Omega^{(2)} + \cdots , \tag{4.2.3}$$

and its matching conditions with the outer solution,

$$\Omega^{(0)} \to 0 , \qquad \Omega^{(1)} \to 0 , \qquad \Omega^{(2)} \to \Omega_b(t, \mathbf{X}) , \tag{4.2.4}$$

and higher order conditions as $\bar{r} \to \infty$. The first two conditions in (4.2.4) are the same as those for a background potential flow and can be replaced by the stronger conditions (2.3.17b) for $n = 0, 1$. Similarly the matching conditions for the leading two terms of the velocity in the inner region remain the same as given by (2.3.17a). Consequently, the leading and the first order solutions $\Omega^{(0)}$ and $\Omega^{(1)}$ for the core structure are independent of the outer vorticity distribution regardless whether it is of order one or identically zero.

The asymptotic analysis will show that the vorticity in the outer region remains order one and the convection of the background vorticity has no effect on the leading order core structure and strength of a filament. These results are now inferred directly from the scalings. In the outer region, the velocity and length scales are order one and hence the vorticity remains order one. In the inner region, the background vorticity in a normal plane of the center line C deviates from the value on C by an order ϵ. The variation of the background vorticity in the inner region remains order one even though the circumferential component can be order ϵ^{-1}. The axial flow along the center line and its motion can only induce an order one variation in the background vorticity. This is true even for a large axial flow so long as C is a closed curve of order one length. We note that an order one variation in the core structure changes the strength of a filament by an order ϵ^2. Here we define the strength of filament by the area integral of the vorticity given by the inner solution minus the local value of the outer solution over the normal plane of C. The order one strength is given by the area integral of $\epsilon^{-2}\Omega^{(0)}$ which is the same as that with a background potential flow and hence is time invariant.

Matched asymptotic analysis was carried out for two dimensional problems by Liu and Ting 1987 and employed to simulate the interaction of decaying trailing vortices in a spanwise nonuniform shear flow. The analysis was carried out for axisymmetric problems by Ishii and Liu 1987 and employed to simulate the interactions of vortex rings submerged in a jet. We note that the background vorticity field will remain undisturbed by the presence of vortex filament(s) if the background flow is a two dimensional simple shear flow (constant vorticity, ζ) or an axisymmetric flow with a parabolic velocity profile (constant effective circumferential vorticity, ϖ/σ).

They carried out the matched asymptotic analysis in the normal time scale and confirmed that the leading order core structure and the velocity of the filament are the same as those in **Chapter 2** for a background potential flow. But there is an order one variation of the background vorticity and hence an order one variation of the outer solution. The motion and decay of the filament is now coupled with the Euler solution of the outer region.

Here we quote a numerical example from Liu and Ting (1987) to demonstrate the interaction of a Lamb vortex defined by (2.2.45) with a two dimensional background rotational flow in the upper half plane (above the ground $y = 0$). Initially, the center of the vortex is located at $X = 0$, $Y = 1$ with core size $\delta_0 = 10^{-3}$ and the background shear flow is $\mathbf{Q}(0, \mathbf{x}) = \hat{\imath}U_0(y) = \hat{\imath}[1 - e^{-y}]$ with vorticity $\zeta_b(x, y) = -e^{-y}$ with $\ell = 1$, $U = 1$ and $4\nu = 10^{-6}$. The background shear flow fulfills the nonslip condition along the wall $y = 0$ but the inviscid outer solution induced by the vortex does not and requires the addition of a boundary layer along the wall. As long as the distance from the vortex center to the wall is much greater than the sum of the core size and the boundary layer thickness, we can consider the outer solution to be an inviscid flow with the boundary condition of $v = 0$ on $y = 0$. The Euler solution of the outer region is coupled with the motion and decay of the inner solution, the Lamb vortex. Shown in Fig. 4.4 are the trajectory of the Lamb vortices with strengths ranging from $\Gamma = -3$ to 3.

The results show that a Lamb vortex with positive circulation, $\Gamma > 0$, drifts downstream (x-direction) and upward (y-direction) while a vortex with negative circulation drifts downstream and downward. The latter eventually ($t > t^*$) turns around and drifts upstream. These contrasting phenomena are more pronounced as $|\Gamma|$ increases. To explain this we consider the case of a vortex with $\Gamma > 0$.

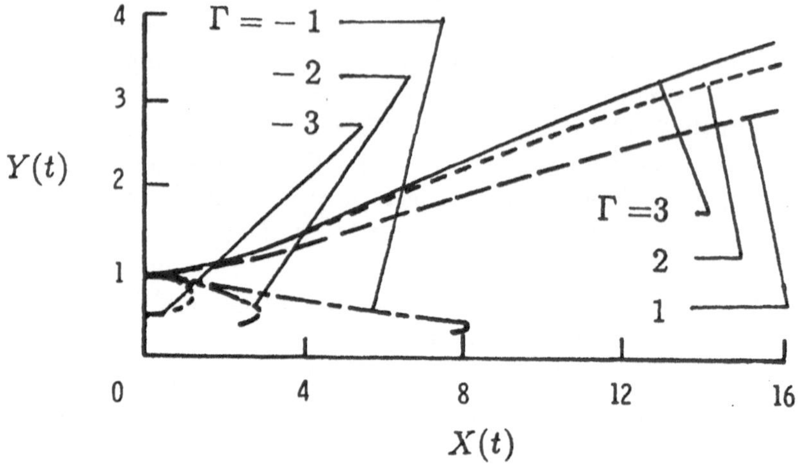

Fig. 4.4. The effect of the strength of a Lamb vortex on its trajectory in a shear flow (Liu and Ting 1987)

The fluid moves downward behind the vortex ($x < X$) and upward ahead of it ($x > X$). For an initial background vorticity ζ_b with $\zeta_b'(y) > 0$, which is the case in Fig. 4.4, the disturbed flow increases the vorticity behind the vortex and decreases the vorticity ahead of it, i.e. $\tilde{\zeta} > 0$ for $x < X$ and $\tilde{\zeta} < 0$ for $x > X$. The background vorticity variation $\tilde{\zeta}$ in turn, induces an upward motion of the vortex. Applying similar arguments for $\Gamma < 0$, we find that the background vorticity variation $\tilde{\zeta}$ induces a downward motion of the vortex. Note that a vortex of negative strength turns around and drifts upstream as it gets closer to the ground because its forward velocity induced by the background shear flow is eventually overcome by the backward velocity induced by its image with respect to the wall as $Y \to 0^+$. It should be pointed out once more that these phenomena will appear only when the background vorticity is nonuniform. When the background flow is either a uniform flow ($\zeta_b = 0$) or a constant shear flow ($\zeta_b =$ constant), the vortex drifts horizontally with no change in the background ($\tilde{\zeta} \equiv 0$).

Additional examples on the motion of a vortex pair in a nonuniform shear flow and examples on the motion and decay of the vortex ring(s) in a axisymmetric jet were presented in the two references quoted above. Those examples show that the trajectories of the (ring) vortices and the variation of the outer solution depend strongly on the signs of the circulation Γ and the (effective) vorticity gradient of the background flow.

4.2.2 Motion and Core Structure of a Geostrophic Vortex

Discrete rectilinear geostrophic (Bessel) vortices were first proposed by Stewart (1943) to model the periodic motion the semipermanent atmospheric pressure systems and by Morikawa (1962) for the prediction of hurricane tracks. Numerical studies of the motion of a geostrophic vortex submerged in a background flow and the interaction of several vortices were made by Morikawa and Swenson (1971) and Bauer and Morikawa (1976). The inviscid theory of a geostrophic vortex has two defects, namely (i) the velocity and the effective height of the atmosphere become infinite at the vortex point and (ii) the velocity of the vortex point is assumed to be moving at the local velocity of the background flow. These defects are the same as those in a potential point vortex, and can be removed by the method of matched asymptotic analysis. We consider the Bessel vortex as the outer solution of a geostrophic vortex with turbulent flow in its core or *eye*. The turbulence is accounted for by introducing an eddy viscosity coefficient ν_e. In **Sec. 4.2.2.1** we explain the reference scales, the small parameters and the equations of geostrophic motion in the tangent plane approximation (see e. g. Pedlosky 1979) and their axisymmetric singular solution (Morikawa 1960), the Bessel vortex. In **Sec. 4.2.2.2** we present the inner solution for the core structure (Ting and Ling 1983).

4.2.2.1 Geostrophic equations of motion

We consider the large scale motion of the atmosphere over the rotating earth with the horizontal scale ℓ much larger that the typical effective thickness H of the atmosphere but much smaller than the radius of the earth. The ratio H/ℓ then defines the small parameter $\epsilon \ll 1$ for the expansion schemes. The equations of motion of the thin atmospheric layer can be approximated by an equivalent two dimensional incompressible flow in the tangent plane, the local horizontal xy plane. Let h and $\mathbf{u}$ denote the effective height and horizontal velocity scaled by H and U respectively. They are governed by the following equations

$$h_t + \nabla \cdot (h\mathbf{u}) = 0 \,, \tag{4.2.5}$$

$$\alpha^2[\partial_t + \mathbf{u} \cdot \nabla]\mathbf{u} + \alpha\aleph\, \hat{k} \times \mathbf{u} = -\nabla h \tag{4.2.6}$$

with

$$\alpha = \frac{U}{\sqrt{gH}}, \qquad \aleph = \frac{2\,\ell\,\omega\, \sin\varphi}{\sqrt{gH}} \,, \tag{4.2.7}$$

where g, ω and φ denote the gravitational acceleration, the angular velocity of the earth and the local latitude respectively. Here $\aleph$ is known as the Coliolis parameter and the independent variables x, y and t are scaled by the length ℓ and time ℓ/U. We assume $\varphi \neq 0$, i. e., to be away from the equator and $\aleph = O(1)$. By applying the curl operator to (4.2.6), we get

$$[\partial_t + \mathbf{u} \cdot \nabla][\alpha\, \zeta - \aleph\, h] = 0 \,, \tag{4.2.8}$$

where ζ denotes the vorticity. Now we consider the flow field in the length scale ℓ to be a small perturbation from equilibrium and identify the reference velocity as

$$U = \epsilon \sqrt{gH} \,, \quad \text{i. e.,} \quad \alpha = \epsilon \,. \tag{4.2.8}$$

The expansion schemes for the outer solutions are:

$$\begin{aligned} h(t,x,y,\epsilon) &= 1 + \epsilon h'(t,x,y) + O(\epsilon^2) \,, \\ \mathbf{u}(t,x,y,\epsilon) &= \mathbf{u}'(t,x,y) + O(\epsilon) \,, \\ \zeta(t,x,y,\epsilon) &= \zeta'(t,x,y) + O(\epsilon) \,. \end{aligned} \tag{4.2.9}$$

From (4.2.5 - 8) we obtain the leading order equations,

$$\nabla \cdot \mathbf{u}' = 0 \,, \tag{4.2.10}$$

$$\aleph \, \hat{k} \times \mathbf{u}' = -\nabla h' \tag{4.2.11}$$

and

$$\partial_t \, [\Delta h' + \aleph^2 h'] = 0 \,. \tag{4.2.12}$$

Equations (4.2.10 - 11) imply that there is a stream function ψ' related to h' by the equation,

$$\Delta[\aleph \, \psi' + h'] = 0 \,, \tag{4.2.13}$$

and hence ψ' can differ from $-h'/\aleph$ by a potential solution. Note that (4.2.12) has an axisymmetric singular solution,

$$h^*(t,r) = -\aleph \, \psi^*(t,r) \quad \text{with} \quad \psi^* = \frac{\Gamma}{2\pi} K_0(\aleph \, r) \,, \tag{4.2.14}$$

where $r^2 = (x-X)^2 + (y-Y)^2$ and $K_0(\aleph \, r)$ denotes the modified Bessel function. The stream function, ψ^*, represents a geostrophic or Bessel vortex with strength Γ and centered at (X, Y). It is the solution of the equation

$$\Delta \psi^* - \aleph^2 \psi^* = -\frac{\Gamma \, \delta(r)}{2\pi r} \,,$$

where δ denotes the delta function. The dependence of ψ^* on t comes from that of $X(t)$ and $Y(t)$. The solution reduces to that of a potential point vortex when $\aleph \to 0$ and has the same logarithmic singularity as $r \to 0$. We can separate the regular parts of h' and ψ' from their singular parts and write

$$\begin{aligned} h'(t,x,y) &= h^*(t,r) + h_1(t,x,y) \,, \\ \psi'(t,x,y) &= \psi^*(t,r) + \psi_1(t,x,y) \end{aligned} \tag{4.2.15}$$

where h_1 and ψ_1 are regular at the vortex point. Note that ψ_1, which can differ from $-h_1 \,/\, \aleph$ by a harmonic function, represents the stream function of the background flow. To close the system, it is assumed that the vortex point moves at the local background velocity, i. e.,

$$\dot{X} = -\partial_y \, \psi_1(t, X, Y) \quad \text{and} \quad \dot{Y} = \partial_x \, \psi_1(t, X, Y) \,. \tag{4.2.16}$$

In the following subsection, we employ the matched asymptotic analysis in the normal time scale to obtain the core structure and remove the singular part of the outer solution (4.2.15) and show that the leading order velocity of the vortex center is given by (4.2.16).

4.2.2.2 Inner solution of a geostrophic vortex

For the inner solution, we use the small length scale $\epsilon\ell$ and the normal time scale ℓ/U (see **Sec. 2.2.2** and **2.2.3**). Let r, θ denote the polar coordinates relative to the vortex center $X(t), Y(t)$ and $\bar{u}, \bar{v}$ the radial and circumferential components of the velocity relative to the moving vortex center. The velocity is written as

$$\mathbf{u} = [\hat{\imath}\dot{X} + \hat{\jmath}\dot{Y}] + [\hat{r}\bar{u} + \hat{\theta}\bar{v}]. \tag{4.2.17}$$

Again we introduce the stretched radial variable $\bar{r} = r/\epsilon$ for the inner solution. We assume that the velocity of the vortex center $(\dot{X}, \dot{Y})$ remains order one while the circumferential velocity will be $O(\epsilon^{-1})$ to match with the singular part of the outer solution. The expansions for the inner solutions are:

$$\bar{u}(t,\bar{r},\theta,\epsilon) = \bar{u}^{(1)}(t,\bar{r},\theta) + O(\epsilon)\ , \tag{4.2.18a}$$
$$\bar{v}(t,\bar{r},\theta,\epsilon) = \epsilon^{-1}[\bar{v}^{(0)}(t,\bar{r},\theta) + O(\epsilon)]\ , \tag{4.2.18b}$$
$$\bar{h}(t,\bar{r},\theta,\epsilon) = \bar{h}^{(0)}(t,\bar{r},\theta) + O(\epsilon)\ , \tag{4.2.18c}$$
$$\bar{\zeta}(t,\bar{r},\theta,\epsilon) = \epsilon^{-2}[\bar{\zeta}^{(0)}(t,\bar{r},\theta) + O(\epsilon)]\ , \tag{4.2.18d}$$
$$X(t,\epsilon) = \bar{X}^{(0)}(t) + O(\epsilon) \quad \text{and} \quad Y(t,\epsilon) = \bar{Y}^{(0)}(t) + O(\epsilon)\ . \tag{4.2.18e}$$

To account for the turbulence in the core structure, we add the eddy diffusion term $\nu_e\Delta\mathbf{u}$ to the right-hand side of the momentum equation (4.2.6). Here ν_e stands for the small eddy viscosity coefficient scaled by $U\ell$. The order of magnitude of ν_e relative to ϵ will be assigned later. We now repeat the procedures of the asymptotic analysis carried out in **Sec. 2.2.2** and **2.2.3** and conclude that the leading order vorticity, circumferential velocity and effective height are axisymmetric, i. e., independent of θ and are related by the following equations,

$$\bar{h}^{(0)}_{\bar{r}}(t,\bar{r}) = [\bar{v}^{(0)}(t,\bar{r})]^2/\,\bar{r} \quad \text{or} \quad \bar{h}^{(0)}(t,\bar{r}) = 1 - \int_{\bar{r}}^{\infty} \frac{[\bar{v}^{(0)}(t,\xi)]^2}{\xi}\, d\xi \tag{4.2.19}$$

and

$$\bar{\zeta}^{(0)}(t,\bar{r}) = [\bar{v}^{(0)}\ \bar{r}]_{\bar{r}}/\,\bar{r}\ . \tag{4.2.20}$$

In (4.2.19) we have made use of the matching condition, $\bar{h}^{(0)} \to 0$ as $\bar{r} \to \infty$.

From the leading order asymmetric part of the momentum equation, the boundary conditions $\bar{u} = \bar{v} = 0$ at $\bar{r} = 0$ and the matching conditions we conclude that the velocity of the vortex center is given by (4.2.16). To obtain the leading order (axisymmetric) solution from the circumferential averages of the higher order equations, we have to specify the order of magnitude of the scaled eddy viscosity ν_e relative to ϵ. We consider two models:

$$\text{Model A}\ , \qquad \bar{\nu}_e = \nu_e/\epsilon^2 = O(1) \tag{4.2.21a}$$

and

$$\text{Model B}\ , \qquad \bar{\nu}_e = \nu_e/\epsilon = O(1)\ . \tag{4.2.21b}$$

For model A, the leading order inner solution is defined by the circumferential averages of the second order equations in the same manner as that in **Sec. 2.2.2**. The vorticity fulfills the simple diffusion equation (2.2.33), which is

$$\bar{\zeta}_t^{(0)} = \frac{\bar{\nu}_e}{\bar{r}}\ [\bar{r}\bar{\zeta}_{\bar{r}}^{(0)}]_{\bar{r}}. \tag{4.2.22}$$

The vorticity and circumferential velocity in the *eye* of the geostrophic vortex is identical to that in the core of a potential vortex with $\bar{\nu}_e$ replacing $\bar{\nu}$. The deviation of the effective height, $\bar{h}^{(0)} - 1$, is proportional to the pressure variation in the latter.

For model B, the eddy viscosity is one order larger than that in model A and hence we have a larger diffusion rate. To account for this fact, we introduce a short time scale $\ell/\sqrt{gH}$ and the corresponding stretched time variable,

$$\tilde{t} = t/\epsilon\ , \tag{4.2.23}$$

for the inner solutions with the math accent bar replaced by tilde. We then obtain from the circumferential averages of the first order equations two conditions relating $\tilde{v}^{(0)}$, $\tilde{h}^{(0)}$ and $\{\tilde{u}^{(1)}\}$, which denotes the axisymmetric part of $\tilde{u}^{(1)}$. These conditions are:

$$[\tilde{r}\tilde{h}^{(0)}]_{\tilde{t}} + [\tilde{r}\tilde{h}^{(0)}\{\tilde{u}^{(1)}\}]_{\tilde{r}} = 0 \tag{4.2.24a}$$

and

$$\tilde{v}_{\tilde{t}}^{(0)} + \{\tilde{u}^{(1)}\}[\tilde{v}_{\tilde{r}}^{(0)} + \frac{\tilde{v}^{(0)}}{\tilde{r}}] = \tilde{\nu}_e[\tilde{v}_{\tilde{r}\tilde{r}}^{(0)} + (\frac{\tilde{v}^{(0)}}{\tilde{r}})_{\tilde{r}}]\ . \tag{4.2.24b}$$

These two equations and (4.2.19) form a closed system for $\tilde{h}^{(0)}, \tilde{v}^{(0)}$ and $\{\tilde{u}^{(1)}\}$. The boundary conditions are $\tilde{v}^{(0)} = 0$ and $\{\tilde{u}^{(1)}\} = 0$ at $\tilde{r} = 0$ and the matching condition is $\tilde{v}^{(0)} \to \Gamma/(2\pi\tilde{r})$, as $\tilde{r} \to \infty$.

Since the system of equations does not have a t-derivative of $\{\tilde{u}^{(1)}\}$, we need only the initial profile of either $\tilde{v}(0)$ or $\tilde{h}^{(0)}$ on account of (4.2.19). Numerical solutions of the system for an initial bell-shaped profile of $\tilde{h}^{(0)}(0, \tilde{r})$ were constructed and the temporal variations of the leading order circumferential velocity and effective height in the *eye* of the geostrophic vortex were presented by Ting and Ling (1983).

Note that for both models, the r^{-1} singularity of the outer solution $\mathbf{u}'$ as $r \to 0$ is removed by the matching condition with the leading order inner circumferential velocity. The logarithmic singularity of the outer solution $\epsilon h'$ will be removed by the matching condition with the next order inner solution $\epsilon\tilde{h}^{(1)}$. The leading order velocity of the vortex center is defined by (4.2.16), i. e., by the local background velocity. But the structure of the *eye* for model A and that for model B are different from each other. In particular, they have different time scales.

In case that the initial velocity of the vortex center disagrees with that given by (4.2.16) or when there is a sudden change of background flow, we have to introduce a two-time analysis similar to that in **Sec. 2.2.1**. Details of the analysis were presented by Ling and Ting (1988). For both models, the additional short time variable τ is scaled by $\epsilon^2\ell/U$ and is equal to t/ϵ^2 or $\bar{t}/\epsilon$. Many results of the two-time analysis in **Sec. 2.2.1** remain valid here. For example, the τ-averaging of the leading order velocity of the vortex center given by the two-time solution remains equal to that of the one-time solution defined by (4.2.16) and the trajectory of the former differs from that of the latter by an oscillation in τ with small period and amplitude of the order of $\epsilon^2\ell/U$ and $\epsilon^2\ell$ respectively. Of course the details of the oscillatory motion do depend on the choice of model A or B. It should be noted that the typical scales ℓ and U are of the order of 100 miles and 10 miles per hour. The normal time scale is of the order of 10 hours. Therefore, the oscillatory motion should be observable. Of course, the two dimensional geostrophic vortex is only a simple model for the dynamics of the thin atmospheric layer. Additional analyses are needed to account for the vertical stratification, the compressibility effect and the energy transfer.

4.3 Interaction of Vortex Filaments with Boundaries

In this section, we will present a digest of the interaction problems. We mentioned before two weak interaction problems, namely the passage of a circular vortex ring around a coaxial sphere (Wang 1970) and the interaction of trailing vortices in spanwise shear wing near ground (Liu and Ting 1987). In a *weak* interaction problem, we assume that the distance of the filament to the rigid surface is much larger than the core size so that the motion and decay of the filament are described by the matched asymptotic solution and the nonslip condition on the surface is fulfilled by the addition of a boundary layer. Because of the presence of the surface, the inviscid flow field outside of the vortical core and the boundary layer is coupled with the motion of the filament even when the inviscid flow is irrotational.

In case that the distance of the filament to the surface is comparable to the core size, we have a *strong* interaction problem. It is well known that vortices can be created due to flow instability, e. g., the generation of Taylor vortices due to the instability of shear flows between two concentric rotating circular cylinders or spherical shells (Taylor 1922) and the generation of Görtler vortices due to boundary layer instability (Görtler 1944). Vortices are also observed in flows over a corner, a cavity and in internal flows (see for example, van Dyke 1975 and Lugt 1983 and references therein). In those examples, strong interaction between a filament and boundary takes place along the entire length of the filament. Simulations of those phenomena by numerical solution of N-S equations are available (see for example, Krause 1980 and Schrauf and Krause 1984).

There are problems in which strong interaction between a filament and surface takes place locally, i. e., the size of the local interaction region is much smaller than the length of the filament and/or the size of the surface. The strong interaction may take place only for a finite duration. Efficient schemes for the analyses of those problems would call for the coupling of the numerical solution of the local strong interaction region(s) with the matched asymptotic solution of the filament(s) and the boundary layer along the surface. To give concrete examples of local strong interactions, we refer to problems in rotorcraft aerodynamics. The importance of interactional aerodynamics of rotorcraft components, namely, the main rotor, fuselage, tail boom and tail rotor relative to the aerodynamics of individual components, has been substantiated by recent experimental investigations (see for example Sheridan and Smith 1980 and Balch 1985). The interactions create anomalies in the flow field resulting in substantial changes in the load distributions, control characteristics and far field acoustic signatures (see for example, Schmitz and Yu 1986 and references therein). The dominant interaction problems will be different for different operation conditions. For example, during hovering the interaction of the main rotor tip vortices with the fuselage will be important. In a near ground operation, the interactions of the tip vortices with the ground and nearby objects have to be included. In a descending motion, the interaction of one tip vortex with another rotor blade becomes important. In a forward motion, the interaction of tip vortices with the tail boom and/or tip rotor blades has to be accounted for. Of course, challenging interaction problems are present in the flow field downstream of the main rotor. They are: the roll up of trailing vortex sheets to helical tip vortices with finite core structures, the interactions of the helical vortex systems among themselves and with the blades. Since the vortical cores are usually turbulent a realistic turbulence model is needed (see for example, Tung et al. 1983).

To show the complexity of the interaction problems, we describe two simple models simulating respectively the glancing impingement and the cutting of a filament by a surface. For example, in an ascending motion a tip vortex may glance the fuselage and then be chopped by the tail boom. While in forward motion, a tip vortex may glance the tail boom and be cut by a tail rotor blade. The glancing and chopping phenomena are illustrated in Fig. 4.5 and Fig. 4.6. Besides the geometries of the filament, namely, its strength Γ, typical core size δ and radius of curvature ℓ, the size L of the moving surface and the velocity V of the filament relative to the surface are important length and velocity scales.

Three stages of glancing impingment are shown in Fig. 4.5. In the first stage, the filament is at finite distance (much larger than the core size δ from the surface and we have a *weak* interaction problem. In the second stage, the *strong* interaction stage, a finite segment of the filament, say of the length S is in close contact or merged with the boundary layer of the surface. In the third stage, a slender filament reappears behind the body. Of course it may not always be the case. The reconnection of the filament may not take place in the third stage if the the length S is large or the duration of the impingement, which is of the order of L/V is too long. For example the relative velocity V can be much smaller than the reference velocity U of the flow field.

Shown in Fig. 4.6 are the stages of the chopping of a filament by a body, say the tail boom. The length of the filament segment chopped by the body is of the order of the size L of the cross section of the body and is much larger than δ. The filament reconnection in the the post chopping stage is unlikely. The process of the break up is still an open question.

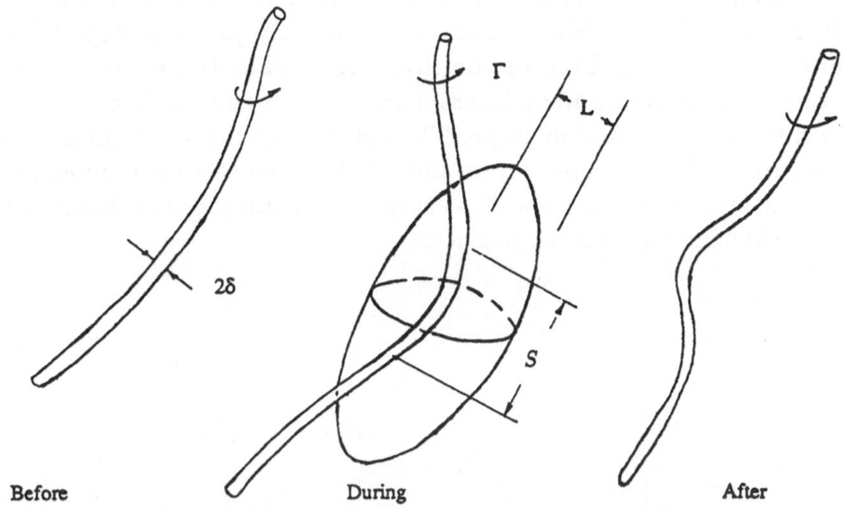

Fig. 4.5. Glancing impingement of a filament by a body

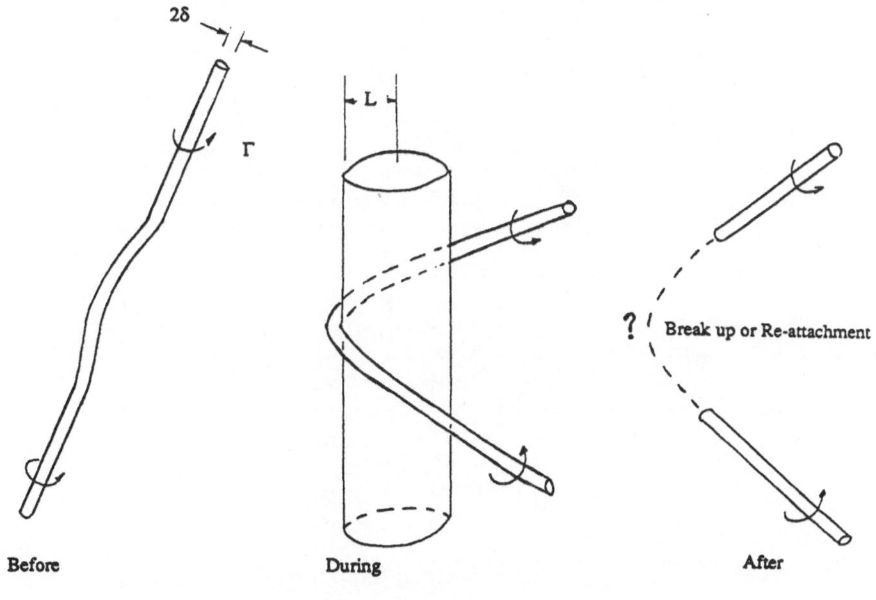

Fig. 4.6. Chopping of a filament by a body

In the chopping of a filament by a thin tail rotor blade we replace the body in Fig. 4.6 by a thin airfoil with its chord c identified as L, which can be of the order of δ. The duration of the chopping stage which is of the order of c/V can be very short since V can be much larger than U. In the limit $U/V \to 0$ and/or $c/\delta \to 0$, the filament remains undisturbed. In general, the reconnection or break up of the filament in the post chopping stage is again an open question.

The strong interaction problem illustrated in either one of the two figures above, has to be treated as a three dimensional unsteady problem with the initial data prescribed at the beginning of the first stage. That is we can create the initial data by the solution of a weak interaction problem and continue the solution until the beginning of the second stage. The solution of the weak interaction problem for a nonsimple body shape is a formidable task. Numerical studies for the strong interaction stage and then the final stage of reconnection or break up should be the next challenging research projects.

Appendix A.1. Governing Equations for Higher Order Solutions

This appendix was first referred to in **Sec.2.2.1**, in which the vorticity and stream function in the inner region of a two-dimensional vortex were expanded in power series of ϵ. The independent variables in the two-time inner solution are: the short time, τ, the normal time t and the polar coordinates $\bar{r}$ and θ. The symbols $\mathcal{M}$ and $\{\,\}$ denote, respectively, the operators for short time and circumferential averaging. Following the terminology introduced in **Sec. 2.2.1** we call the θ-average of a function its "symmetric part", while the remainder its "asymmetric part". These two parts carry the subscripts c and a, respectively.

It was pointed out in **Sec.2.2.1** that among all the governing equations for the nth order two-time solution only the vorticity evolution equation is nonlinear for $n \geq 2$, because of the inhomogeneous term, $\tilde{F}_n$. The term is the sum of products of jth and $(n-j)$th order terms for $j = 1, \dots, (n-1)$. In this appendix, we explain how to decompose the nth order vorticity equation into separate equations for the asymmetric and symmetric parts of the nth order solution and how to decompose the τ-average of the nth order equation into one equation for the τ-average of the asymmetric part of the nth order solution and one equation for that of the symmetric part of the $(n-2)$nd order solution.

The second decomposition implies that the dependence of the symmetric parts of the $(n-1)$th and nth order solutions on the normal time, t, remains undetermined at this stage. Since the nonlinear inhomogeneous term $\tilde{F}_n$ in general involves $(n-1)$th order solution, we need to show that our scheme is consistent, i. e., our scheme provides a closed system of equations to determine the solution to any order. In **Sec.2.2.1** we did show how to determine the leading and first order solutions except for the dependence of the symmetric first order solution on the normal time, t. Hence, by the method of induction, it is sufficient to show that

(I) *If the jth order solutions for $j \leq n-1$ are defined except for the τ-average of the symmetric part of the $(n-1)$st order solution, then one can determine the same with $n-1$ repaced by n.*

We begin by extending the first and second order equations (2.2.27a) and (2.2.28a) to the following equation for the nth order solution with $n \geq 2$:

$$\tilde{\zeta}_\tau^{(n)} + \mathcal{L}[\tilde{\psi}^{(n)}, \tilde{\zeta}^{(n)}] = \tilde{F}_n(\tau, t, \bar{r}, \theta) + \mathcal{G}\tilde{\zeta}^{(n-2)} \,, \qquad (A.1.1)$$

where

$$\tilde{F}_n = -\sum_{j=1}^{n-1} \frac{1}{\bar{r}} [\tilde{\psi}_\theta^{(j)} \tilde{\zeta}_{\bar{r}}^{(n-j)} - \tilde{\zeta}_\theta^{(j)} \tilde{\psi}_{\bar{r}}^{(n-j)}] \qquad (A.1.2)$$

and

$$\mathcal{G} = -\frac{\partial}{\partial t} + \bar{\nu}\bar{\Delta} \, . \qquad (A.1.3)$$

Here the linear operator $\mathcal{L}$ is defined in (2.2.27*b*). Let us consider a product, say $\tilde{\psi}_\theta^{(j)}\tilde{\zeta}_{\bar{r}}^{(n-j)}$, in $\tilde{F}_n$. By using the definition of the symmetric and asymmetric parts, we evaluate the θ-average of the product,

$$\begin{aligned}\{\tilde{\psi}_\theta^{(j)}\tilde{\zeta}_{\bar{r}}^{(n-j)}\} &= \{[\tilde{\psi}_a^{(j)}]_\theta\}[\tilde{\zeta}_c^{(n-j)}]_{\bar{r}} + \{[\tilde{\psi}_a^{(j)}]_\theta[\tilde{\zeta}_a^{(n-j)}]_{\bar{r}}\} \\ &= \{[\tilde{\psi}_a^{(j)}]_\theta[\tilde{\zeta}_a^{(n-j)}]_{\bar{r}}\} \, .\end{aligned} \qquad (A.1.4)$$

This equation says that

(II) *The θ-average of $\tilde{F}_n$ contains only the asymmetric parts of jth order solution for $j = 1, \ldots (n-1)$.*

Now the θ-average of (A.1.1) becomes an equation for the symmetric part

$$[\tilde{\zeta}_c^{(n)}]_\tau = \{\tilde{F}_n\} + \mathcal{G}\tilde{\zeta}_c^{(n-2)} \, . \qquad (A.1.5)$$

The difference between (A.1.1) and (A.1.5) then yields an equation for the asymmetric part,

$$[\tilde{\zeta}_a^{(n)}]_\tau + \mathcal{L}[\tilde{\psi}_a^{(n)}, \tilde{\zeta}_a^{(n)}] = \tilde{F}_n - \{\tilde{F}_n\} + \mathcal{G}\tilde{\zeta}_a^{(n-2)} \, . \qquad (A.1.6)$$

By taking the τ-average of the above equation, we obtain an equation for the τ-average of the asymmetric part, namely

$$\mathcal{L}[\mathcal{M}\tilde{\psi}_a^{(n)}, \mathcal{M}\tilde{\zeta}_a^{(n)}] = \mathcal{M}[\tilde{F}_n - \{\tilde{F}_n\}] + \mathcal{G}\mathcal{M}\tilde{\zeta}_a^{(n-2)} \, . \qquad (A.1.7)$$

From statement II and the assumption in statement I, we see that inhomogeneous terms in (A.1.5) are defined because they depend only on the asymmetric parts of the jth order solutions for $j = 1, \ldots, (n-1)$ and the symmetric part of $(n-2)$nd order solution. The symmetric part to nth order may thus be written as

$$\tilde{\zeta}_c^{(n)}(\tau, t, \bar{r}) = \tilde{\zeta}_p^{(n)}(\tau, t, \bar{r}) + \tilde{\zeta}_h^{(n)}(t, \bar{r}) \, , \qquad (A.1.8)$$

where the subscripts p and h stand for the particular integral of (A.1.5) which vanishes at $\tau = 0$ and for the homogeneous solution. The former is defined by quadrature, while the latter is any function independent of τ. Since the τ-average of (A.1.8) yields

$$\mathcal{M}\tilde{\zeta}_c^{(n)} = \mathcal{M}\tilde{\zeta}_p^{(n)} + \tilde{\zeta}_h^{(n)} \, , \qquad (A.1.9)$$

we conclude that the dependence of the τ-average of the symmetric nth order solution on t, is yet undefined. In fact, the assumption in statement (I) implies that even the τ-average of the symmetric $(n-1)$st order solution is still unknown. To find an equation for the latter, $\mathcal{M}\tilde{\zeta}_c^{(n-1)}$, we take the τ-average of (A.1.5) with n replaced by $n+1$. The result is

$$\{\frac{\partial}{\partial t} - \bar{\nu}\bar{\Delta}_0\}\mathcal{M}\tilde{\zeta}_c^{(n-1)} = \mathcal{M}\{\tilde{F}_{n+1}\} \, . \qquad (A.1.10)$$

Using statement (II) and the assumption in statement (I), we draw the following conclusions:

(1) The unknowns on the right-hand side of (A.1.10) are the asymmetric parts of the nth order solution.
(2) The right-hand sides of of (A.1.6) and (A.1.7) involve the unknown, the τ-average of the symmetric part of the $(n-1)$st order solution, via $\mathcal{M}\{\tilde{F}_n\}$.
(3) The right-hand side of (A.1.5) is defined because because of the assumption in statement (I).

Equations (A.1.5), (A.1.6), (A.1.7) and (A.1.10) thereby form a closed system that determines the τ-average of the symmetric part of the $(n-1)$st order solution and the nth order solution except for the τ-average of its symmetric part. Thus statement (I) is confirmed.

Note that for $n = 1$ the system is uncoupled because (A.1.5), (A.1.6), (A.1.7) and (A.1.10) are all homogeneous. This enabled us to solve first for the leading order core structure and then determine the asymmetric part of the first order solution in **Sec.2.2.1**.

Also note that the operator,

$$\mathcal{L} = \frac{1}{\bar{r}}[\bar{\psi}_{\theta}^{(0)}\frac{\partial}{\partial\bar{r}} - \bar{\psi}_{\bar{r}}^{(0)}\frac{\partial}{\partial\theta}] \ , \tag{A.1.11}$$

involves only spatial derivatives so that only (A.1.10) contains a t-derivative of the short time average of the symmetric part to $(n-1)$st order. As an extension of a result obtained for $n = 1$ in **Sec. 2.2.1**, we infer that the normal time, t, plays the role of a parameter in the asymmetric part of the nth order solution for all n.

Appendix A.2. Second Order Two-time Solutions

In this appendix we derive the results stated at the end of **Sec. 2.2.4** regarding the τ-average of the second order two-time solution.

We first summarize the results on the first order asymmetric solutions derived in **Sec. 2.2.3** and **2.2.4**. Equations (2.2.56) and (2.2.57) imply that the τ-average of the two-time asymmetric solution is a trivial solution,

$$\bar{\psi}_a^{(1)}(t,\bar{r},\theta,s) = \mathcal{M}\tilde{\psi}_a^{(1)}(\tau,t,\bar{r},\theta,s) = 0. \qquad (A.2.1)$$

If the initial data happen to be such that not only the leading but also the first order core structures are symmetric and that the initial velocity of the vortex center agrees with the local background velocity, (2.2.58), then (A.2.1) defines a trivial first order asymmetric solution independent of the short time variable. However, in reality, the first order initial core structure is not available and the leading order velocity of the vortex center can clearly differ from the local background velocity. Thus, in general it is necessary to construct asymmetric two-time solutions. They obey (2.2.64) and (2.2.69a, b), which imply that the solutions do not have second and higher harmonics in θ and that their dependence on the two independent variables, τ and θ, appears only in the linear combination, $\theta - \bar{\omega}\tau$. Therefore, the partial derivatives in τ and θ are related by

$$\frac{\partial}{\partial\theta}\tilde{\psi}_a^{(1)}(\tau,t,\bar{r},\theta) = -\frac{1}{\bar{\omega}}\frac{\partial}{\partial\tau}\tilde{\psi}_a^{(1)}(\tau,t,\bar{r},\theta)\ . \qquad (A.2.2)$$

This relationship enables us to express the inhomogeneous term $\tilde{F}_2$ in the second order equation (2.2.28) as a derivative of τ. We then replace (2.2.32a) by

$$\tilde{F}_2 = \frac{1}{\bar{\omega}}\frac{\partial}{\partial\tau}\left[\frac{1}{2}[(\bar{\omega}\bar{a} - \bar{b})(\tilde{\zeta}^{(1)})^2]_{\bar{r}} + \tilde{\zeta}^{(1)}\tilde{\psi}_{\bar{r}}^{(1)}\right]\ , \qquad (A.2.3a)$$

which in turn yields

$$\mathcal{M}\tilde{F}_2 = 0\ . \qquad (A.2.3b)$$

From (A.2.3) and the compatibility condition (2.2.33) the τ-average of the second order equation (2.2.28a) becomes a homogeneous equation. This equation and the τ-average of (2.2.23) for $n = 2$ yield the following equation for the one-time asymmetric stream function $\bar{\psi}_a^{(2)}$,

$$[\bar{\psi}_{\bar{r}}^{(0)}\bar{\Delta} + \bar{\zeta}_{\bar{r}}^{(0)}](\bar{\psi}_a^{(2)})_\theta = 0\ , \qquad (A.2.4)$$

where

$$\psi^{(2)}(t,\bar{r},\theta) = \mathcal{M}\bar{\psi}^{(2)}(\tau,t,\bar{r},\theta)\ . \tag{A.2.5}$$

Equation (A.2.4) is identical to (2.2.49) for the corresponding first order solution. The two homogeneous boundary conditions (2.2.24) at $\bar{r}=0$ hold for both $n=1$ and 2. Only the the matching conditions are different. The condition for the second order solution as $\bar{r}\to\infty$ is

$$\begin{aligned}\frac{1}{\bar{r}}\bar{\psi}^{(2)}_{\theta}\hat{r} - \bar{\psi}^{(2)}_{\bar{r}}\hat{\theta} \to &\\ &-\{\dot{X}^{(1)}\hat{\imath}+\dot{Y}^{(1)}\hat{\jmath}\}\\ &+\{[X^{(1)}\Psi_{yx}(X^{(0)},Y^{(0)})+Y^{(1)}\Psi_{yy}(X^{(0)},Y^{(0)})]\,\hat{\imath}\\ &\quad-[X^{(1)}\Psi_{xx}(X^{(0)},Y^{(0)})+Y^{(1)}\Psi xy(X^{(0)},Y^{(0)})]\,\hat{\jmath}\}\\ &+\{[\bar{r}\cos\theta\,\Psi_{yx}(X^{(0)},Y^{(0)})+\bar{r}\sin\theta\,\Psi_{yy}(X^{(0)},Y^{(0)})]\,\hat{\imath}\\ &\quad-[\bar{r}\cos\theta\,\Psi_{xx}(X^{(0)},Y^{(0)})+\bar{r}\sin\theta\,\Psi_{xy}(X^{(0)},Y^{(0)})]\,\hat{\jmath}\}\ .\end{aligned} \tag{A.2.6}$$

On the right-hand side of (A.2.6), the vector inside the first pair of curly brackets is the unknown first order velocity of the vortex center. The second vector accounts for the change of the background velocity due to the first order displacement of the vortex center and the third vector represents the contribution of local background velocity gradients.

Again we note that the solution of the system, (A.2.4), (2.2.24) and (A.2.6), is quasi-steady, i. e., that the normal time, t, can be treated as a parameter. If we assume that the inner solution depends only on the normal time and carry the one-time analysis to the second order†, we would find that the second order solution is governed by the same system of equations, the differential equation (A.2.4), two boundary conditions (2.2.24) at $\bar{r}=0$ and the matching condition (A.2.6). Thus we extend the equivalence relationship stated below (2.2.39) from the first to the second order solutions:

- *The short time average of the second order two-time solution is equivalent to the solution of one-time analysis, provided the first order asymmetric two-time solution fulfills condition (A.2.2) or (2.2.69a).*

We then construct the one-time solution of the system (A.2.4), (2.2.24) and (A.2.6) by repeating the steps from (2.2.49) to (2.2.59) in **Sec. 2.2.3**.

The Fourier coefficients of the asymmetric part of the second order stream function, $\bar{\psi}^{(2)}_{jk}$, $j=1,2,\ldots$, $k=1,2$, have to fulfill (2.2.51) and the boundary conditions (2.2.52) at $\bar{r}=0$. With the superscript suppressed, these equations read

$$\{\frac{\partial^2}{\partial\bar{r}^2}+\frac{1}{\bar{r}}\frac{\partial}{\partial\bar{r}}+[\frac{\bar{\zeta}^{(0)}_{\bar{r}}}{\bar{\psi}^{(0)}_{\bar{r}}}-\frac{j^2}{\bar{r}^2}]\}\,\bar{\psi}_{jk}=0\ , \tag{A.2.7}$$

and

$$\bar{\psi}_{jk}=0\ ,\qquad [\bar{\psi}_{jk}]_{\bar{r}}=0\qquad \text{at}\quad \bar{r}=0\ , \tag{A.2.8}$$

† The one-time analysis was performed by Ting and Tung in 1965.

for $j = 1, 2, \ldots$ and $k = 1, 2$. Here the ratio $\bar{\zeta}_{\bar{r}}^{(0)}/\bar{\psi}_{\bar{r}}^{(0)}$ is finite at $\bar{r} = 0$ and decays rapidly as $\bar{r} \to \infty$. Therefore, (A.2.7) behaves as the Laplace equation for the jth harmonics, when $\bar{r} \to 0$ and also when $\bar{r} \to \infty$.

The matching condition (A.2.6) yields the following conditions on the Fourier coefficients,

$$\bar{\psi}_{11} \to [\dot{Y}^{(1)} + X^{(1)}\Psi_{xx}(X^{(0)}, Y^{(0)}) + Y^{(1)}\Psi_{xy}(X^{(0)}, Y^{(0)})]\,\bar{r}\,, \quad (A.2.9a)$$

$$\bar{\psi}_{12} \to [-\dot{X}^{(1)} + X^{(1)}\Psi_{yx}(X^{(0)}, Y^{(0)}) + Y^{(1)}\Psi_{yy}(X^{(0)}, Y^{(0)})]\,\bar{r}\,, \quad (A.2.9b)$$

$$\bar{\psi}_{21} \to [\Psi_{xx} - \Psi_{yy}](t, X^{(0)}, Y^{(0)})\,\bar{r}^2/4 = \Psi_{xx}(t, X^{(0)}, Y^{(0)})\,\bar{r}^2/2\,, \quad (A.2.10a)$$

$$\bar{\psi}_{22} \to \Psi_{xy}(t, X^{(0)}, Y^{(0)})\,\bar{r}^2/2\,, \quad (A.2.10b)$$

and

$$\bar{\psi}_{jk} \to 0 \qquad \text{for} \quad j = 3, 4, \ldots,\ k = 1, 2\,. \quad (A.2.11)$$

Note that the contributions of the local background velocity gradients appear in the inhomogeneous terms in (A.2.10a, b) for the second harmonics of the inner solution.

By repeating the arguments used in **Sec. 2.2.3** for the Fourier coefficients of $\bar{\psi}^{(1)}$, we obtain the following results for those of $\bar{\psi}^{(2)}$:

$$\bar{\psi}_{jk} \equiv 0\,, \qquad \text{for} \quad j = 1\,,\ j = 3, 4, \ldots \text{ and } k = 1, 2\,. \quad (A.2.12)$$

Conditions (A.2.9a, b) in turn yield

$$\begin{aligned} \dot{X}^{(1)} &= X^{(1)}\Psi_{yx}(X^{(0)}, Y^{(0)}) + Y^{(1)}\Psi_{yy}(X^{(0)}, Y^{(0)})\,, \\ \dot{Y}^{(1)} &= -X^{(1)}\Psi_{xx}(X^{(0)}, Y^{(0)}) - Y^{(1)}\Psi_{xy}(X^{(0)}, Y^{(0)})\,. \end{aligned} \quad (A.2.13)$$

By combining this equation for the first order velocity with (2.2.58) for the leading order velocity, we get

$$\hat{\imath}\dot{X} + \hat{\jmath}\dot{Y} \;=\; \hat{\imath}\Psi_y(X, Y) - \hat{\jmath}\Psi_x(X, Y) + O(\epsilon^2)\,. \quad (A.2.14)$$

Thus we conclude :

- *In the normal time scale, the velocity of the vortex center defined by the asymptotic analysis differs from the local background velocity by no more than $O(\epsilon^2)$. That is to say that the trajectory of the vortex center deviates by at most $O(\epsilon^2)$ from the stream line of the background potential flow passing through the initial position of the vortex center.*

The two nontrivial Fourier coefficients $\bar{\psi}_{2k}$, $k = 1, 2$, are governed by the differential equation (A.2.7) in $\bar{r}$ for $j = 2$, the boundary condition $\bar{\psi}_{2k} = 0$ at $\bar{r} = 0$ and the matching conditions (A.2.10a and b). These coefficients can be related to one canonical solution $\bar{\psi}_{2c}(t, \bar{r})$ as follows:

$$\bar{\psi}_{21} = \frac{1}{2}[\Psi_{xx} - \Psi_{yy}](X^{(0)}, Y^{(0)})\,\bar{\psi}_{2c} \qquad \text{and} \qquad \bar{\psi}_{22} = \Psi_{xy}(X^{(0)}, Y^{(0)})\,\bar{\psi}_{2c}\,. \quad (A.2.15)$$

The canonical solution $\bar{\psi}_{2c}$ obeys the differential equation (A.2.7) with $j = 2$ and the boundary conditions

$$\bar{\psi}_{2c} = 0 \quad \text{at} \quad \bar{r} = 0 \qquad \text{and} \qquad \bar{\psi}_{2c} \to \bar{r}^2/2 \quad \text{as} \quad \bar{r} \to \infty\,. \tag{A.2.16}$$

Thus we can determine the canonical solution for a given leading order core structure in (A.2.7) with $j = 2$. Since the equation behaves as the Laplace equation for the second harmonics as $\bar{r} \to \infty$, the solution $\bar{\psi}_{2c}$ behaves as

$$\bar{\psi}_{2c} = \bar{r}^2/2 + c(t)\,\bar{r}^{-2} + o(\bar{r}^{-m}) \tag{A.2.17}$$

for a large m and the function $c(t)$ is defined by the canonical solution.

For example, when $\zeta^{(0)}$ is represented by the similarity solution (2.2.45), we have

$$\frac{\bar{\zeta}^{(0)}_{\bar{r}}}{\bar{\psi}^{(0)}_{\bar{r}}} = \frac{\beta(\eta)}{4\bar{\nu}\bar{t}} \tag{A.2.18a}$$

with

$$\beta(\eta) = 4\eta^2/(e^{\eta^2} - 1)\,, \tag{A.2.18b}$$

where $\bar{t} = t + t_0^*$ and $\eta^2 = \bar{r}^2/(4\bar{\nu}\bar{t})$. We can then express the canonical solution in terms of $\bar{t}$ and η with $\bar{\psi}_{2c} = 2\bar{\nu}\bar{t}g(\eta)$. Equation (A.2.7) becomes an equation for $g(\eta)$ with the independent variable $\bar{r}$ replaced by η and the ratio in (A.2.18a) by $\beta(\eta)$. The boundary conditions (A.2.16) and (A.2.17) become $g(0) = 0$ and $g(\eta) \sim \eta^2 + C\eta^{-2}$ as $\eta \to \infty$. A numerical solution of $g(\eta)$ was constructed and the constant C was found to be $C = 4.37$. The coefficient $c(t)$ in (A.2.17) then becomes $8C(\bar{\nu}\bar{t})^2$. *

From the behavior (A.2.17) of the canonical solution and with the asymmetric part of the second order inner solution, $\epsilon^2[\bar{\psi}_{21}\cos 2\theta + \bar{\psi}_{22}\sin 2\theta]$, related to the canonical solution by (A.2.15), we see from (A.2.10a, b) and (A.2.17) that as $\bar{r} \to \infty$ the $\bar{r}^2$ terms in the inner solution match with the contributions of the local velocity gradients of the outer solution. The $\bar{r}^{-2}$ terms in the inner solution then induce $\epsilon^4 r^{-2}$ terms , quadrupoles, in the outer solution. Thus we have shown that

- *The leading contribution of the vortical core to the outer stream function other than that of a classical point vortex is a fourth order quadrupole,* $\epsilon^4 c(t) r^{-2}[\Psi_{xx}\cos 2\theta + \Psi_{xy}\sin 2\theta]_{\text{at } X^{(0)},Y^{(0)}}$. *Its strength depends on the local velocity gradient of the background flow and the global effect of the leading order core structure through the function* $c(t)$.

As mentioned before, the asymmetric part of a one-time inner solution is a quasi-steady solution because it is governed by a partial differential equation in $\bar{r}, \theta$ with time t acting as a parameter. In particular, the first order asymmetric

* Details of the analysis were described in Ting and Tung (1965). However, their function α is in error by a factor of 1/2 which results in an incorrect value for the constant C.

part of the one-time solution is identically zero and the velocity of the vortex center is defined by the background velocity, (2.2.66). Thus the one time solution does represent the complete first order solution if it happens to be in agreement with the initial data. In other words, a first order two-time solution will be needed only if the initial vorticity distribution has an $O(\epsilon^{-1})$ asymmetric contribution and/or if the initial velocity of the vortex disagrees with the local background velocity. The latter case arises for example when a vortex that is held fixed for $t < 0$ is suddenly released at $t = 0$.

Using condition (A.2.2) on the first order two-time solution, we can express the inhomogeneous term $\tilde{F}_2$ as a θ-derivative, i. e., replace (A.2.3) by

$$\tilde{F}_2 = -\frac{\partial}{\partial\theta}\left[\frac{1}{2}[(\bar{\omega}\bar{a} - \bar{b})(\tilde{\zeta}^{(1)})^2]_{\bar{r}} + \tilde{\zeta}^{(1)}\tilde{\psi}^{(1)}_{\bar{r}}\right] , \qquad (A.2.19a)$$

and, therefore, not only its short time average but also its θ-average vanish:

$$\{\tilde{F}_2\} = 0 . \qquad (A.2.19b)$$

Consequently, the second order equation (2.2.40a) for the symmetric two-time solution becomes:

$$(\tilde{\zeta}^{(2)}_c)_\tau = 0 \qquad \text{or} \qquad \tilde{\zeta}^{(2)}_c = \tilde{\zeta}^{(2)}_c(t, \bar{r}) , \qquad (A.2.20)$$

which implies that the symmetric part of the second order inner solution does not depend on the short time.

Note that (2.2.36a) or (2.2.37) state that, quite generally, the symmetric part of the first order inner solution is varying only on the normal time, because the leading order inner solution is symmetric and τ-independent and thus, $\tilde{F}_1 \equiv 0$. In contrast, (A.2.20) holds only if the first order two-time solution satisfies the special condition (A.2.2). With this understanding, we may conclude that :

- *The symmetric parts of the zeroth, the first and the second order vorticity distributions have only a normal time dependence*

Finally we point out that it was shown in **Sec. 2.2.1**, following (2.2.34), that the leading order inner solution in the two-time analysis coincides with the symmetric one-time solution. We can not draw the same conclusion even for the symmetric first order two-time solution let alone the second order solution, although they are both independent of τ. The reason is provided in Appendix A.1. We recall that in the two-time analysis the governing equation for the symmetric second order solution is given by (A.1.10) for $n = 2$. This equation is obtained from the τ- and θ-average of the third order equation. The inhomogeneous term $\mathcal{M}\{\tilde{F}_3\}$ in (A.1.10) contains the average of products of the first and second order asymmetric two-time solutions and does not, in general, vanish identically. This inhomogeneous term is absent in the one-time analysis.

Appendix A.3. Equations of Motion of Filaments

In **Sec. 2.3.4** we uncoupled the equations for the core structure of a filament from those for the motion of its centerline and expressed the global contributions of the core structure to the motion in terms of the initial core structure and the length of the filament. Consequently, a closed system of equations of motion for the filament was obtained. As mentioned in **Sec.2.3.4.3**, we list here the system of equations and define all the symbols so that we do not have to refer back to the main text.

Let us consider the flow field of N vortex filaments in a background flow with velocity potential $\Phi(\mathbf{x})$. Let Γ_i and $\mathbf{X}_i(s,t)$ denote the circulation and position vector of the centerline of the ith filament, for $i = 1, 2, \ldots, N$. The parameter s_i increases in the direction of positive Γ_i. For each filament, its centerline and its inner structure, i.e., the large axial and circumferential velocity distribution, are prescribed through appropriate initial conditions. We collect now all formulae that describe, say, the ith filament embedded in an outer flow that is the superposition of the outer potential field, $\nabla\Phi$, and the velocities induced by the other filaments. In these formulae we suppress the subscript i unless necessary. Also, we omit the superscript (0), denoting the leading order entities.

At $t = 0$, the vector function defining the centerline as a simple closed curve C is prescribed as

$$\mathbf{X}(0,s) \quad \text{and} \quad 0 \le s \le 2\pi \qquad \text{with} \quad \mathbf{X}(0,s+2\pi) = \mathbf{X}(0,s)\ . \tag{A.3.1}$$

The initial profiles of axial vorticity and velocity, ζ and w, are

$$\zeta(0,\mathbf{x}) = \epsilon^{-2}\zeta_0(\bar{r}) \quad \text{and} \quad w(0,\mathbf{x}) = \epsilon^{-1}w_0(\bar{r}) \tag{A.3.2}$$

where $\bar{r} = r/\epsilon$ and $\epsilon = \delta^*/\ell$. Here ℓ denotes a typical length scale of the flow field such as a typical length of a filament and δ^* a is typical core size. The initial profiles are independent of the circumferential coordinate, θ, and of s and they decay exponentially in $\bar{r}$. This is consistent with conditions (2.3.18) and (2.3.19a) for the asymptotic analysis. The variables r, θ and s are the curvilinear coordinates relative to C used to describe the core structure. They are related to a position vector, $\mathbf{x}$, in space by

$$\mathbf{x} = \mathbf{X}(t,s) + r\hat{r}(t,s)\ , \tag{A.3.2}$$

where r denotes the minimum distance from $\mathbf{x}$ to C and $\hat{r}$ the unit radial vector.

We denote the unit tangent, normal and binormal vectors of the centerline by $\hat{\tau}$, $\hat{n}$ and $\hat{b}$ and its curvature and torsion by κ, and T, respectively. These quantities are related to $X(t,s)$ by the Serrett-Frenet formulae (2.1.45).

The equation for the velocity of the ith filament is

$$\mathbf{X}_t(t,s) = \mathbf{Q}^*(t,\mathbf{X}) + \frac{\Gamma\kappa}{4\pi}\left[\ln\frac{1}{\epsilon} + C_v(t)\right]\hat{b} + \frac{\Gamma\kappa}{4\pi}C_w(t)\,\hat{b}\ , \tag{A.3.3}$$

where

$$\mathbf{Q}^* = \mathbf{Q}_2 + \mathbf{Q}^f - \hat{\tau}[(\mathbf{Q}_2 + \mathbf{Q}^f)\cdot\hat{\tau}]\ . \tag{A.3.4}$$

The removal of the tangential component insures that $\hat{\tau}\cdot\mathbf{X}_t = 0$. In (A.3.4), $\mathbf{Q}_2(t,\mathbf{X})$ denotes the velocity of the flow field at $\mathbf{X}$ in the absence of the ith filament,

$$\mathbf{Q}_2 = \nabla\Phi(\mathbf{X}) + \frac{1}{4\pi}\sum_{\substack{j=1\\ j\neq i}}^{N}\Gamma_j\int_{C_j}\frac{\mathbf{X}_j - \mathbf{X}}{|\mathbf{X}_j - \mathbf{X}|^3}\times d\mathbf{X}_j\ , \tag{A.3.5}$$

and $\mathbf{Q}^f(t,s)$ is the finite part of the Biot-Savart integral for the ith filament at $\mathbf{X}$. In the matched asymptotic analysis, the singular terms in the Biot-Savart integral are cancelled analytically and $\mathbf{Q}^f$ is given by

$$\mathbf{Q}^f(t,s) = \frac{\Gamma}{4\pi}\left\{\int_{-\pi}^{\pi}\mathbf{G}(t,s+\bar{s},s)\,d\bar{s} + \left[\ln(2\sqrt{S_+S_-}) - 1\right]\kappa\hat{b}\right\}\ , \tag{A.3.6}$$

where

$$\mathbf{G}(t,s',s) = \mathbf{F}(t,s',s) - 2\kappa\hat{b}/|\lambda|\ , \qquad (s'\neq s)\ , \tag{A.3.7a}$$

$$= \frac{\hat{\tau}\times\mathbf{B}}{3}\,\mathrm{sgn}(s'-s), \qquad (s' = s\pm 0)\ , \tag{A.3.7b}$$

$$\mathbf{F}(t,s',s) = \frac{\mathbf{X}(t,s') - \mathbf{X}(t,s)}{|\mathbf{X}(t,s') - \mathbf{X}(t,s)|^3}\times\mathbf{X}_s(t,s)\ , \tag{A.3.8}$$

with

$$\lambda(t,s',s) = \int_s^{s'}\sigma(t,s^*)\,ds^*\ , \tag{A.3.9}$$

$$\sigma(t,s) = |\mathbf{X}_s(t,s)|\ , \tag{A.3.10}$$

$$\mathbf{B}(t,s) = \frac{\partial\kappa}{\partial s}\hat{n} + \kappa T\sigma\hat{b}\ , \tag{A.3.11}$$

$$S_\pm = \lambda(t,s\pm\pi,s) \tag{A.3.12}$$

and

$$S(t) = S_+(t) + S_-(t) = \int_0^{2\pi}\sigma(t,s)\,ds\ . \tag{A.3.13}$$

Here $S(t)$ denotes the length of C. The integrand $\mathbf{G}$ in (A.3.6) is a piecewise continuous function of $\bar{s} = s' - s$, with a jump discontinuity at $\bar{s} = 0$ as defined in (A.3.7b). Therefore, $\mathbf{Q}^f$ given by (A.3.6) can be evaluated numerically at each instant. The remaining unknowns in (A.3.4) are $C_v(t)$ and $C_w(t)$. They denote, respectively, the global contributions of the large circumferential and axial flow and are related to the inner structure of the filament. The core size is defined by

$$\delta(t) = [4\nu\tau_1(t)/S(t)]^{1/2} \tag{A.3.14}$$

$$\text{where} \quad \tau_1 = \int_0^t S(t')\,dt' + \tau_{10} \tag{A.3.15}$$

$$\text{with} \quad \tau_{10} = \frac{\pi S_0}{2\Gamma\bar{\nu}} \int_0^\infty \zeta_0(\bar{r})\,\bar{r}^3\,d\bar{r} \tag{A.3.16}$$

The improper integrals in the definition (2.3.73) of the global contribution C_v due to the larger circumferential flow were evaluated analytically and C_v is found to be

$$C_v(t) = \ln\left(\frac{1}{\bar{\delta}(t)}\right) + \frac{1}{2}\{1+\gamma-\ell n\,2\} + 4\pi\sum_{n=1}^{N}\alpha_n\tau_1^{-n}(t) + 8\pi^2\sum_{n=2}^{N}\gamma_n\tau_1^{-n}(t) \tag{A.3.17}$$

with

$$\alpha_n = \frac{\bar{\nu}}{\Gamma}D_n\sum_{j=1}^{N}p_{n,j}\left(1-\frac{1}{2^j}\right)(j-1)!$$

$$\gamma_h = \sum_{m=1}^{h}\left(\frac{\bar{\nu}}{\Gamma}\right)^2 D_m D_{h-m}\beta_{m,h-m}\,, \tag{A.2.18}$$

$$\beta_{n,h} = \sum_{i=1}^{n}(p_{n-1,i}-p_{n,i})\sum_{j=1}^{h}\frac{(i+j-1)!}{2^{i+j}}(p_{h-1,j}-p_{h,j}).$$

To compute the global contribution $C_w(t)$ due to the large axial flow, we introduce $\delta^w(t), \tau_1^w(t), \tau_{10}^w$ in analogy to (A.3.14 - 16),

$$\delta^w(t) = [4\nu\tau_1^w(t)/S(t)]^{1/2}, \tag{A.3.19a}$$

$$\tau_1^w = \int_0^t S(t')\,dt' + \tau_{10}^w \tag{A.3.19b}$$

and

$$\tau_{10}^w = \frac{\pi S_0}{2\bar{\nu}m(0)}\int_0^\infty w_0(\bar{r})\,\bar{r}^3\,d\bar{r} \tag{A.3.19c}$$

and obtain

$$C_w = -\frac{16\pi^2}{[S^{(0)}]^3\tau_1^w}\sum_{n=0}^{N}\omega_n[\tau_1^w]^{-n} \tag{A.3.20}$$

with

$$\omega_n = \sum_{m=0}^{n}C_m C_{n-m}P_{m,n-m}, \quad P_{n,i} = \sum_{j=0}^{N}p_{i,j}\sum_{k=0}^{n}p_{n,k}(j+k)!/2^{j+k+1}\,. \tag{A.3.21}$$

Here γ is the Euler number, $p_{n,k}$ is the coefficient of x^k in the Laguerre polynomial $L_n(x)$ and $(N+1)$ is the number of terms used in the series representations

(2.3.80a,b) of w and ζ to fit the initial data (A.3.2) which in turn define the coefficients in the series. They are

$$C_n = 2S_0\left(\frac{\tau_{20}^w}{S_0}\right)^{n-1} \int_0^\infty \bar{w}_0(\bar{r}) L_n(\eta^2)\, \eta \, d\eta \,, \tag{A.3.22}$$

$$D_n = \frac{2}{S_0}\left(\frac{\tau_{20}}{S_0}\right)^{n+1} \int_0^\infty \bar{\zeta}_0(\bar{r}) L_n(\eta^2)\, \eta \, d\eta \,, \tag{A.3.23}$$

where $\bar{r} = \eta\sqrt{4\bar{\nu}\tau_{10}/S_0}$, $S_0 = S(0)$, $\delta_0 = \sqrt{4\nu\tau_{10}/S_0}$ in (A.3.23). For the C_n's in (A.3.22) we replace τ_{10} by τ_{10}^w. Note that $C_1 = 0$ and $D_1 = 0$ because of the choice of τ_{10} and τ_{10}^w (see (2.3.82)). If the initial profiles are similar and $\tau_{10}^w = \tau_{10}$, we have $C_n = D_n = 0$ for $n \geq 1$.

Equations (A.3.4 and 15), supplemented by the remaining equations in this Appendix applied to the ith filament for $i = 1, \ldots, N$, form a closed system of equations of motion for the N filaments.

Appendix A.4. Formulae for the Coefficients in (3.2.60 and 61)

This appendix was first referred to in **Sec.3.2.2.** We defined

$$a_{km} = \Gamma_k \Gamma_m a^*_{km} \ , \ b_{km} = \Gamma_k \Gamma_m b^*_{km} \ \text{ and } \ e_{km} = \Gamma_k \Gamma_m e^*_{km} \ , \qquad (A.4.1)$$

with

$$a^*_{km} = \langle \partial_x \zeta^*_k \ \partial_x \zeta^*_m \rangle \ , \quad b^*_{km} = \langle \partial_y \zeta^*_k \ \partial_y \zeta^*_m \rangle \quad \text{and} \quad e^*_{km} = \langle \partial_x \zeta^*_k \ \partial_y \zeta^*_m \rangle \ , \quad (A.4.2)$$

so that the (*) quantities are independent of the strengths Γ_k, Γ_m and are functions of $X_k - X_m$, $Y_k - Y_m$, δ_k and δ_m. By converting the integrand to that for an error function, we obtain

$$a^*_{km} = \frac{2}{\pi \delta^4_{km}} e^{-R^2_{km}/\delta^2_{km}} \left[1 - \frac{2(X_k - X_m)^2}{\delta^2_{km}}\right] \qquad (A.4.3)$$

and

$$e^*_{km} = e^*_{mk} = \frac{-4(X_k - X_m)(Y_k - Y_m)}{\pi \delta^6_{km}} e^{-R^2_{km}/\delta^2_{km}} \ , \qquad (A.4.4)$$

where $\delta^2_{km} = \delta^2_m$ and $R^2_{km} = (X_k - X_m)^2 + (Y_k - Y_m)^2$.

We note that R_{km} denotes the distance between the k-th and m-th center. The coefficient b^*_{km} is given by (A.4.3) with $(X_k - X_m)^2$ replaced by $(Y_k - Y_m)^2$. For the constants on the right side of (3.2.50), we define

$$C_k = \Gamma_k \sum_m \Gamma_m \sum_{\ell \neq m} \Gamma_\ell c^*_{km,\ell} \ , \qquad (A.4.5)$$

with

$$\begin{aligned} c_{km,\ell} &= \langle (\partial_x \zeta^*_k (\boldsymbol{V}^*_\ell \cdot \nabla \zeta^*_m) \rangle \\ &= \langle (\partial_x \zeta^*_k) [-(y - Y_\ell)\, \partial_x \zeta^*_m + (x - X_\ell)\, \partial_y \zeta^*_m][1 - e^{-r^2_\ell/\delta^2_\ell}]/r^2_\ell \rangle \ . \end{aligned}$$

By using the polar coordinates relative to the ℓ-th vortex center and the integral representations of the modified Bessel functions, we identify the above area integral as the sum of integrals of Bessel functions, see formula 6.618-4 and 6.631-4 of Gradshteyn (1965). The final result is

$$\begin{aligned} c^*_{km,\ell} = \frac{2}{\pi^3} e^{-R^2_{km}/\delta^2_{km}} & \left[-X^\ell_k (X^\ell_m \sin\tau - Y^\ell_m \cos\tau) I^\ell_{km} \right. , \\ & \left. - \frac{Y^\ell_m}{2} II^\ell_{km} + \frac{1}{2}(X^\ell_m \sin 2\tau - Y^\ell_m \cos 2\tau) III^\ell_{km}\right] \ , \end{aligned} \qquad (A.4.6)$$

where

$$X_j^\ell = X_j - X_\ell\,,\quad Y_j^\ell = Y_j - Y_\ell\,,\qquad j = k, m\,,$$
$$X_{km} = (X_k\delta_k^2 + X_m\delta_m^2)\,/\,(\delta_k^2 + \delta_m^2)\,,\quad Y_{km} = (Y_k\delta_k^2 + Y_m\delta_m^2)\,/\,(\delta_k^2 + \delta_m^2)\,,$$
$$R_{km,\ell}^2 = (X_{km} - X_\ell)^2 + (Y_{km} - Y_\ell)^2\,,$$
$$X_{km} - X_\ell = X_{km,\ell}\cos\tau\,,\quad Y_{km} - Y_\ell = R_{km,\ell}\sin\tau\,.$$

We note that the point (X_{km}, Y_{km}) is the center of gravity of the k-th and m-th core weighted by the inverse of the square of their core size and R_{km}^ℓ denotes the distance between (X_{km}, Y_{km}) and the ℓ-th center. The functions I, II and III in (A.4.6) are

$$I_{km}^\ell = \beta_{km,\ell}[1 - e^{-R_{km,\ell}^2/\delta_{km,\ell}^2}]\quad II_{km}^\ell = \frac{1}{2}[\alpha_{km} - \gamma_{km,\ell}\, e^{-R_{km,\ell}^2/\delta_{km,\ell}^2}]$$

and

$$III_{km}^\ell = \frac{1}{2}(\alpha_{km} - \gamma_{km,\ell}) + \frac{1}{2}(\gamma_{km,\ell} - 4\beta_{km,\ell}^2)(1 - e^{-R_{km,\ell}^2/\delta_{km,\ell}^2}]\,,$$

where

$$\alpha_{km} = \delta_k^2\delta_m^2/(\delta_k^2 + \delta_m^2)\,,\quad \beta_{km,\ell} = \alpha_{km}/R_{km,\ell}\,,$$
$$\gamma_{km,\ell} = (\delta_k^2\delta_m^2 + \delta_m^2\delta_\ell^2 + \delta_\ell^2\delta_k^2)/(\delta_k\delta_m\delta_\ell)^2\quad\text{and}$$
$$\delta_{km,\ell}^2 = (\delta_k^2\delta_m^2 + \delta_m^2\delta_\ell^2 + \delta_\ell^2\delta_k^2)/(\delta_k^2 + \delta_m^2)\,.$$

Likewise, we define

$$D_k = \Gamma_k \sum_m \Gamma_m \sum_{\ell\neq m} \Gamma_\ell d_{km,\ell}^*\,,\qquad \text{with}$$

$$\begin{aligned} d_{km,\ell}^* &= \langle(\partial_y\zeta_k^*)(\boldsymbol{V}_\ell^*\cdot\nabla\zeta_m^*)\rangle \\ &= \frac{2}{\pi^3}e^{-R_{km,\ell}^2/\delta_{km}^2}\left[-Y_k^\ell(X_m^\ell\sin\tau - Y_m^\ell\cos\tau)I_{km}^\ell\right. \\ &\qquad \left. + \frac{1}{2}X_m^\ell II_{km}^\ell - \frac{1}{2}(X_m^\ell\cos 2\tau + Y_m^\ell\sin 2\tau)III_{km}^\ell\right]\,. \end{aligned}$$

Thus all the coefficients $a_{km}, \ldots$, in (3.2.50), which are defined by area integrals in the xy plane, are reduced to elementary functions of the strengths and core sizes and the coordinates of one center relative to the other.

References

Balch, D. T. (1985): "Experimental Study of Main Rotor - Tail Rotor - Airframe Interaction in Hover" J. Amer. Helicopter Soc. **30**, 49-56.

Batchelor, G. K. (1967): *An Introduction io Fluid Dynamics*, Cambridge University Press.

Bauer, L., Morikawa, G.K. (1976): "Stability of Rectilinear Geostrophic Vortices in Stationary Equilibrium", Phy. Fluid **19**, 929-942.

Berezovskii, A. A., Kaplanskii, F. B. (1988) "Diffusion of a Ring Vortex", Fluid Dynamics, **22**, 832-836, Plenum Publ., Translated from Izvestiya Akademii Nauk USSR, Mekhanika Zhidkosti i Gaza, **6**, 10-15, 1987.

Breuer M., Haenel D. (1989): "Solution of the 3-D Navier Stokes Equations for the Simulation of Vortex Breakdown", 8th GAMM Conf. on Numerical Fluid Mechanics, to appear in Lecture Notes on Numerical Fluid Dynamics, Vieweg.

Callegari, A., Ting, L. (1978): "Motion of a Curved Vortex Filament with Decaying Vortical Core and Axial Velocity", SIAM J. Appl. Math., **35**, 148-175.

Chorin, A.J. (1988): "Scaling Laws in the Lattice Vortex Model of Turbulence", Comm. Math. Phys., **4**, 167-176.

Chorin, A.J. (1989): "Constraint Random Walks and Vortex Filaments in Turbulence Theory", preprint, Sept1989.

Courant, R., Hilbert, D. (1953): *Methods of Mathematical Physics*, Interscience Publishers, New York.

Chamberlain, J. P., Liu, C. H., (1985): "Navier-Stokes Calculations for Three Dimensional Vortical Flows in Unbounded Domain", AIAA J., **23** , 868-874.

Chamberlain, J. P., Weston, R. P. (1984): "Three-Dimensional Navier-Stokes Calculations of Multiple Interacting Vortex Rings", AIAA Paper 84-1545.

Crow, S. C. (1970): "Aerodynamic Sound Emission as a Singular Perturbation Problem", Studies in Appl. Math. **XLIX**, 21-44.

Doligalski, T. L., Walker, J. D. (1984): "The Boundary Layer Induced by a Convected Two-dimensional Vortex", J. Fluid Mech. **139**, 1-28.

van Dyke, M. (1975): *Perturbation Methods in Fluid Dynamics*, Parabolic Press, Stanford, CA.

van Dyke, M. (1982): *An Album of Fluid Mechanics*, Parabolic Press, Stanford CA.

Flaschka H., Newell A. C. (1975): "Integrable Systems of Integrable Evolution Equations", in *Dynamical Systems, Theory and Applications*, Lecture Notes in Physics, **38**, 355-440.

Fohl, T. and Turner, J. S. (1975): "Colliding Vortex Rings", Phys. Fluids, **18**, 433-436.

Fung, Y. T., Liu, C. H. and Gunzburger, M. D. (1979): "Simulation of the Pressure Field near a Jet by Randomly Distributed Vortex Rings", AIAA J., **17**, 553-557.

Görtler, H. (1944): "Einige Bemerkungen über Strömungen in rotierenden Flüssigkeiten", ZAMM **24**, 210.

Gradshteyn, I. S. and Ryzhik, I. M. (1965): *Table of Integrals, Series and Products*, Academic Press, New York.

Guiraud, J. P., Zeytounian, R. Kh. (1977): "A Double-Scale Investigation of the Asymptotic Structure of Rolled-Up Vortex Sheets", J. Fluid Mech., **79**, 93-112.

Guiraud, J. P., Zeytounian, R. Kh. (1979): "Rotational Compressible Inviscid Flow with Rolled Vortex Sheets. An Analytical Algorithm for the Computation of the Core", J. Fluid Mech., **101**, 393-401.

Guiraud, J. P., Zeytounian, R. Kh. (1982): "Vortex Sheets and Concentrated Vorticity. A Variation on the Theme of Asymptotic Modeling in Fluid Mechanics", ONERA T.P. No. 1982-124.

Gunzburger, M. D. (l972): " Motion of Decaying Vortex Rings with Nonsimilar Vorticity Distribution", J. Engineering Math., **6**, 53-61.

Gunzburger, M. D. (l973): "Long Time Behavior of a Decaying Vortex", ZAMM, **53**, 751-760.

Gunzburger, M. D. (1989): *Finite Element Methods for Viscous Incompressible Flows*, Academic Press, New York.

Hall, M. G. (1961): "A Theory for the Core of a Leading Edge Vortex", J. Fluid Mech., **11**, 209-228.

Howard, L., (1967): Arch. Rat. Mech. Anal., **1**, 113-123.

Ishii, K., Liu, C. H. (1987):"Motion and Decay of Vortex Rings Submerged in a Rotational Flow" AIAA paper 87-0043.

Ishii, K., Hussain, F., Kuwahara, K., Liu, C. H. (1989): "The Dynamics of Vortex Rings in an Unbounded Domain", in *Advances in Turbulence 2*, Eds.: Fernholz, H. H. and Fiedler, H. E., 51-56, Springer-Verlag.

von Kármán, Th. and Burgers, J. M. (1963): "General Aerodynamic Theory-Perfect Fluids", in *Aerodynamic Theory* **II**, Editor Durand, W. H., Dover Publ., NY.

Keller J. (1980): "Darcy's Law for Flow in Porous Media", in *Nonlinear Partial Differential Equations in Engineering and Applied Sciences*, Eds.: R. L. Swirnberg, A. J. Katimowski, J. S. Papadkis, Dekker Publications, 429-443.

Kellog, O. D., (1967): *Fundations of Potential Theory*, 143, 236, Spring-Verlag.

Klein, R., Majda, A.J. (1990): "Self-Stretching of a Perturbed Vortex Filament I: The Asymptotic Equation for Deviations from a Straight Line", accepted for publication in Physica D, Sept. 1990.

Klein, R., Ting, L. (1990): "Far Field Potential Flow Induced by a Rapidly Decaying Vorticity Distribution", ZAMP, **41**, 395-418.

Kleinstein, G., Ting, L. (1971): "Optimum Solution for Heat Conduction Problems", ZaMM, **51**, 1-16.

Krause, E. (1980): " Strive for Accuracy-improvements of Predictions", Comp. Fluids, **8**, 31-57.

Krause E. (1990): "Vortex Breakdown: Physical Issue and Computational Simulation", 3rd Int. Conf. of Fluid Mechanics, Cairo Egypt, Jan. 1990.

Krause, E., Liu, C. H., Ting, L. (1985): "Vortex Dominated Flow with Diffusive Core Structure", AIAA paper 85-1556.

Krause E., Liu C. H. (1989): "Numerical Studies in Compressible Flows around Delta- and Double Delta Wings", Z. Flugwiss. Weltraumforschung, **12**, 291-301.

Lamb, H. (1932): *Hydrodynamics*, Dover Publ., New York.

Leonard, A. (1974): "Numerical Simulation of Interacting Three-Dimensional Vortex Filaments", Lecture Notes in Physics, **35**, 245-250, Springer-Verlag.

Lighthill, M. J. (1952): "On Sound Generated Aerodynamically: I. General Theory", Proc. Roy. Soc. **A211**, 564-587.

Ling, G. C., Ting, L. (1988):"Two time Scales Inner Solutions and Motion of a Geostrophic Vortex", Scientia Sinica, Ser. A, **31**, 806-817.

Liu, C. H., Ting, L. (1982): "Numerical Solutions of Viscous Flow in Unbounded Fluid", Lecture Notes in Physics **170** , 357-363, Springer-Verlag, 357-363.

Liu, C. H., Tavantzis, J., Ting, L. (1986): "Numerical Studies of Motion of Vortex Filaments – Implementing the Asymptotic Analysis", AIAA J., **24** , 1290-1297.

Liu, C. H., Ting, L., Weston, R. P. (1986): "Boundary conditions for N-S Solutions of Merging of Vortex Filaments", in *Numerical Methods in Fluid Dynamics*, Ed. Oshima, K. (Japan Soc. Compu. Fluid Dynamics) Vol. 2, 255-264.

Liu, C. H., Ting, L. (1987): "Interaction of Decaying Trailing Vortices in Spanwise Shear Flow", Computers and Fluids, **15**, 77-92.

Liu, G. C., Hsu C. H. (1984): "Numerical Studies of Interacting Vortices", in *Proceedings of 4th International Conference on Applied Numerical Modelling*, Eds. Hsia, H. M. et al (National Cheng Kung University, Taiwan), 656-665.

Lo, K. C. R. and Ting L. (1975): "Studies of the merging of Vortices", Division of Applied Sciences, New York University Report AA-75-10.

Lo, K. C. R. and Ting L. (1976): "Studies of the merging of Vortices", Phys. Fluids, **19**, 912-913.

Lugt, H. J. (1983): *Vortex Flow in Nature and Technology*, John Wiley, New York.

Majda, A. (1984): *Compressible Fluid Flow and Systems of Conservative Laws in Several Space Variables*, Springer-Verlag.

Maxworthy, T. (1977): "Some Experimental Studies of Vortex Rings", J. Fluid Mech., **81**, 465-495.

Milne-Thompson, L. M. (1973): *Theoretical Hydrodynamics* Dover Publ., New York.

Möhring, W. (l978): "On Vortex Sound at Low Mach Number", J. Fluid Mech., **85**, 685-691.

Moore, D. W., Saffman, P. G. (1972): "The Motion of a Vortex Filament with Axial Flow", Phil. Trans. Roy. Soc., London, Series **A 272**, 403-429.

Morikawa, G.K. (1960): "Geostrophic Vortex Motion", J. Meteorol., **17**, 148-158.

Morikawa, G.K. (1962): "On the Prediction of Hurricane Tracks Using a Geostrophic Point Vortex", in *Proceedings of the International Symposium on Numerical Weather Prediction in Tokyo*, Edited by S. Syono (Meteorol. Soc. Japan, Tokyo), 349-354.

Morikawa, G.K. and Swenson, E.V. (1971): "Interacting Motion of Rectilinear Geostrophic Vortices", Phys. Fluid **14**, 1058-1073.

Magnus, W., Oberhettinger, F., Soni, R. P. (1966): *Formulas and Theorems for the Special Functions of Mathematical Physics*, Springer-Verlag.

Moore, D. W. (1975): "The Rolling up of a Semi-Infinite Vortex Sheet", Proc. Roy. Soc., **A 345**, 417-430.

Moreau, J. J. (1948): "Sur Deux Theoremes Generaux de la Dynamique d'un Milieu Incompressible Illimite", C. R. Acad. Sci. Paris, **226**, 1420-1422.

Moreau, J. J. (1949): "Sur la Dynamique d'un Ecoulement Rotationnel", C. R. Acad. Sci. Paris, **229**, 100-102.

Obermeier, F., (1985): "Aerodynamic Sound Generation Caused by Viscous Processes", J. Sound Vibration, **99**, 111-120.

Olsen, J. H., Goldburg, A., Rogers, M. (Editors) (1971): *Aircraft Wake Turbulence*, Plenum Publ., New York.

Oshima, Y. and Asaka, S. (1975), "Interaction of Two Vortex Rings Moving Side by Side", Natural Science Report, **26**, 31-37.

Oshima, Y. and Asaka, S. (1977) "Interaction of Two Vortex Rings Along Parallel Axes in Air", J. Phys. Soc. Japan, **42**, 708-713.

Oshima, Y. and Izutsu, N. (1988): "Cross-linking of Two Vortex Rings", Phys. Fluids **31**, 2401-2404.

Payne, F. M., Ng, T. T., Nelson, R. C., Schiff, L. B. (1986): "Visualization and Flow Surveys of the Leading Edge Vortex Structure on Delta Wing Planforms", AIAA paper 86-0330.

Peace, A. J. and Riley, N. (1983): "A Viscous Vortex Pair in Ground Effect", J. Fluid Mech. **129**, 409-426.

Pedlosky, J. (1979): *Geophysical Fluid Dynamics*, Springer-Verlag.

Poincaré, H. (1893): *Théorie des Tourbillons*, Ed. G. Carré, chap. IV, Deslis Frères.

Prandtl, L. and Tietjens, O. G. (1957): *Applied Hydro- and Aeromechanics*, Dover Publ., New York.

Reyna, L., Menne, S. (1988): "Numerical Prediction of Flow in Slender Vortices", Comp. & Fluids, **16**, 239-256.

Ribner, H. S. (1962): "Aerodynamic Sound from Fluid Dilatation", Univ. Toronto, Inst. Aerophysics Report No. 86.

Saffman, P. G. (1970): "The Velocity of Viscous Vortex Rings", Studies in Appl. Math., **49**, 371-380.

Schneider, W. (1978): *Mathematische Methoden in der Strömungsmechanik*, Vieweg.

Schlichting, H. (1978): *Boundary Layer Theory*, McGraw-Hill.

Schmitz, F. H. and Yu, Y. H. (1986): "Helicopter Impulsive Noise: Theoretical and Experimental Status", J. Sound Vib. **109**, 361-422.

Schrauf, G. and Krause, E. (1985): " Symmetric and asymmetric Taylor Vortices in a Spherical Gap", IUTAM Symposium Novosibirsk 1984. Ed. V. V. Kozlov, Springer Verlag.

Sears, W. R. (Editor) (1954), *General Theory of High Speed Aerodynamics*, Princeton University Press.

Sheridan, P. F. and Smith, R. P. (1980): "Interactional Aerodynamics - A New Challenge to Helicopter Technology", J. Amer. Helicopter Soc. **25**, 3-21.

Staufenbiel, R. (1985): Chairman, Special Research program 25, *Colloquium on Vortex Breakdown*, RWTH Aachen.

Stewart, H. J. (1943): "Periodic Properties of the Semi-Permanent Atmospheric Pressure Systems", Q. Appl. Math. **1**, 262-267.

Struik, D. J. (1961): *Lectures on Classical Differential Geometry*, Addison - Wesley, Reading, MA.

Swarztrauber, P. N. and Sweet, R. A. (1979): "Alogorithm 541, Efficient FORTRAN Subprograms for the solution of Separable Elliptic Partial Differential Equations [D3]", ACM Transactions on Mathematical Software, **5**, 352-364.

Taylor, G. I. (1921): " Experiments with Rotating Fluids", Proc. Roy. Soc. London, *A* **100**, 114.

Taylor, G. I. (1922): "The Motion of a Sphere in a Rotating Liquid" Proc. Roy. Soc. London, *A* **102**, 180.

Ting, L., Chen, S. (1967): "Perturbation Solutions and Asymptotic Solutions in Boundary Layer Theory", J. Engineering Math., **1**, 327-340.

Ting, L., Tung, C. (1965): "Motion and Decay of a Vortex in Nonuniform Stream", Phys. Fluids, **8**, 1309-1051.

Ting, L. (1971): "Studies in the Motion and Decay of Vortices", in *Aircraft Wake Turbulence*, Ed. by Olsen, J. H., Goldburg, A., Rogers, M., Plenum Publ., New York, 11-39.

Ting, L. (1981): "Studies on the Motion and Decay of Vortex Filaments", Lecture Notes in Physics, **148**, 67-105 Springer-Verlag.

Ting, L. (1983): "On the Application of the Integral Invariants and Decay Laws Of Vorticity Distributions", J. Fluid Mech., **127**, 497-506.

Ting, L., Ling, G. C. (1983): "Studies of the Motion and Core Structure of a Geostrophic Vortex", Proc. 2nd Asian Congress of Fluid Mechanics, 900-905, Science Press, Beijing, China.

Ting, L. (1986): "Theoretical and Numerical Studies of Vortex Interaction and Merging", in *Numerical Methods in Fluid Dynamics I*, Ed.: K. Oshima, Japan Soc. Compu. Fluid Dynamics, 218-229.

Ting, L., Liu, G. C. (1986): "Merging of Vortices with Decaying Cores and Numerical Solutions of Navier-Stokes Equations", Lecture Notes in Physics **264**, 612-616, Springer Verlag.

Ting, L., Miksis, M. J. (1990): "On Vortical Flow and Sound Generation", SIAM J. Appl. Math., **50**, 521-536.

Truesdell, C. (1951): "Vorticity Averages", Can. J. Math., **3**, 69-86.

Truesdell, C. (1954): *The Kinematics of Vorticity*, Indiana University Press.

Tung, C., Ting, L. (1966): "Motion and Decay of a Vortex Ring", Phys. Fluids, **10**, 901-910.

Tung, C., Pucel, S. L., Caradonna, F. X. and Morse, H. A. (1983): "The Structure of Trailing Vortices Generated by Model Rotor Blades", Vertica **7**, 33-43.

Wang, H. C. (1970): " The Motion of a Vortex Ring in the Presence of a Rigid Sphere", Annual Report of the Institute of Physics, Academia Sinica, Taiwan, 85-93.

Wendt, J. F. (1982): Lecture Series Director, *High-Angle of Attack Aerodynamics*, AGARD lecture Series Preprint No. 342.

Wentz, W. H., Kohlmann, D. L. (1971): "Vortex Breakdown on Slender Sharp-Edged Wings", Journal of Aircraft, **8**, 150-161.

Wentz W. H., Kohlmann D. L. (1971): "Vortex Breakdown on Slender Sharp-Edged Wings", J. Aircraft, **8**, 156-161.

Weston, R. P. , Ting, L., Liu, C. H. (1986): "Numerical Studies of the Merging of Vortices", AIAA Paper 86-0557.

Wu, J. C. and Thompson, J. F. (1973): "Numerical Solutions of Time-Dependent Incompressible Navier-Stokes Equations Using an Integro-Differential Formulation", Comput. Fluids, **1**, 197-215.

Yih, C. S. (1979): *Fluid Mechanics*, West River Press, Ann Arbor.

Young, A. D. (1983): Symposium Chairman, *Aerodynamics of Vortical Type flows in Three Dimensions*, AGARD Conference Proceedings No. 342.

Subject Index